Breeding in a World of Scarcity

Isabel Roldán-Ruiz • Joost Baert
Dirk Reheul
Editors

Breeding in a World of Scarcity

Proceedings of the 2015 Meeting of the Section "Forage Crops and Amenity Grasses" of Eucarpia

Springer
EUCARPIA

Editors
Isabel Roldán-Ruiz
Institute for Agricultural and Fisheries Research (ILVO)
Melle
Belgium

Dirk Reheul
Ghent University
Ghent
Belgium

Joost Baert
Institute for Agricultural and Fisheries Research (ILVO)
Melle
Belgium

ISBN 978-3-319-28930-4 ISBN 978-3-319-28932-8 (eBook)
DOI 10.1007/978-3-319-28932-8

Library of Congress Control Number: 2016939624

Printed on acid-free paper

This Springer imprint is published by Springer Nature
The registered company is Springer International Publishing AG Switzerland

Preface

The 31st Eucarpia Fodder Crops and Amenity Grasses Congress took place in Ghent, Belgium, on 13–17 September 2015. Attendance was good, with 124 scientists and breeders from 25 countries. The Institute for Agricultural and Fisheries Research (ILVO) and Ghent University were co-organizers; it was the third time that Belgium had the honor of hosting this congress.

The theme of the meeting was "breeding in a world of scarcity." Scientific presentations and discussions were divided into four sessions: (1) scarcity of natural resources, (2) scarcity of breeders, (3) scarcity of land, and (4) scarcity of focus. Session 1 refers to the consequences of climate change, reduced access to natural resources, and increased pressure for adopting more sustainable agricultural practices. Global warming results in more extreme weather conditions. Biodiversity and genetic resources are under pressure as a consequence of climatic change and anthropogenic actions. High-yielding crops require high doses of nutrients with shrinking availabilities. Plant breeding may help by developing varieties with a more efficient use of water and nutrients and a better tolerance to biotic and abiotic stress. Session 2 refers to the shrinking number of breeders. Field breeders are becoming a rare breed, and modern plant breeders are expected to combine knowledge from different disciplines far more than in the past. There is a need for a mutual empathy between field-oriented and lab-oriented breeding activities. New methods of phenotyping and genotyping need to be integrated in breeding and bridge the gap between lab and field. Session 3 deals with the scarcity of agricultural land. Agricultural land has to be optimally used. Forage needs to be intensively produced in a sustainable way, at a competitive cost while still meeting the energy, protein, and health requirements of livestock. Well-adapted varieties, species, and mixtures of grasses and legumes are needed, not only to use as feed but also to use as turf and bioenergy and to provide ecosystem services. Session 4 refers to the fading of focus in primary production triggered by a range of societal demands. There are few farmers left, and they are asked to meet many consumer demands. Various crops and management systems are involved. Both large-scale, multi-purpose species and varieties and specialized niche crops are required to fulfill all these diverse needs and expectations.

This book contains the invited and submitted papers presented at the conference, whose Parts I, II, III, and IV correspond to the four sessions described above. Part V summarizes the conclusions of the debates, working groups, and workshops held during the meeting. Two open debates were organized: one on the future of grass and fodder crop breeding and a second one on feed quality breeding and testing. The content of these debates was determined on the basis of a survey in which several breeding companies and institutes participated. Different, and sometimes contrasting, views of these topics were presented and discussed in plenary sessions. The 31st section conference hosted meetings of two working groups, namely, "Multisite rust evaluation" and "Festulolium." During the "genomic selection and association mapping" workshop, participants shared experiences about the use of genetic and genomics tools in forage crop breeding. In the "phenotyping" workshop, current applications of noninvasive phenotyping tools in forage crop breeding research were presented, with a focus on the implementation in practical breeding. Part V contains also short sketches of breeding ideas presented as short communications by conference participants meant to help create progress in forage crop breeding.

We gratefully acknowledge the efforts of the members of the scientific committee (Ulf Feuerstein, Roland Kölliker, Paolo Annicchiarico, Philippe Barre, Susanne Barth, Johan De Boever, Alex De Vliegher, Trevor Gilliland, Mike Humphreys, Bernhard Ingwersen, Bernadette Julier, Petter Marum, Jan Nedelnik, Ulrich Posselt, Niels Roulund, Daniele Rosellini, Dejan Sokolovic, Leif Skot), for the critical review of the offered papers. Their pivotal contribution ensured the high quality of the chapters included in this book. Special thanks go to Miriam Levenson (ILVO) for the careful English-language editing of all submissions.

The local organizing committee (Jonas Aper, Mathias Cougnon, Johan De Boever, Alex De Vliegher, An Ghesquiere, Geert Lejeune, Katrien Liebaut, Nancy Mergan, Hilde Muylle, Tom Ruttink, Ariane Staelens Kristiaan Van Laecke, and Tim Vleugels) did a great job before, during, and after the meeting. It was a pleasure to organize the conference with such a good, tight team.

Finally, we express our gratitude to all the participants for daring to attend a meeting peppered with unconventional formats and for their valuable contributions to the presentations, debates, and discussions.

Melle and Ghent, Belgium

Isabel Roldán-Ruiz
Dirk Reheul
Joost Baert

Contents

Part I

Scarcity of Natural Resources

1 Breeding Forages to Cope with Environmental Challenges in the Light of Climate Change and Resource Limitations

Á. Helgadóttir, L. Østrem, R.P. Collins, M. Humphreys, A. Marshall, B. Julier, F. Gastal, Ph. Barre, and G. Louarn

Abstract

Global climate change and increased pressure for adopting more sustainable agricultural practices call for new approaches in breeding forage crops. In the cool temperate regions of Europe these crops may benefit from a warmer and prolonged growing season, but new stresses may emerge during autumn and winter, whereas further south risk of drought will increase. In addition, future forage crops have to use both nutrients and water more efficiently maximize production per unit area. This paper presents examples of how perennial forage crops can be adapted to the projected European environmental conditions through breeding. In the Nordic region, the focus is on identifying traits that are important for high yields under changed overwintering conditions and management practices. In temperate maritime Europe, the breeding focus is on forage grass and legume root systems for ecosystem service, nutrient and water use, as well as the advantages and potential for Festulolium, including its role in ruminant nutrition. In temperate and southern Europe, breeders aim to develop varieties that can survive long drought periods and recover rapidly following autumn rains, as well as improving adapted legume species with the following aims: reducing use of synthetic fertilizers, mitigating the environmental impacts of ruminant production systems; and reducing

Á. Helgadóttir (✉)
Department of Land and Animal Resources
Agricultural University of Iceland,
Árleyni 22, IS-112 Reykjavík, Iceland
e-mail: aslaug@lbhi.is

L. Østrem
Food and Agriculture Division
NIBIO, Fureneset, Fure, 6967
Hellevik i fjaler, Norway

R.P. Collins • M. Humphreys • A. Marshall
IBERS, Aberystwyth University,
Gogerddan Campus, Aberystwyth, UK

B. Julier
Unité de Recherche Pluridisciplinaire Prairies et Plantes Fourragères, Institut National de la Recherche Agronomique (INRA), Lusignan, France

F. Gastal
Unité de Recherche Pluridisciplinaire Prairies et Plantes Fourragères, Unité Experimentale Fourrages Ruminants Environnement, INRA, Lusignan, France

Ph. Barre • G. Louarn
Unité de Recherche Pluridisciplinaire Prairies et Plantes Fourragères, INRA, Lusignan, France

I. Roldán-Ruiz et al. (eds.), *Breeding in a World of Scarcity: Proceedings of the 2015 Meeting of the Section "Forage Crops and Amenity Grasses" of Eucarpia*,
DOI 10.1007/978-3-319-28932-8_1

their dependency on external protein-rich feeds. Forage production systems, which are commonly found in areas less suited to grain production, can contribute significantly to future food security but only if forage crops can be successfully adapted to meet future environmental challenges.

Keywords
Breeding • Climate change • Environmental sustainability • Forage crops

Introduction

The rising world population and a rapidly growing middle class are projected to result in increased demand for high quality food products from ruminants, both meat and milk. This demand is projected to nearly double by 2050 (FAO 2006). The corresponding increases in emissions of biogenic greenhouse gases (GHG) are expected to be the main driver for the projected temperature increase towards the end of this century (IPCC 2013). Agriculture is responsible for around one-third of global GHG emissions, of which 37 % of the methane production comes from livestock (FAO *loc. sit.*). Agrochemicals have polluted waterways, biodiversity is being threatened by intensive cultivation, and groundwater use is unsustainable in many places. Strategies must be found to meet the expected demand for animal products without further compromising the environment (Tilman et al. 2011). Ruminant products can be obtained from either grain or pasture-based systems. Grain-based feed competes directly with food for people, whereas pasture systems convert plant material unsuitable for human consumption into valuable food products. Such systems are commonly found in areas that are less suitable for grain production; they are dominated by grasslands and forage crop production. They can therefore contribute significantly to human food security by adding both calories and protein (Foley et al. 2011).

Changing climatic conditions and the increased pressure for adopting more sustainable agricultural practices will alter the suitability and profitability of the pasture systems currently employed in different parts of the world. In the cool temperate regions of northern Europe, global warming is expected to increase productivity primarily because of an extended growing season and longer frost-free periods (Olesen et al. 2011). At the same time, new stresses may emerge, especially in relation to overwintering of perennial species (Höglind et al. 2013). In large parts of southern Europe, conditions for summer crop production are expected to deteriorate, primarily due to increased risk of drought, whereas the growth period will be extended in late autumn and early spring (Olesen et al., *loc. sit.*). In all European regions, forage crops will have to use both nutrients and water more efficiently to maximize production per unit area. The use of N-fixing species should be emphasized to reduce the energy required to produce artificial fertilizers and to improve the protein autonomy of animal fodder in Europe. In addition, forage production systems will need to contribute to additional ecosystem services (soil, air and water quality, reduction of GHG from agriculture, and biodiversity conservation). Various options are available for adapting pasture systems to a changing climate and limited resources, from modifying or making substantial changes to current production systems to developing completely new systems (Lee et al. 2013). Breeding of forage crops has an important role to play in this evolution. One route is to develop new varieties of currently-grown species by identifying traits that improve their ability to cope with the expected environmental challenges. Exotic material can also be introduced and adapted to a new environment. Selecting the best approach will require information on various factors, such as the potential impact of climate change on the environment in question, the available genetic variation within gene pools of the relevant species, and the potential for introducing exotic material into existing breeding programs without sacrificing already-obtained improvements. Desired traits will be improved forage quality, increased resistance to biotic and abiotic factors, fine-tuned plant development processes and maximized resource use efficiency (Kingston-Smith et al. 2013).

This paper gives examples of how perennial forage crops can be adapted to the projected environmental conditions in three main European climates: Nordic, temperate maritime, and temperate or Mediterranean.

The Nordic Environment

Environmental Conditions and Projected Future Climate

The Nordic region of Europe (55–70 °N) has mild sea currents that result in average temperatures several degrees higher than comparable latitudes anywhere in the world. Within the region climatic conditions can vary considerably. The main limiting factors for forage crops are a short and moderately cool growing season and various winter stresses such as frost, ice encasement, low temperature fungi, prolonged snow cover, water logging and low light intensity. A rise in average temperature is projected to be similar to the global mean in the south and west part of the region (2.6–4.8 °C), and nearly double that in the north and east (Nordic Council of Ministers 2014). The highest increases will be seen during winter and in areas with a continental climate. Precipitation is predicted to increase in most parts of the region, especially in winter, but summer precipitation may be more evenly distributed and may even decrease in the southernmost part of the region (<60 °N). More extreme precipitation events are expected. The expected changes in climate at northern latitudes will result in a longer growing season with higher temperatures during the growing season, and increased biomass production potential. A modeling approach based on currently-grown timothy (*Phleum pratense* L.) cultivars across Norway demonstrated that increased biomass production was primarily attributed to more cuts per growing season (Persson and Höglind 2014). However, disease pressure and complex interactions between various environmental factors influencing overwintering may offset the potential gain (Rapacz et al. 2014). In continental sites with reduced snow cover, frost injury during winter for timothy may increase while more extensive damage is expected for the less adaptive perennial ryegrass (*Lolium perenne* L.) (Thorsen and Höglind 2010).

Adaptive Strategies in Response to Climate Change

The interaction of temperature and photoperiod controls the local adaptation of perennial forage crops by governing important physiological processes such as vernalization, flowering and cold acclimation. This interaction in turn determines winter survival and seasonal yield distribution. The projected climatic changes will lead to increased temperatures while seasonal photoperiod remains unchanged. Plants need to be adapted to different combinations of climatic variables and possess higher biomass production potential. We should therefore consider factors such as growth cessation in autumn in relation to acclimation, deacclimation and reacclimation, carbohydrate dynamics during autumn and winter, and photosynthetic activity and respiration at low temperature and light intensity.

For the foreseeable future, timothy is likely to remain the most common forage grass species grown in cold temperate regions at higher latitudes (Østrem et al. 2013). With a prolonged growing season, worthy breeding goals would be to improve its regrowth capacity and to make better use of favorable conditions in spring/early summer for elongation growth and production, while simultaneously avoiding the risk of frost damage in late winter (Rapacz et al. 2014). Perennial ryegrass and Festulolium (×*Festulolium* Aschers. et Graebn.) are currently grown in maritime regions but cannot be reliably grown inland and north of 60 °N. These high yielding species with good regrowth capacity and superior feed quality will become a promising option for a prolonged growing season and milder winters. Studies of their growth cessation and photosynthetic acclimation have shown a clear relationship between leaf elongation rate and photosynthetic acclimation, thus implying an inadequate autumnal growth cessation for sufficient cold acclimation (Østrem et al. 2014). Significant G×E interactions for biomass production, regrowth capacity and winter survival in recent cultivar trials reflect the considerable variation in climatic conditions in the Nordic region (Østrem et al. 2015), showing the importance of regional testing to find the optimal cultivars for persistence and production at each location.

The onset of acclimation is expected to start later in the autumn at lower solar radiation levels and thus less optimal conditions. Studies in controlled environments have shown reduced cold acclimation efficiency for northern varieties of perennial ryegrass and timothy compared with more southern ones when the temperature is higher than normal during preacclimation (Dalmannsdóttir et al. 2015). An increased use of non-native species in future conditions will therefore depend on plants being adapted to unique combinations of photoperiod and temperature during autumn. A sensible strategy for future breeding programs should be to select genotypes where growth cessation is controlled by photoperiod rather than temperature, while still maintaining a certain level of photosynthetic activity in autumn in order to build up sufficient carbohydrate reserves for winter survival (Østrem et al. 2014). At warmer winter temperatures, warm spells in mid-winter may trigger deacclimation and cause premature elongation growth. In perennial ryegrass it will probably be more efficient to increase the maximum frost tolerance rather than improve the ability to reacclimate after warm spells during unstable winters (Nordic Council of Ministers 2014). Additionally, waterlogged soil in autumn is expected due to increased precipitation (IPCC 2013). Freezing tests have indicated that increased abiotic stress caused by higher levels of waterlogged soil may enhance frost tolerance of timothy and red clover at low temperatures. At higher temperatures, hardening of red clover was reduced because of higher respiration and reduced assimilation rates, whereas timothy was less affected (Dalmannsdóttir et al. 2012).

The current genetic diversity in perennial ryegrass, timothy and meadow fescue (*Festuca pratensis* Huds.) found in the Nordic region is rather restricted, as it probably originates from a limited number of introductions (Rognli et al. 2013). Future breeding, irrespective of crop species, requires efficient ways to incorporate wild adapted genetic resources and exotic material into the current breeding base. This long-term task requires public support and active knowledge transfer from public-sector scientists into real-life crop improvement (Helgadóttir 2014). A good example is the Nordic Public-Private Partnership (Nordgen 2015) on pre-breeding of perennial ryegrass. Approximately 400 diverse accessions have been phenotyped, recombined to introgress exotic germplasms, and genotyped by sequencing to estimate relatedness and develop tools for associating genome regions with traits. These efforts will result in locally adapted germplasm for specific climates/sites and germplasm with wide adaptations (phenotypically stable) (Nordic Council of Ministers 2014).

Forage legumes play a vital role in future sustainable agriculture, and in the Nordic region both red (*Trifolium pratense* L.) and white clover (*T. repens* L.) cultivars are sold. In white clover, adequate genetic variation for cold tolerance and rapid adaptational changes have been manifested (Helgadóttir et al. 2001). Generally, persistent types are small-leaved with low production potential but simultaneous selection for yield and winter hardiness has been successful following hybridization between northern and southern adapted populations (Helgadóttir et al. 2008). Red clover shows extensive variation for yield but the basis for genetic variation of cold tolerance has been less investigated than in white clover (Annicchiarico et al. 2015). A recent study using molecular markers has demonstrated rapid genetic change of diverse populations when grown in contrasting environments (Collins et al. 2012). When developing red clover cultivars for the current and future climate in the north, similar to perennial ryegrass, it is important to broaden the genetic base and identify valuable material for future pre-breeding and breeding projects. One such project is currently being undertaken where wild material and landraces in the Nordic collection of red clover are evaluated at four locations across the Nordic region.

Temperate Maritime Environments

The UK climate is predicted to include warmer, wetter winters and hotter, drier summers with increasing frequency of extreme weather events (Harrison et al. 2001). Incidents of soil waterlogging, floods, or droughts will all have significant effects when occurring separately, and when they occur in sequence, ever-increasing incremental effects are expected that will reduce grassland production and crop persistence in the UK. Soil hardening after summer droughts will result in heavily compacted soils, which in turn will exacerbate the effects of high rainfall events in autumn. We propose that improvements in plant traits affecting waterlogging and drought tolerance can be delivered through breeding for combined deeper and more extensive root systems of major grassland species, and will pave the way for improved, agronomically superior forage varieties that also deliver environmental services including improved soil hydrology and carbon sequestration.

In European temperate maritime environments, the need for grassland swards that produce above-ground biomass of sufficient quality to sustain high levels of animal performance while also delivering ecosystem services is well-recognized (e.g., Isselstein and Kayser 2014). In temperate forage species, breeding for above-ground sustainability traits such as stress resistance and forage quality has made rapid progress (Abberton et al. 2008) but the design of root systems has received considerably less attention. The technical difficulties associated with investigating the functioning of plant root systems *in situ* have diminished full appreciation of their potential to deliver multiple environmental benefits. However, with the development of improved phenotyping technologies and increased emphasis on sustainable grassland systems, the links between root system traits and environmental functioning (Bardgett et al. 2014) are now a major research focus.

Plant roots have a greater impact on soil water status than different capacities for water extraction alone (Macleod et al. 2007). Hydrologists have described a pivotal role for vegetation in the regulation of soil water content, including root activity that can initiate biophysical changes in soil hydraulic properties (Whalley et al. 2005), that occur in function of the porosity of the soil and the rooting depth of the vegetation. Developments in dynamic root imaging, combined with an increased understanding of the genetic base of variation in root architecture, could bring about a step change in our awareness of how forage plant root systems may be designed to balance above-ground biomass productivity and below-ground biotic and abiotic interactions. Plant breeding approaches at IBERS (Wales, UK) now involve novel phenotyping systems that can be used at different scales (from genotype to farm system) to enable quantification of the impact of plant root systems on important aspects of soil quality. As proof of principle, rainfall run-off was measured over a 2-year period using hydrologically-isolated field plots, each containing varieties of different *Lolium* and *Festuca* species. The Festulolium variety, cv. Prior (a hybrid of *L. perenne* and *F. pratensis*), had a 51 % lower run-off than a current *L. perenne* variety and 43 % lower than *F. pratensis*. In a detailed phenotyping study, cv. Prior produced very large root systems which subsequently degenerated extensively, especially at depth, leading to significantly enhanced soil porosity. This trait assisted soil water retention and reduced over-land flow compared to its parent species and the other grasses assessed (Macleod et al. 2013). The outcome of these observations led to a 5-year project, SUREROOT (http://www.sureroot.uk/), that includes use of two UK National Capability Facilities, the National Plant Phenotyping Centre (NPPC) at IBERS (Aberystwyth) and the Farm Platform at North Wyke (Devon). Detailed analyses of root ontogeny involving both grasses and clovers are being obtained in the NPPC and changes measured over time. The impacts of root architecture and ontogeny on soil structure and hydrology under different field conditions and livestock management systems is being compared to assess the potential for a combined strategy for grassland that provides high agronomic value and flood mitigation.

Exploiting interspecific and intergeneric hybrids in forage grasses and legumes

1. Festulolium: The research carried out in SUREROOT uses novel Festulolium hybrids that include the attributes of both *Lolium* and *Festuca* species, but may also outperform their parent species in a number of growth and performance traits. Predicted climate change scenarios open the possibility for including novel grasses previously considered suitable for livestock production systems in temperate maritime environments (Kingston-Smith et al. 2013). Examples include hybrids involving either *L. multiflorum* or *L. perenne* together with *F. arundinacea var glaucescens,* a drought and heat-tolerant grass species found in Mediterranean regions. This has potential to restrict plant mediated proteolysis in the rumen, and reduce greenhouse gas emissions and N losses by livestock (O'Donovan 2015). This germplasm is also deep rooting, which has enhanced its drought resistance (Durand et al. 2007), consistent with alternative hybrids involving *Lolium* spp. and *F. mairei* (Atlas fescue). All of this germplasm has high forage yields and excellent nutritive value (Humphreys et al. 2014) and will be employed in the SUREROOT project in addition to two diploid drought-tolerant introgression-lines formed following transfers of genes for drought resistance from chromosome three of two *Festuca* species into *Lolium* (Humphreys et al. 2012). Natural Festulolium hybrid combinations also exist and contain potentially useful adaptations for use in plant breeding such as novel genetic variants for flooding tolerance. For example, *Festulolium loliaceum* (derived from hybridization of *L. perenne* and *F. pratensis*) is found in flood- and water-meadows, and appears better adapted to waterlogging than either of its parental species (Humphreys and Harper 2008).
2. *Trifolium repens* × *T. ambiguum* hybrids: Wide genetic variation within the gene pool of white clover has been used successfully in the production of new varieties with improvements in many traits. However, less variation has been identified for traits such as drought tolerance, which is difficult to improve significantly by conventional selection (Abberton and Marshall 2005). Caucasian or Kura Clover (*Trifolium ambiguum* M. Bieb.) is a rhizomatous perennial legume species with good drought tolerance and persistence (Coolbear et al. 1994). Hybrids have been developed between white clover and Caucasian clover (Marshall et al. 2001) to introgress the rhizomatous trait from the latter as a strategy for improving drought tolerance while retaining the desirable agronomic traits associated with white clover. Fertile backcross (BC) hybrids (derived from backcrossing to white clover) have been produced that are essentially like white clover, but which have rhizomes and stolons. A drought experiment carried out in deep soil bins showed that the hybrids maintained lower values of leaf relative water content (RWC) and leaf water potential than Caucasian clover, but higher levels than white clover at the same level of drought. The mechanism by which Caucasian clover maintains a higher leaf RWC is not known, but it has been established that this species and both generations of the BC hybrids allocate a higher proportion of their total DM yield to roots than white clover, *i.e.,* they maintain a higher root weight ratio (Marshall et al., *loc. sit.*). A recent experiment (Marshall et al. 2015) compared the below-ground morphology of plants of Caucasian clover, white clover, and the BC hybrids grown in 1 m long plastic 'root pipes' (Fig. 1.1). Caucasian clover had a greater root weight than white clover at depths below 0.2 m. The BC1 and BC2 hybrids also had a greater root weight than white clover at depths below 0.2 m, which suggests that introgression increased the allocation of resources to roots at lower depths, thereby contributing to the improved drought tolerance of the BC hybrids compared with white clover. This is

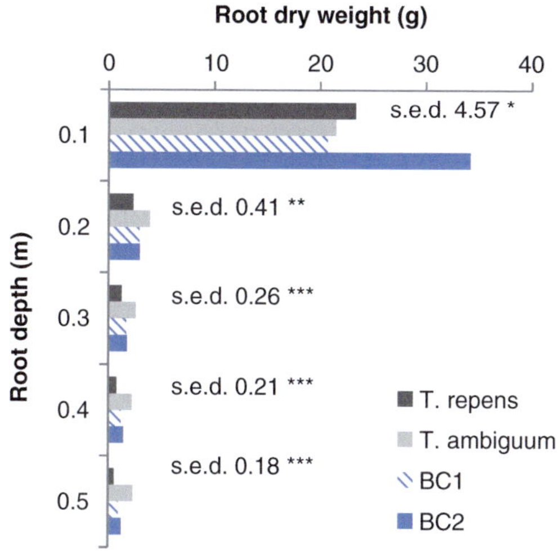

Fig. 1.1 Root dry weight in 0.1 m sections of soil columns containing *T. repens*, *T. ambiguum*, BC1 and BC2 hybrids

likely to be advantageous in environments where water reserves exist at depth. Conversely, where moisture reserves are confined to the upper layers of soil, then root depth becomes less important than the ability of plants to produce an efficient extraction system in these surface layers. Although the depth of root systems depends on soil type, cultivar and management, white clover is generally considered to be a shallow-rooted species, with most roots distributed in the top 0.1–0.2 m of soil (Caradus 1990). This distribution pattern is likely to reduce the ability of the species to persist under drought, and suggests a mechanism through which the BC hybrids, with greater root weight than white clover below 0.2 m, are more drought tolerant than their parent.

The Southern European Environment

Environmental Conditions and Projected Future Climate

Southern Europe is characterized either by a temperate climate or (along the Mediterranean Sea) a Mediterranean climate. The average annual temperature of southern Europe is expected to increase: more during summer (3–4 °C across the Iberian Peninsula) than during winter (1.5–2 °C) (IPCC 2013). More frequent hot summers and fewer cold winters are therefore expected to occur. Annual rainfall is expected to decrease between 0 and 10 % in southern areas, particularly during the summer (IPCC 2013). As evapotranspiration is also expected to increase due to the rise in temperature and solar radiation, soil water deficit should increase more than the projected decrease in summer rainfall. This trend is exacerbated by the extremely limited potential for irrigation. Crop simulation models show that rises in CO_2 concentration and in autumn-winter-spring temperatures are both favorable to forage production but would not compensate for the decrease in summer production (Durand et al. 2010). To cope with these new constraints, several key actions could be considered. First, new varieties can be created which are better adapted to new climatic conditions. Second, forage legumes in forage production systems can be increased. Legume species can be used in pure or mixed swards. They can also be used in rotations to accomplish three aims: (1) reduce the use of mineral N fertilizers, (2) reduce environmental impacts of ruminant systems (nitrate leaching, N_2O and methane emissions) (Lüscher et al. 2014) and (3) to increase the persistency of perennial grasslands to support severe drought periods, favor carbon storage and biodiversity and avoid soil erosion. Forage for the driest areas can be based on self-reseeding winter annuals.

Strategies to Adapt Forage Species to Climate Change in Southern Europe

In southern Europe, lucerne (*Medicago sativa* L.), tall fescue (*Festuca arundinacia* Schreb.) and cocksfoot (*Dactylis glomerata* L.) are the most cultivated perennial species. Interest in these species is reinforced by the expected impacts of climate change. The large diversity of ecotypes, landraces and varieties adapted to various Mediterranean conditions have been little used in breeding programs thus far (Lelièvre and Volaire

2009; Annicchiarico et al. 2013; Poirier et al. 2012). A few annual species are also used as annually self-regenerating pasture (e.g., subterranean clover, annual medics, annual ryegrass) in Mediterranean areas (Grashaw et al. 1989; Texeira et al. 2014). These forage species rely on seed production and survival in the seed bank to avoid summer droughts and may become more important to overcome prolonged periods of stress in the driest areas.

Lucerne, a major legume species traditionally cultivated in many areas of southern (and eastern) Europe, has a high potential for biomass and protein production. It is well suited to provide ecosystem services in forage-crop rotations, because it can supply large amounts of N to subsequent crops, improve soil structure and capture nitrate in soils. Considering climate change, the choice of the best adapted lucerne varieties should focus on the autumn dormancy range. This dormancy, driven by both low temperature and short photoperiod, induces a reduction of vegetative growth that is related to winter survival. With mild temperatures in autumn and spring and less severe frost in winter, dormancy could skip from 4–5 (Flemish types) to 5–6 (known as Mediterranean types in France) in temperate Europe and from 5–6 to 7–8 in Mediterranean Europe, on a dormancy scale ranging from 1 (winter-dormant) to 11 (winter-active). The varieties of temperate and Mediterranean origin have a similar developmental response to temperature but the Mediterranean cultivars have a lower productivity potential and photosynthesis at high temperatures (G. Louarn, pers. comm.). Part of the breeding effort currently devoted to Flemish types should now be transferred to Mediterranean types by using classical breeding criteria (forage yield, disease and pest resistance, feeding value, lodging resistance, seed production). For Mediterranean climatic conditions in the context of climate change, development of lucerne varieties adapted to rain-fed conditions is an objective (http://reforma.entecra.it/). Landraces traditionally grown in rain-fed conditions have shown interesting agronomic performance in a multi-site trial in northern and southern Mediterranean regions (Annicchiarico et al. 2011a). Combining adapted autumn dormancy and tolerance to summer drought would create lucerne varieties tolerant to Mediterranean climates in a water-restricted sustainable agriculture.

Mixtures of lucerne and perennial grasses can be used to meet the requirement of an energy and input-limited agriculture. Mixtures grown with no or little N fertilization or herbicides have to produce at least as much forage as the best yielding species grown in pure stand and/or to improve annual yield distribution, and the forage must have a well-balanced N/energy ratio. Usually varieties are bred for pure stand performance, considering that a good variety in pure stand is also good in mixture. However, in recent experiments, we have found that forage production and quality of lucerne varieties in mixture was only partly related to their performance in pure stands (Julier et al. 2014).

To cope with dry conditions, perennial grasses have developed different strategies either to produce (drought resistance) or survive (drought survival) during drought (Volaire et al. 2009, 2014). For forage species, drought resistance is targeted in areas subjected to moderate droughts, while drought survival is the main objective for areas with severe and chronic droughts. To achieve drought resistance or survival, the main physiological mechanisms are classified as dehydration delay (or avoidance) of either lamina or meristems, dehydration tolerance of either lamina or meristems and summer dormancy (Volaire et al. 2009). However, there is a general trade-off between summer dormancy and productivity under favorable conditions of autumn and spring (Shaimi et al. 2009). An ideotype has been defined that would be highly drought tolerant through summer dormancy but nevertheless productive at other seasons.

There is a gradient of winter "dormancy" (i.e., reduction of plant growth) with latitude from Mediterranean ecotypes to Nordic ecotypes. Mediterranean ecotypes grow continuously from autumn to spring when water is not a limiting factor. In contrast, Nordic ecotypes have a long growth cessation from autumn to spring that confers cold tolerance, but they are characterized by a vigorous spring-summer growth (Annicchiarico et al. 2011b; 2013). New plant material could be created with high leaf growth rates (related to forage productivity) and different levels of summer and winter dormancy, depending on the climatic region targeted. It is possible to select for productivity in Mediterranean material or to gather favorable traits from Mediterranean material (dehydration tolerance, growth during mild and humid seasons, i.e. autumn, winter, spring) and from elite temperate material (forage productivity, cold and disease resistances and seed productivity) (Barre et al. 2014).

The use of self-regenerating annual species has been reported to increase forage production, soil fertility and carbon sequestration, and they are more resilient than perennials to extreme climatic events (Wolfe and Dear 2001; Texeira et al. 2014). The adaptability of self-regenerating annual species is governed by traits enabling them to complete seed production over the duration of the local growing season (determined by winter rainfall and summer drought). The flowering and maturity dates are critical. The ability to overcome low or zero seed production during extremely stressful years or management has been obtained by selecting varieties with hard-seededness, ensuring persistence in the seed bank and long-term regeneration (Reed et al. 1989).

Adaptation of grass and legume species could also be obtained by mixing Mediterranean and temperate types into mono- or pluri-specific mixtures. Mixing such types of tall fescue, cocksfoot and lucerne in monospecific mixtures may help to regulate forage production during the year and to reduce the disequilibrium between spring and summer production (http://www.animalchange.eu/). This approach was particularly interesting in the case of grass-lucerne associations, with both seasonal growth complementarities between the species and the plant types, and positive N facilitation effects brought by the legume to the grass species. In another design, in which both the number of species and genotypes varied, multispecies mixtures were more productive than monocultures in dry conditions but it was shown that the temporal stability of production increased with the number of genotypes only (Prieto et al. 2015). Use of species and population diversity could be beneficial to forage productivity in constrained environments but requires adapted cultivars that can resist drought, cold and N-limited conditions.

Conclusion

Forage production systems, commonly found in marginal areas less suited to grain production, can contribute significantly to future food security – but only if forage crops can be successfully adapted to future environmental challenges.

References

Abberton MT, Marshall AH (2005) Progress in breeding perennial clovers for temperate agriculture. Journal of Agricultural Science 143: 117–135.

Abberton MT, Marshall AH, Humphreys MW, MacDuff JH, Collins RP, Marley CL (2008) Genetic improvement of forage species to reduce the environmental impact of temperate livestock grazing systems. Advances in Agronomy 98: 311–355.

Annicchiarico P, Pecetti L, Abdelguerfi A, Bouizgaren A, Carroni AM, Hayek T, Bouzina MM, Mezni M (2011a) Adaptation of landrace and variety germplasm and selection strategies for lucerne in the Mediterranean basin. Field Crops Research 120: 283–291.

Annicchiarico P, Pecetti L, Bouzerzour H, Kallida R, Khedim A, Porqueddu C, Simoes NM, Volaire F, Lelievre F (2011b) Adaptation of contrasting cocksfoot plant types to agricultural environments across the Mediterranean basin. Environmental and Experimental Botany 74: 82–89.

Annicchiarico P, Pecetti L, Abdelguerfi A, Bouzerzour H, Kallida R, Porqueddu C, Simoes NM, Volaire F (2013) Optimal forage grass germplasm for drought-prone Mediterranean environments. Field Crops Research 148: 9–14.

Annicchiarico P, Barrett B, Charles Brummer E, Julier B, Marshall AH (2015). Achievements and challenges in improving temperate perennial forage legumes. Critical Reviews in Plant Sciences 34: 327–380.

Bardgett R, Mommer L, De Vries FT (2014) Going underground: root traits as drivers of ecological processes. Trends in Ecology and Evolution 29: 692–699.

Barre P, Zhouri L, Kallida R, Julier B, Shaimi N, Fakiri M, Volaire F (2014) Towards productive summer dormant cocksfoot for Mediterranean climates. In: Sokolovic D, Huyghe C, Radovic J (eds) Quantitative traits breeding for multifunctional grasslands and turf. 30th Meeting of the EUCARPIA Fodder Crops and

Amenity Grasses Section, Vrnjačka Banja, Serbia, 2013. Springer, Heidelberg, p 145.

Caradus JR (1990) The structure and function of white clover root systems. Advances in Agronomy 43: 1–46.

Collins RP, Helgadóttir Á, Frankow-Lindberg BE, Skot L, Jones C, Skot KP (2012) Temporal changes in population genetic diversity and structure in red and white clover grown in three contrasting environments in northern Europe. Annals of Botany 110: 1341–1350.

Coolbear P, Hill MJ, Efendi F (1994) Relationships between vegetative and reproductive growth in a four year old stand of Caucasian clover (*Trifolium ambiguum* M Bieb.) cv. Monaro. Proceedings of the Agronomy Society of New Zealand 24: 77–82.

Dalmannsdóttir S, Østrem L, Larsen A (2012) Effects of water saturated soil on winter survival of red clover (*Trifolium pratense*) and timothy (*Phleum pratense*) in Norway. Grassland Science in Europe 17: 103–105.

Dalmannsdóttir S, Rapacz M, Jørgensen M, Østrem L, Larsen A, Rødven R, Rognli O.A (2015) Temperature before cold acclimation affects cold tolerance and photoacclimation in timothy (*Phleum pratense* L.), perennial ryegrass (*Lolium perenne* L.) and red clover (*Trifolium pratense* L.). Journal of Agronomy and Crop Science doi:10.1111/jac.12149.

Durand JL, Bariac T, Ghesquiere M, Brion P, Richard P, Humphreys MW and Zwierzykowski Z (2007) Ranking of the depth of water extraction by individual grass plants using natural ^{18}O isotope abundance. Environmental and Experimental Botany 60: 137–144.

Durand JL, Bernard F, Lardy R, Graux I (2010) Changement climatique et prairie: l'essentiel des impacts. In : Brisson N, Levrault F (eds) Livre vert du projet climator. Ademe Editions, 334 p

FAO (Food and Agriculture Organization) (2006) Livestock's long shadow: environmental issues and options. FAO, Rome. ftp://ftp.fao.org/docrep/fao/010/a0701e/A0701E00.pdf

Foley JA, Rmankutty N, Brauman KA, Cassidy ES, Gerber JS, Johnston M, Mueller ND, O'Connell C, Ray DK, West PC, Balzer C, Bennett EM, Carpenter SR, Hill J, Monfreda C, Polasky S, Rockstrom J, Sheehan J, Siebert S, Tilman D, Zaks DPM (2011) Solutions for a cultivated planet. Nature 478: 337–342.

Grashaw D, Read JW, Collins WJ, Carter ED (1989) Sown pastures and legume persistence: An Australian overview. In: Marten GC *et al.* (eds.) Persistence of forage legumes. American Society of Agronomy, pp 1–22.

Harrison PA, Berry PM, Dawson TP (2001) Climate change and nature conservation in Britain and Ireland: Modelling natural resource responses to climate change (the MONARCH project). Technical Report. Oxford, UK: UKCIP.

Helgadóttir Á (2014) Why is public plant breeding important in the Nordic region? Journal of the Swedish Seed Association 123: 23–28.

Helgadóttir Á, Dalmannsdóttir S, Collins RP (2001) Adaptational changes in white clover populations selected under marginal conditions. Annals of Botany 88: 771–780.

Helgadóttir Á, Marum P, Dalmannsdóttir S, Daugstad K, Kristjánsdóttir TA, Lunnan T (2008) Combining winter hardiness and forage yield in white clover (*Trifolium repens*) cultivated in northern environments. Annals of Botany 102: 825–834.

Höglind M, Thorsen SM, Semenov MA (2013) Assessing uncertainties in impact of climate change on grass production in Northern Europe using ensembles of global climate models. Agricultural and Forest Meteorology 170: 103–113.

Humphreys MW, Harper JA (2008) *Festulolium loliaceum*, an understudied natural UK grass hybrid species that may provide benefits to UK grasslands withstanding the onsets of climate change. European Crop Wild Relative Newsletter 6.

Humphreys MW, Macleod CJA, Whalley WR, Turner LB, Farrell MS, Ghesquiere M, Haygarth PM (2012). Designing grass cultivars for droughts and floods. In: (Chapter 21) Breeding Strategies for Sustainable Forage and Turf Improvement. Barth S., Milbourne D. (Eds). Springer, pp 171–180.

Humphreys MW, O'Donovan SA, Farrell MS, Gay A, Kingston-Smith AL (2014) The potential of novel *Festulolium* (2n = 4x = 28) hybrids as productive, nutrient-use-efficient fodder for ruminants. Food and Energy Security; DOI: 10.1002/fes3.50.

IPCC (2013) The Physical Science Basis. Contribution of Working Group I to the Fifth Assessment Report of the Intergovernmental Panel on Climate Change. Cambridge, United Kingdom and New York, NY, USA, Cambridge University Press.

Isselstein J, Kayser M (2014) Functions of grassland and their potential in delivering ecosystem services. Grassland Science in Europe 19: 199–214.

Julier B, Louarn G, Gastal F, Surault F, Sampoux J.P, Maamouri A, Fernandez L. (2014) Les associations graminées – légumineuses prairiales. Comment sélectionner des variétés adaptées pour accroitre leur productivité et faciliter leur conduite? Innovations Agronomiques 40: 61–72.

Kingston-Smith AH, Marshall AH, Moorby JM (2013) Breeding for genetic improvement of forage plants in relation to increasing animal production with reduced environmental footprint. Animal 7, Supplement 1: 79–88.

Lee JM, Clark, AJ, Toche JR (2013) Climate-change effects and adaptation options for temperate pasture-based dairy farming systems: a review. Grass and Forage Science 68: 485–503.

Lelièvre F, Volaire F (2009) Current and potential development of perennial grasses in rainfed Mediterranean farming systems. Crop Science 49: 2371–2378.

Lüscher A, Müller-Harvey I, Soussana JF, Rees RM, Peyraud JL (2014) Potential of legume-based grassland-livestock systems in Europe: a review. Grass and Forage Science 69: 206–228.

Macleod CJA, Binley A, Clark LJ, Hawkins SL, Humphreys MW, Turner LB, Whalley WR, Haygarth PM (2007) Genetically modified hydrographs: what can grass genetics do for temperate catchment hydrology? Hydrological Processes 21(16): 2217–2221.

Macleod CJA, Humphreys MW, Whalley WR, Turner LB, Binley A, Watts CW, Skot L, Joynes A, Hawkins SL, King IP, O'Donovan SA, Haygarth PM (2013) A novel grass hybrid to reduce flood generation in temperate regions. Nature Scientific Reports; DOI: 10.1038/srep01683.

Marshall AH, Rascle C, Abberton MT, Michaelson-Yeates TPT, Rhodes I (2001) Introgression as a route to improved drought tolerance in white clover (*Trifolium repens* L.). Journal of Agronomy and Crop Science 187: 11–18.

Marshall AH, Lowe M, Collins RP (2015) Variation in response to moisture stress of young plants of interspecific hybrids between white clover (*T. repens* L.) and Caucasian clover (T. *ambiguum* M. Bieb.). Agriculture 5: 353–366.

Nordic Council of Ministers (2014) Climate change and primary industries: Impacts, adaptation and mitigation in the Nordic countries. TemaNord 2014:552. http://norden.diva-portal.org/smash/get/diva2:737875/FULLTEXT03.pdf

Nordgen (2015). Public Private Partnership on pre-breeding. Accessed on 15 October 2015 at http://www.nordgen.org/index.php/en/content/view/full/1907.

O'Donovan SA (2015) Transmission and expression of *Festuca arundinacea var glaucescens* traits in *Lolium* spp. for improved efficiency of protein use by ruminants. PhD Thesis Aberystwyth University, 2015.

Olesen JE, Trnka M, Kersebaum KC, Skjelvåg AO, Seguin B, Peltonen-Sainio P, Rossi F, Kozyra J, Micale F (2011) Impacts and adaptation of European crop production systems to climate change. European Journal of Agronomy 34: 96–112.

Persson T, Höglind M (2014) Impact of climate change on harvest security and biomass yield of two timothy ley harvesting systems in Norway. Journal of Agricultural Science 152: 205–216.

Poirier M, Durand JL, Volaire F (2012) Persistence and production of perennial grasses under water deficits and extreme temperatures: importance of intraspecific vs. interspecific variability. Global Change Biology 18: 3632–3646.

Prieto I, Violle C, Barre P, Durand JL, Ghesquière M, Litrico I (2015) Complementary effects of species and genetic diversity on productivity and stability of sown grasslands. Nature Plants; DOI: 10.1038/NPLANTS.2015.33

Rapacz M, Ergon Å, Höglind M, Jørgensen M, Jurczyk B, Østrem L, Rognli OA, Tronsmo AM (2014) Overwintering of herbaceous plants in a changing climate – still more questions than answers. Plant Science 225: 34–44.

Reed KFM, Mathison MJ, Crawford EJ (1989) Adaptation, regeneration and persistence of annual legumes in temperate pasture. In: Marten GC *et al.* (eds) Persistence of forage legumes. American Society of Agronomy, pp. 69–89.

Rognli OA, Fjellheim S, Pecetti L, Boller B (2013) Semi-natural grasslands as a source of genetic diversity. Grassland Science in Europe 18: 303–313.

Shaimi N, Kallida R, Volaire F, Al Faiz C (2009) Summer dormancy in orchardgrass: evaluation and characterization through ecophysiological and genetic studies. Crop Science 49: 2353–2358.

Texeira R, Proença V, Valada T, Crespo D, Domingos T (2014) Sown biodiverse pastures as a win-win approach to reverse the degradation of Mediterranean ecosystems. Grassland Science in Europe 19: 258–260.

Thorsen SM, Höglind M (2010) Assessing winter survival of forage grasses in Norway under future climate scenarios by simulating potential frost tolerance in combination with simple agroclimatic indices. Agricultural and Forest Meteorology 150: 1272–1282.

Tilman D, Balzer C, Hill J, Befort BL (2011) Global food demand and the sustainable intensification of agriculture. Proceedings of the National Academy of Sciences of the United States of America 108: 20260–20264.

Volaire F, Barkaoui K, Norton M (2014) Designing resilient and sustainable grasslands for a drier future: Adaptive strategies, functional traits and biotic interactions. European Journal of Agronomy 52: 81–89.

Volaire F, Norton MR, Lelièvre F (2009) Summer drought survival strategies and sustainability of perennial temperate forage grasses in Mediterranean areas. Crop Science 49: 2386–2392.

Whalley WR, Riseley B, Leeds-Harrison PB, Bird NRA, Leech PK, Adderley WP (2005) Structural differences between bulk and rhizosphere soil. European Journal of Soil Science 56: 353–360.

Wolfe EC, Dear BS (2001) The population dynamics of pastures, with particular references to southern Australia. *In*: Tow PG, Lazenby A (eds.) Competition and succession in pastures. CABI Publishing, pp 119–148.

Østrem L, Volden B, Larsen A (2013) Morphology, dry matter yield and phenological characters at different maturity stages of × *Festulolium* compared with othergrass species. Acta Agriculturae Scandinavica Section B: Soil and Plant Science 63: 531–542.

Østrem L, Rapacz M, Larsen A, Dalmannsdottir S, Jørgensen M (2014) Influences of growth cessation and photoacclimation on winter survival of non-native Lolium-Festuca grasses in high-latitude regions. Environmental and Experimental Botany 111: 21–31.

Østrem L, Helgadóttir Á, Niemeläinen O, Seppänen M, Greve Pedersen M, Halling M, Rognli OA (2015) Environmental impact on winter survival and production in non-native grasses in the Nordic countries. *In*: Zeverte-Rivza S (ed.) Proceedings of the 25th NJF Congress, Nordic View to Sustainable Rural Development, Riga, Latvia 16–18 June 2015, pp. 127–133 (ISBN 978-9934-14-548-3).

Performance of Diploid and Tetraploid Perennial Ryegrass Synthetics with Variable Numbers of Parents

2

B. Boller, C. Grieder, and F.X. Schubiger

Abstract

When building synthetic varieties by crossing a fixed number of elite genotypes in a polycross, breeders are faced with a tradeoff between selection intensity and inbreeding depression. We investigated the relationship between synthetic size and herbage dry matter (DM) yield by analyzing replicated field data of 93 diploid and 73 tetraploid perennial ryegrass (*Lolium perenne* L.) synthetics of the Agroscope breeding program. In line with theoretical expectations, DM yield of diploid synthetics declined when the number of parents dropped below seven. However, no such decline was observed with tetraploid synthetics. We conclude that in tetraploid perennial ryegrass, selection intensity can be increased to result in synthetics with only five parents without hampering DM yield.

Keywords

Parent number • *Lolium perenne* • Polycross • Synthetic • Breeding • Dry matter yield

Introduction

Synthetics built by cloning and polycrossing a fixed number of elite parent genotypes are the most common type of cultivar in forage grass breeding. When selecting the parental elite genotypes of a candidate cultivar, breeders must consider the tradeoff between selection intensity and inbreeding depression occurring during seed increase of the synthetic. This decrease in performance of the syn-2 and following generations, caused by mating of related individuals, will be the more pronounced the fewer parents form the initial polycross. The theoretical foundation of this relationship, mathematically captured by the so-called Sewall Wright formula, was comprehensively described by Posselt (2010) for the case of perennial ryegrass (*Lolium perenne* L.). He modeled performance of synthetics based on experimental

B. Boller (✉) • C. Grieder • F.X. Schubiger
Agroscope, Institute for Sustainability Sciences, Zurich, Switzerland
e-mail: beat.boller@agroscope.admin.ch

I. Roldán-Ruiz et al. (eds.), *Breeding in a World of Scarcity: Proceedings of the 2015 Meeting of the Section "Forage Crops and Amenity Grasses" of Eucarpia*,
DOI 10.1007/978-3-319-28932-8_2

data of diploid perennial ryegrass and found that the optimum number of parents for maximum yield was between 7 and 11. Recently, attempts were made to reduce the effect of inbreeding in few-parent-synthetics by choosing, with the aid of molecular markers, genetically more distant genotypes (Kölliker et al. 2005). Ghesquiere and Baert (2007) and Baert et al. (2007), in model experiments, investigated the possibility of narrowing a perennial ryegrass synthetic down to as little as two parents. They concluded that inbreeding depression was less important in tetraploid than in diploid material, and less important for herbage than for seed yield. In this study, we test (i) if the size-performance-relation as observed in practical breeding follows theoretical expectations and (ii) if there is a difference in optimum synthetic size between diploid and tetraploid plants. The results suggest that when using tetraploid breeding material of perennial ryegrass, the risk of a decrease in performance with too-few constituents of a synthetic is minimal.

Materials and Methods

Plant Material and Experimental Data

The study is based on results obtained in Agroscope's perennial ryegrass breeding program. Data encompass a total of 93 diploid and 73 tetraploid synthetics, created between 1987 and 2009 and composed of 5–68 parents. Out of these, 7 diploid and 12 tetraploid synthetics reached the status of a cultivar listed in at least one country.

In Agroscope's breeding scheme (see Humphreys et al. 2010), after cloning and polycrossing a number of genotypes (Ng), the progeny of each genotype is sown in replicated rows and a selected number of half-sib progenies (Np) is allowed to inter-pollinate and set seed. Bulk seed harvested on these progenies constitutes the synthetic. We describe the size of the synthetic as Np + (Ng − Np)/2, considering that Np genotypes deliver their full and (Ng − Np) half their genetic makeup into the synthetic. All synthetics of this study were tested in field trials, carried out between 2000 and 2014 at Zürich-Reckenholz (47.43°N, 8.52°E), Ellighausen (47.61°N, 9.14°E) and Oensingen (47.28°N, 7.73°E), Switzerland. Field trials consisted of 1.5 × 6 m plots with three replications arranged in Latin rectangles and were harvested five times in both the first (H1) and second (H2) main harvest year. Total dry matter (DM) yield was determined by weighing fresh herbage with a plot harvester followed by DM determination via oven drying. Most entries were sown in at least 2 years at all three locations, resulting in yield data from six "environments" (combination of location and year of sowing). However, deviations were possible with some entries being sown in only one or in more than 2 years (especially those reaching status as a standard cultivar). On average, each entry was tested in 6 (2–32) environments.

Statistical Analysis

In a first step, least squares means of total DM yield of H1 and H2 were estimated for each synthetic using general linear models. Combined analysis over all series of trials was possible through overlapping entries between trial series, even though there was no common set of standard varieties in all trials. The following statistical model was used:

$$y_{ijn} = \mu + s_i + e_j + r_{jn} + \varepsilon_{ijn} \quad (2.1)$$

where μ is the general mean, s_i is the effect of the ith synthetic, e_j the jth environment, r_{jn} the nth replication within the jth environment and ε_{ijn} the residual. Mean performance of each synthetic was then regressed on the two covariates "year of polycross creation" (p) and "time of inflorescence emergence" (f) to correct for the general breeding progress achieved over the years or any effect of earliness, respectively:

$$y_i = \mu + xp_i + yf_i + \varepsilon_i \quad (2.2)$$

Finally, the residuals (εi) from Eq. (2.2), i.e. the deviation from the predicted yield of synthetic i, were regressed on synthetic size to evaluate the size-performance-relation. Since tetraploid synthetics clearly outyielded diploid ones irrespective of synthetic size, separate regression analyses were performed for these two groups. SAS® procedures were used for all statistical analyses.

Results and Discussion

DM yield of tetraploid synthetics exceeded that of diploids by 8.5 % in H1 and by 12.9 % in H2 (Table 2.1). Overall, early synthetics yielded significantly higher in H1 but the influence of earliness in H2 was not significant. General breeding progress was rather slow but significant for both H1 and H2. These differences justify considering ploidy, earliness and year of polycross in the subsequent analysis of the influence of synthetic size on performance.

Diploid and tetraploid synthetics differed in the relationship between synthetic size and yield in H1 (Fig. 2.1). A significant decline of DM yield with lower number of parents was observed in the diploid, but not in the tetraploid synthetics. The decline of yield with lower numbers of parents in the diploid synthetics was best described by a highly significant ($p = 0.0006$) hyperbolic regression (Table 2.2).

Table 2.1 Influence of ploidy level, earliness and year of polycross on herbage dry matter yield of 166 perennial ryegrass synthetics

Factor level / coefficient	Herbage yield H1 (dt ha^{-1})	Herbage yield H2 (dt ha^{-1})
Ploidy		
2x (n=93)	100.3	84.4
4x (n=73)	108.8	95.3
p-Value	<0.0001	<0.0001
Time of inflorescence emergence (dt ha^{-1} d^{-1})	−0.254 (p<0.001)	0.079 (p=0.250)
Year of polycross (dt ha^{-1} yr^{-1})	0.233 (p=0.026)	0.250 (p=0.026)

p probability of error (F-test)

The analysis of DM yield data of the second harvest year (H2) showed the same trends (Fig. 2.2), but the relationship between size of the synthetic and performance was somewhat looser than that observed in H1. Irrespective of the regression model and the ploidy level, p (probability of error) values for H2 were always higher than those for H1, indicating a less good fit of the observed values to the model (Table 2.2). However, for the diploid synthetics, again, the hyperbolic regression yielded the lowest probability of error, hence the best fit to the model. Again, a large difference in behavior between diploid and tetraploid synthetics was observed, with the latter always being far away from statistical significance of the regression. The decline of performance with low numbers of parents of the diploid synthetics was consistent with the predictions of Posselt (2010), who found a marked decline below seven parents. When we analyzed subsets of our diploid data, the linear regression of H1 and H2 DM-yield on synthetic size became insignificant when synthetics with less than 7 (H1) or 6.5 (H2) parents were omitted (data not shown).

Contrary to expectations (Posselt 2010), no optimum number of parents for synthetics was apparent. This was mainly due to the lack of decline of DM yield with very high numbers of parents. Obviously, in the case of model experiments like those reported by Posselt (2010), the maximum number of parents possible is the number of genotypes actually present in the model set, which will include the poorest performing ones. In our practical plant breeding data set, even large numbers of parents still represent a restricted selected fraction of the total variation available.

The absence of a decline in performance with low numbers of parents at the tetraploid level is consistent with the data presented by Baert et al. (2007) who found a lesser importance of inbreeding depression with tetraploid than with diploid breeding material of perennial ryegrass. In tetraploids, the chance of having at least one positive (dominant) allele is

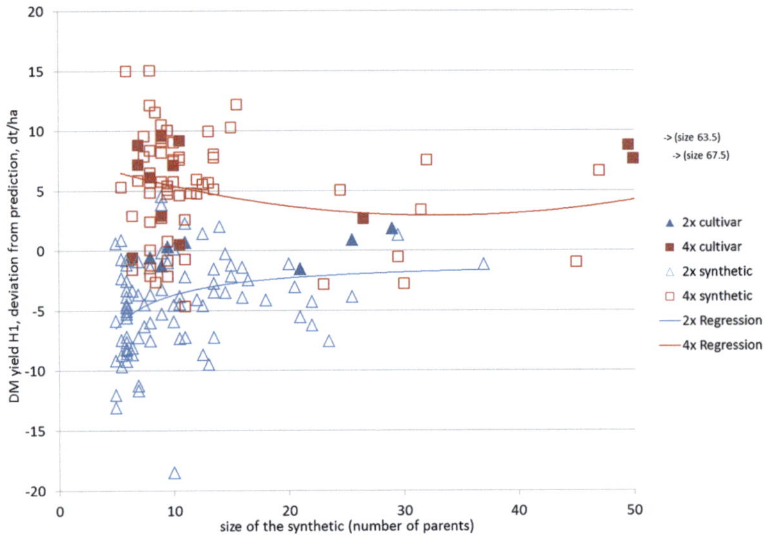

Fig. 2.1 Influence of synthetic size on total first harvest year (H1) herbage dry matter yield (deviation from prediction based on earliness and year of polycross) of 93 diploid (2x) and 73 tetraploid (4x) perennial ryegrass synthetics. Regressions according to the best solution of models summarized in Table 2.2 are shown

Table 2.2 Coefficients and statistical significance (*p* probability of error (F-test)) for different regression models describing the relationship between synthetic size (*n*) and total herbage dry matter yield (*Y*) of 93 diploid and 73 tetraploid perennial ryegrass synthetics

		Diploid (2x) synthetics				Tetraploid (4x) synthetics			
Model	Yr	a	b	c	p	a	b	c	p
Linear: Y = a + b × n	H1	−6.24	.189		.0029	5.58	−.019		.672
	H2	−6.46	.189		.0145	5.52	.006		.897
Quadratic: Y = a + b × n + c × n²	H1	−7.53	.405	−.007	.0077	8.04	−.310	.0047	.166
	H2	−8.40	−.010	.512	.0257	6.49	−.109	.0018	.755
Hyperbolic: Y = a + b × (1/n)	H1	−.88	−28.1		.0006	4.06	12.7		.370
	H2	−.76	−31.1		.0018	4.86	7.44		.590
Logarithmic Y = a + b × log(n)	H1	−10.19	2.67		.0011	7.21	−.781		.418
	H2	−10.75	2.82		.0045	6.00	−.167		.859

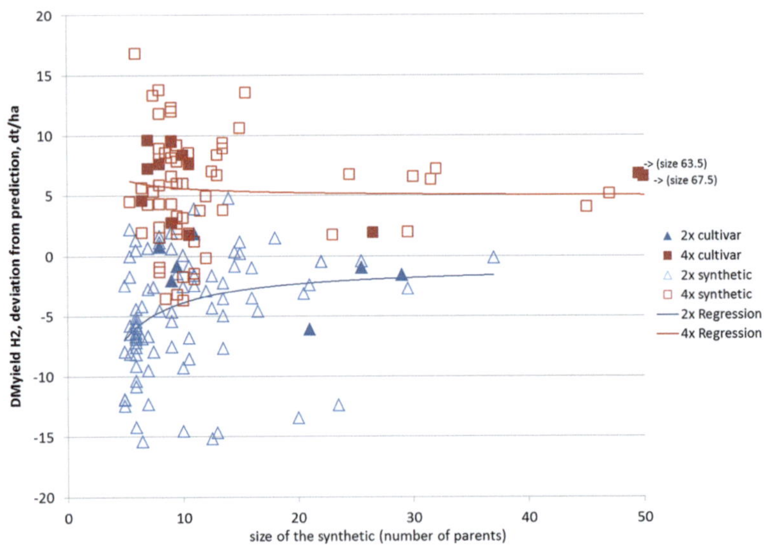

Fig. 2.2 Influence of synthetic size on total second harvest year (H2) herbage dry matter yield (deviation from prediction based on earliness and year of polycross) of 93 diploid (2x) and 73 tetraploid (4x) perennial ryegrass synthetics. Lines represent the hyperpolic regressions that showed best fit to the data (Table 2.2)

much higher than in diploids and, under a preponderance of digenic interactions, the observed lower inbreeding depression can be expected. Apparently, tri- and tetragenic interactions, which would lead to an increased inbreeding depression in tetraploids, are of low importance. This is in line with Gallais (2003), who reports a lower importance of tri- and tetragenic interactions for induced compared to natural tetraploids.

Conclusions

Based on the analysis of data from an actual breeding program, this study shows a clearly different behavior of diploid and tetraploid perennial ryegrass synthetics when considering the influence of synthetic size on performance. While the expected decline of herbage dry matter yield due to inbreeding depression was apparent when the number of diploid parents dropped below seven, no such decline was observed with tetraploid material. Indeed, selection intensity in tetraploid perennial ryegrass can be increased, at least down to five parents, without hampering dry matter yield of synthetics.

References

Baert J, Ghesquiere A, Muylle H (2007) Comparison between two breeding methods in tetraploid Lolium perenne: polycross vs. F2. In: Lübberstedt T, Studer B, Graugaard S (eds) Proceedings of the 26th Meeting of the Fodder Crops and Amenity Grasses Section of EUCARPIA, Perugia, Italy, 3–7 September 2006, Perugia, pp 100–103

Gallais A (2003) Quantitative genetics and breeding methods in autopolyploid plants. INRA, Paris

Ghesquiere A, Baert J (2007) Comparison between two breeding methods in perennial ryegrass: polycross vs. F2. In: Lübberstedt T, Studer B, Graugaard S (eds) Proceedings XXVIIth EUCARPIA Symposium on Improvement of Fodder Crops and Amenity Grasses, Copenhagen (Denmark), 19th to 23th August 2007, Copenhagen, pp 153–155

Humphreys M, Feuerstein U, Vandewalle M, Baert J (2010) Ryegrasses. In: Boller B, Posselt UK, Veronesi F (eds) Fodder Crops and Amenity Grasses. Handbook of Plant Breeding, vol 5. Springer Science+Business Media, New York, pp 211–260

Kölliker R, Boller B, Widmer F (2005) Marker assisted polycross breeding to increase diversity and yield in perennial ryegrass (*Lolium perenne* L.). Euphytica 146: 55–65

Posselt UK (2010) Breeding methods in cross-pollinated species. In: Boller B, Posselt UK, Veronesi F (eds) Fodder Crops and Amenity Grasses. Handbook of Plant Breeding, vol 5. Springer Science+Business Media, New York, pp 39–87

3

Phenotyping Genetic Diversity of Perennial Ryegrass Ecotypes (*Lolium perenne* L.)

A. Bothe, S. Nehrlich, E. Willner, and K.J. Dehmer

Abstract

Perennial ryegrass (*Lolium perenne* L.) is the most valued temperate grass species in Europe. Between 2002 and 2010, a total of 352 *L. perenne* accessions originating from four European countries (Bulgaria, Croatia, Ireland and Spain) were evaluated in four phenotyping experiments at the Malchow experimental station of the Leibniz Institute of Plant Genetics and Crop Plant Research (IPK). Ten plant parameters were visually scored, including development before and after winter, spring growth, plant biomass and incidence of crown rust and disease symptoms (leaf spot symptoms). Accessions from Croatia, Ireland and Spain were characterized by higher plant biomass in the 2nd scoring year than Bulgarian accessions. Considering the incidence of rust and diseases in perennial ryegrass accessions, the present study showed that the variation for scored crown rust and disease symptoms was higher in Irish and Bulgarian accessions. Correspondence analysis (CA) was carried out to examine the relationship between scored traits and 352 *L. perenne* accessions. The results presented in this study provide a brief description of *L. perenne* accessions maintained at the Satellite Collections North of the IPK Genebank.

Keywords

Perennial ryegrass accessions • Geographic origin • Genebank • Phenotyping • Correspondence analysis (CA)

Introduction

Perennial ryegrass (Lolium perenne) is one of the most valued temperate grass species in Europe. Especially its high productivity, nutritive value, digestibility and grazing tolerance makes it a valuable source for feeding livestock

A. Bothe (✉) • S. Nehrlich • E. Willner
K.J. Dehmer
Leibniz Institute of Plant Genetics and Crop Plant Research (IPK), Genebank, Satellite Collections North, Malchow/Gross Luesewitz, Sanitz, Germany
e-mail: bothe@ipk-gatersleben.de

I. Roldán-Ruiz et al. (eds.), *Breeding in a World of Scarcity: Proceedings of the 2015 Meeting of the Section "Forage Crops and Amenity Grasses" of Eucarpia*,
DOI 10.1007/978-3-319-28932-8_3

(Cunningham et al. 1994; Bolaric et al. 2005). With regard to the production demand for feeding livestock and forecast impacts of climate change on agricultural systems, intensive research and adapted breeding strategies are required to breed perennial ryegrass cultivars with an increased total and seasonal dry matter yield, as well as an improved tolerance to biotic and abiotic stress factors. Because progress in developing new forage cultivars depends on the creation of new allele combinations and the selection of desirable genotypes, perennial ryegrass ecotypes represent a vital source of genes for crop improvement and can make a valuable contribution to increasing the genetic variation for traits associated with high dry matter yield (Bolaric et al. 2005). Presently, the efficient and successful use of perennial ryegrass accessions from genetic resources collections in breeding programs relies on the availability of evaluation and characterization data. In light of these considerations, the aim of this study was primarily to explore the genetic diversity of a diverse set of perennial ryegrass accessions for plant productivity and associated traits.

Materials and Methods

Plant Material and Experimental Setup

To capture a representative range of genetic variability within European perennial ryegrass accessions, 352 diverse *L. perenne* accessions were evaluated. The accessions originated from Bulgaria (1998), Croatia (1996/97), Ireland (2002) and Spain (1999). The accessions were evaluated in four field experiments conducted between 2002 and 2010 at the Malchow/Poel station of research group Satellite Collections North of the Leibniz Institute of Plant Genetics and Crop Plant Research (IPK) in Northern Germany (Longitude 11°28′26″ E, Latitude 53°59′40″ N, 10 m ASL, mean annual precipitation 562 mm, annual temperature 8.9 °C, sandy loam soil). Information concerning the origin of 352 *L. perenne* accessions and the timing of the phenotyping experiments is given in Table 3.1.

Table 3.1 List of accessions which were collected in four European countries during 1996 and 2002 and used for phenotyping experiments at the Malchow/Poel station of the IPK

Phenotyping experiment	Country of origin	No. of accessions	Evaluation period (calendar year)
I	Bulgaria	62	2002–2004
II	Croatia	46	2003–2005
III	Spain	91	2004–2006
IV	Ireland	153	2008–2010

Briefly, accessions and eight standard varieties (Barclay, Fennema, Juwel, Sambin, Gladio, Respect, Cancan and Aberelf) were characterized over a 3 year period in four replicates of ten plants, which were grown as rows (2 m long and 50 cm apart). In each experiment, accessions and standard varieties (checks) were arranged in complete randomized blocks. Fertilization and cutting were applied according to common agricultural practice.

Scorings

Ten parameters were investigated. Principal growth stages were scored using the extended BBCH scale of Hess et al. (1997). Over a 3 year period the following traits were visually scored on an ordinal scale from 1 to 9, with scorings of 9 indicating a superior performance: growth before winter (GBW), growth after winter (GAW), spring growth (GIS), above plant biomass (PBM) and flowering tendency (FLW). The term "flowering tendency" is a comprehensive term, including the flowering tendency of accessions without vernalisation in the sowing year and the re-heading of accessions after cuts in the years after sowing. Under natural climatic conditions, rust and disease incidence were assessed by using a progressive scale of 1–9, with: 1 = more than 75 % of the foliage was covered with rust or disease symptoms and 9 = no rust or disease symptoms. The rust evaluation was based on the percentage of crown rust pustules on the foliage. The evaluation of disease symptoms represents a percentage estimate of leaf area infected by

Helminthosporium spp., *Drechslera* spp. and/or *Rhynchosporium orthosporum*.

Statistical Analysis

Statistical analysis of examined traits was performed using the statistical software package R version 3.1.2 and the software package Paleontological Statistics (PAST). Since climatic conditions blur differences between evaluated accessions, standard varieties (checks) were included in each phenotyping experiment and replicated four times. The current study considered the four phenotyping experiments as four environments with varying climatic conditions. Data of the eight standard varieties were analyzed by using the restricted maximum likelihood (REML) approach for fitting a linear mixed model with checks (genotypes) having a fixed effect. The phenotyping environments as well as replications (blocks) within each environment were considered as random effects. The 'lme4' package for R was applied to estimate the genotypic variance of checks and the residual variance component which originated from year effects of different phenotyping environments. According to the mixed model, significant differences between phenotyping environments were detected (data not shown). Hence, accessions were compared based on adjusted accessions means (lsmeans). In general, adjusted accession means were computed by subtracting mean values of each accession in the particular phenotyping experiment (environment) by its corresponding year effect. To compute the year effect, the following equation was used:

$$C_i = \overline{X_i} - \sum_{i=1}^{n} \frac{\overline{x_i}}{n}$$

Where C_i is the effect of the calendar year, $\overline{X_i}$ is the mean of standard checks in the *i*-th year and n is the number of phenotyping environments.

Based on adjusted means, relationships between scored traits were investigated by correspondence analysis (CA).

Results and Discussion

Multiple Mean Comparisons of Accessions

Perennial ryegrass accessions collected from different European countries and check varieties were adjusted for the year effect, allowing 352 perennial ryegrass accessions to be compared to the eight "check" varieties (Tables 3.2 and 3.3).

Accessions collected from Spain achieved high means for growth before winter scored in the 1st year (GBW1), growth in spring assessed in the 2nd scoring year (GIS2), flowering tendency scored in the 2nd year (FLW2), plant biomass scored in the 2nd year (PBM2), absence of rust symptoms scored in the 2nd year (RUST2) and growth before winter scored in the 2nd scoring year (GBW2). Considering the development of Spanish ryegrass accessions (GIS2) and their flowering tendency (FLW2) in the 2nd scoring year, adjusted accession means exceeded average means of check varieties. Accessions collected in Croatia showed high scoring means for growth before winter scored in the 1st year (GBW1), growth in spring in the 2nd scoring year (GIS2), plant biomass in 2nd scoring year (PBM2) and growth before winter assessed in the 2nd scoring year (GBW2). When comparing the means of accessions originating from Croatia with the means of the check varieties, in this study the Croatian accessions showed a similar performance level to standard varieties. The same observation was made when comparing means of Irish accessions with the corresponding means of check varieties. In Irish accessions the genetic variability (CV) for rust and disease scorings was higher than 10 % indicating the presence of a reasonable genetic variance for these traits. Accessions collected in Bulgaria were characterized by higher crown rust susceptibility (RUST2). At the same time, these accessions showed a low susceptibility to diseases (DIS2). Scores of the plant biomass in the 2nd growing year (PBM2) were the lowest for Bulgarian accessions, with PBM2 scores ranging from 1 to 6. Nevertheless, with respect

Table 3.2 Adjusted mean, minimum, maximum values and coefficient of variation for ten visually scored traits of 352 *L. perenne* accessions originating in Bulgaria, Spain, Croatia and Ireland

Year of scoring	Trait	Bulgaria[a]			Spain[b]			Croatia[c]			Ireland[d]		
		Mean	Min/Max	CV (%)	Mean	Min/Max	CV (%)	Mean	Min/Max	CV (%)	Mean	Min/Max	CV (%)
1	FLW	1.2	1.1/2.9	27.0	1.4	1.1/4.9	37.3	1.4	1.1/5.9	53.8	1.1	0.6/3.9	56.9
1	GBW	6.5	5.3/8.1	10.4	7.7	3.1/7.8	9.1	7.9	5.4/9.1	12.1	7.2	5.8/8.0	5.4
2	GIS	7.2	2.8/9.0	13.0	8.0	2.8/9.0	10.7	7.0	5.3/9.0	10.6	6.1	3.5/7.5	13.7
2	FLW	2.9	1.9/5.1	26.7	4.0	2.5/5.7	16.9	3.5	2.4/5.4	28.4	1.6	0.5/3.0	34.4
2	PBM	3.6	0.7/6.3	30.1	5.9	3.0/6.7	10.8	5.7	5.1/6.6	7.7	5.0	3.5/6.2	8.8
2	RUST	1.5	1.3/2.1	16.4	3.7	3.2/4.4	8.2	3.4	2.7/4.4	11.3	3.5	2.1/5.1	12.6
2	DIS	5.0	3.1/7.7	18.1	3.2	2.5/3.8	8.2	3.1	2.6/3.9	8.6	2.6	1.8/4.6	19.4
2	GBW	5.9	1.8/8.0	18.4	6.5	4.7/8.2	9.0	6.6	5.6/7.1	5.8	6.7	5.1/7.8	8.5
3	GAW	3.3	2.3/4.1	12.5	4.0	3.3/4.6	9.2	4.5	3.6/5.3	9.6	5.4	3.2/7.2	15.2
3	GIS	5.5	4.3/7.3	11.6	4.6	2.5/7.0	22.9	5.6	4.7/6.7	7.6	6.0	3.5/8.0	13.9

CV coefficient of variation, *GBW* growth before winter, *GAW* growth after winter, *GIS* growth in spring, *FLW* flowering tendency, *PBM* biomass, RUST absence of rust symptoms, DIS absence of disease symptoms

[a]Scoring of 62 accessions from Bulgaria between 2002 and 2004

[b]Scoring of 91 accessions from Spain between 2004 and 2006

[c]Scoring of 46 accessions from Croatia between 2003 and 2005

[d]Scoring of 153 accessions from Ireland between 2008 and 2010

Table 3.3 Adjusted mean, minimum, maximum values and coefficient of variation for ten visually scored traits of eight check varieties which were included in phenotyping experiments

Year of scoring	Trait	Check varieties scored with accessions from Bulgaria[a]			Check varieties scored with accessions from Spain[b]			Check varieties scored with accessions from Croatia[c]			Check varieties scored with accessions from Ireland[d]		
		Mean	Min/Max	CV (%)	Mean	Min/Max	CV (%)	Mean	Min/Max	CV (%)	Mean	Min/Max	CV (%)
1	FLW	1.1	1.1/1.1	0	1.1	1.1/1.1	0	1.1	1.1/1.1	0	1.1	0.6/1.9	43.2
1	GBW	7.8	7.3/8.3	4.8	7.8	7.6/7.8	1.1	7.8	5.9/9.6	18.0	7.8	7.0/8.3	5.7
2	GIS	7.0	4.4/8.2	17.3	7.0	4.9/8.2	14.8	7.0	5.8/8.1	12.1	7.0	5.1/8.1	16.0
2	FLW	2.9	2.3/3.5	18.3	2.9	2.2/4.2	21.9	2.9	2.4/3.6	14.4	2.9	2.3/3.3	11.2
2	PBM	5.8	3.7/7.5	18.7	5.8	4.0/6.5	14.9	5.8	5.1/7.1	10.7	5.8	3.7/7.7	20.4
2	RUST	3.5	2.5/5.1	23.3	3.5	3.4/3.7	3.7	3.5	3.2/4.2	10.7	3.5	3.1/3.8	7.6
2	DIS	3.0	2.3/3.7	16.3	3.0	2.8/3.3	5.4	3.0	2.9/3.1	4.3	3.0	2.6/3.3	8.9
2	GBW	6.7	6.0/7.8	9.1	6.7	6.0/7.5	8.5	6.7	5.6/7.4	8.2	6.7	5.6/7.3	9.9
3	GAW	4.4	2.7/5.3	18.5	4.4	4.1/4.8	5.2	4.4	3.6/4.8	8.5	4.4	2.7/5.5	19.2
3	GIS	5.9	3.3/6.7	20.1	5.9	5.3/6.3	6.8	5.9	3.7/6.7	18.0	5.9	3.2/7.0	22.0

CV coefficient of variation, *GBW* growth before winter, *GAW* growth after winter, *GIS* growth in spring, *FLW* flowering tendency, *PBM* biomass, RUST absence of rust symptoms, DIS absence of disease symptom

[a]Scoring of eight check varieties between 2002 and 2004

[b]Scoring of eight check varieties between 2004 and 2006

[c]Scoring of eight check varieties between 2003 and 2005

[d]Scoring of eight check varieties between 2008 and 2010

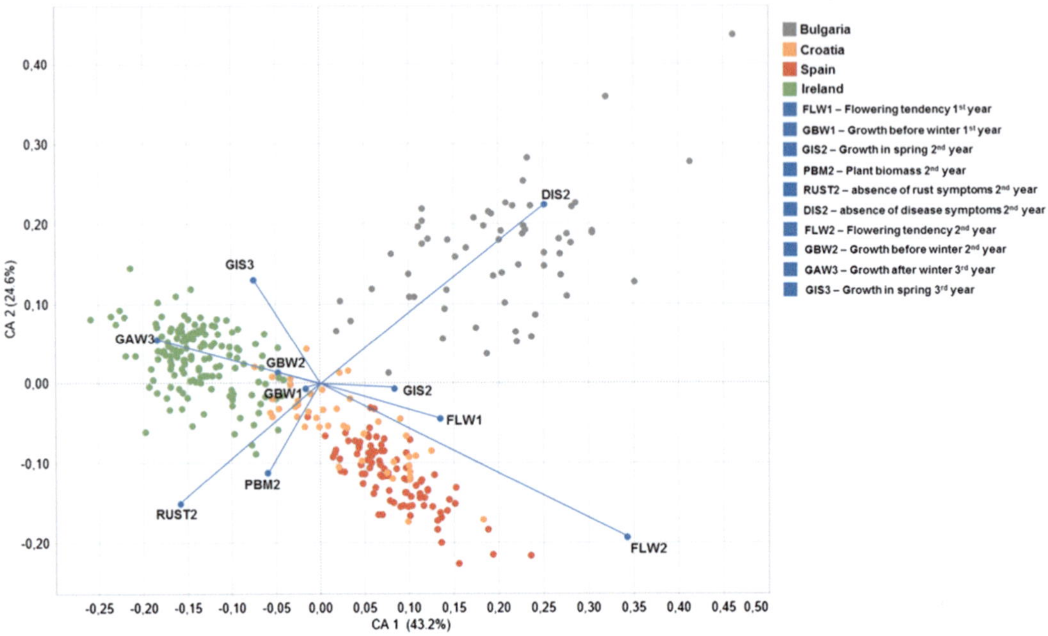

Fig. 3.1 Correspondence analysis (CA) scatter plot of *Lolium perenne* accessions collected from Bulgaria, Croatia, Spain and Ireland and ten traits

to calculated coefficient of variation (CV), a large variation for flowering tendency and plant biomass production was found in Bulgarian accessions. Overall, the accessions investigated in this study displayed a wide variability for flowering tendency in the 1st and 2nd scoring year (FLW1 and FLW2). Accessions collected in Ireland and Bulgaria showed a higher variability (CV) for crown rust and disease symptoms indicating the higher possibilities for selection under the environmental conditions of the study site. Accessions originating from Spain and Croatia were characterized by higher plant biomasses similar to the examined check varieties. Nevertheless, all ecotypes were moderately to highly susceptible to rust and diseases.

Correspondence Analysis (CA)

Relationships between visually scored traits and 352 *Lolium perenne* accessions were investigated by correspondence analysis (CA). Traits included in the CA were accession means of ten evaluated parameters adjusted for the year effect. As displayed in Fig. 3.1, the first and second component (CA1 and CA2) explained 43.2 % and 24.6 % of the total variation, respectively. The most important traits contributing to the first component were flowering tendency in the 1st scoring year (FLW1), flowering tendency in the 2nd scoring year (FLW2) and absence of disease symptoms in the 2nd scoring year (DIS2). The second component was mainly explained by plant biomass (PBM2) and absence of rust symptoms (RUST2) in the 2nd scoring year. Accessions collected in Bulgaria were mainly associated with a higher insensitivity to diseases, except for crown rust. However, accessions collected in Spain and Croatia were preliminary associated with a higher flowering tendency in the 1st and 2nd scoring year (FLW1 and FLW2) as well as a favorable growth in spring in the 2nd year (GIS). Finally, accessions from Ireland were preliminary related to a better development after winter (GAW3) and a higher spring growth (GIS3) in the 3rd scoring year. Accessions collected from Bulgaria form a separate cluster. Additionally, for accessions collected in Ireland,

Croatia and Spain a partial overlap was observed indicating on average a similar performance level of these accessions.

Conclusion

The precise characterisation and evaluation of plant genetic resources stored in the IPK Genebank provide users with valuable information concerning desirable variation for specific traits (e.g. rust and disease susceptibility as well as plant productivity). The present study showed that perennial ryegrass accessions originating from Bulgaria, Croatia, Ireland and Spain differ in their development in spring, flowering tendency, incidence of rust and disease symptoms and plant productivity. In this respect, data gathered in the phenotyping experiments presented here provide additional insight into the wide natural diversity of perennial ryegrass ecotypes.

References

Bolaric S, Barth S, Melchinger AE, Posselt UK (2005) Molecular genetic diversity within and among German ecotypes in comparison to European perennial ryegrass cultivars. Plant Breed 124:257–262. doi: 10.1111/j.1439-0523.2005.01108.x

Cunningham PJ, Blumenthal MJ, Anderson MW, et al (1994) Perennial ryegrass improvement in Australia. New Zeal J Agric Res 37:295–310. doi: 10.1080/00288233.1994.9513068.

Hess M, Barralis G, Bleiholder H (1997) Use of the extended BBCH scale—general for the descriptions of the growth stages of mono; and dicotyledonous weed species.

4 An Assessment of Phenotypic Diversity Within a Collection of Natural *Lolium perenne* Ecotypes and Cultivars from Across Europe

D. Gasior, J. Harper, M. Humphreys, A. Lovatt, A. Armstead, and R. Sanderson

Abstract

Ninety-one *Lolium perenne* ecotypes, collected from across Europe, and six varieties were grown in a randomized field trial over 2 years. The plants were analyzed for a range of morphological characters. The obtained results indicate that high variability actually exists among and within natural perennial ryegrass ecotypes and the cultivars and the differences observed between populations were based more on genetics than geography.

Keywords

Perennial ryegrass • Geographic origin • Principal components analysis

Introduction

Grasses are important in Europe both for their value as a healthy and sustainable fodder for livestock and, especially in semi-natural grasslands. They also provide environmental, social and cultural ecosystem services. These cultural benefits are mainly due to the persistency of grasses and the large amounts of grassland (European Union: >40 % of total agricultural land; UK >70 %).

Within the agricultural grass species, perennial ryegrass (*Lolium perenne* L.) is considered to be the optimal major source of forage for temperate regions. It is widely cultivated due to its high yield, good digestibility, and tolerance to grazing pressures; it is also often chosen for amenity purposes (Humphreys 2010). Perennial ryegrass is a long-lived perennial grass native to Europe, North Africa and temperate Asia but due to human activity it has also been widely distributed into United States, Japan, Australia and New Zealand (Thorogood 2003). Intensive ryegrass breeding for animal feed has concentrated mainly on factors affecting desirable agronomic characteristics, such as selection based on the time of flowering. That tight breeding focus may limit

D. Gasior (✉) • J. Harper • M. Humphreys
A. Lovatt • A. Armstead • R. Sanderson
IBERS, Aberystwyth University, Gogerddan, Aberystwyth, Ceredigion SY23 3EE, UK
e-mail: dkg@aber.ac.uk

I. Roldán-Ruiz et al. (eds.), *Breeding in a World of Scarcity: Proceedings of the 2015 Meeting of the Section "Forage Crops and Amenity Grasses" of Eucarpia*,
DOI 10.1007/978-3-319-28932-8_4

the available genetic diversity to be utilised within their breeding programs (Hu et al. 2011; Statkevičiūtė et al. 2012). The range of selection criteria is now increasing in response to political and public demands. The effects of a changing climate are increasing the pressures on grass breeders as environmentally extreme conditions are becoming more commonplace. These environmental changes have also shown existing varieties to be susceptible to biotic and abiotic stresses. Therefore, breeders must consider creating new gene pools by intercrossing divergent genetic resources (Humphreys et al. 2003). The success of breeding strategies depends on the quality of the basic material. Using wild ecotypes collected in their natural habitats where there has been selective pressure for particular traits can provide valuable new germplasm for breeding purposes. This would minimize genetic erosion and enlarge the gene pool available for selection. In general, natural ecotypes are not as productive as commercial varieties but natural ecotypes are a resource of novel genetic variation that is a potential source of genes for improved resistance and or durability to a range of biotic and abiotic stresses (Lemažienė and Kanapeckas 2008). The aim of this study is to assess the phenotypic variation within natural perennial ryegrass ecotypes collected from across Europe and to evaluate the diversity of their morphological traits.

Materials and Methods

Plant Material

Based on a range of eco-geographic habitats and in relation to latitude and longitude, 91 natural or semi-natural ecotypes and six perennial ryegrass cultivars included as controls were selected from across Europe. The perennial ryegrass cultivars were selected on the basis of heading date and divided into three groups: (1) early heading: Moy, Irleand, Liprior, England, (2) intermediate heading: AberDart, Wales, Premium, Netherlands, (3) late heading: Melle, Belgium, AberAvon, Wales. Ecotypes were collected from 21 countries: Austria, Belgium, Bulgaria, Czech Republic, England, France, Greece, Hungary, Italy, Ireland, Netherlands, Norway, Poland, Portugal, Romania, Scotland, Slovakia, Spain, Switzerland, Turkey, and Wales. These represent a number of different site physiologies (grasslands, wasteland, forest, orchard, ditch, amenity, agricultural abandoned or agricultural mixed). All seed was obtained from the IBERS Lolium gene bank. The accessions were sown in a greenhouse in summer and transplanted to the field in September 2012. The field trial consisted of three completely randomized blocks of 97 accessions represented by 10 genotypes within each accession for a total of 970 plants per block. Genotypes of each ecotype were planted using a spacing of 50 cm. The plants were allowed to establish during the autumn/winter of 2012 and scoring of traits began in the spring of 2013. The experimental block was maintained for 2 years with weeding as required (inter-row cultivation by mechanical rotary cultivators and hand weeding) and applications of Growhow Multicut Sulphur fertilizer 23-4-13-7 (N-P-K-S) in spring and post-harvest. From spring 2013 the following parameters were measured on each plant: winter survival, vegetative growth habit before vernalization (growth habit) and after vernalization, leaf width at vegetative state, heading date, plant width after vernalization, biomass score, fresh weight, dry weight, phenotypic variability within ecotype, regrowth after harvesting as scored 5 weeks after cutting, secondary heading scored 7 weeks after first cut, autumn growth scored 4 weeks after second cut, disease score (crown rust, mildew and stem rust), digestible dry matter, water soluble carbohydrate, plant height – natural height at emergence, flag leaf length, width, head length, number of spikelet per head and number of flowering heads. After measuring, all plants were cut back.

Statistical Analyses

Coefficient of variation (CV), standard error deviation (SD) and principal component analysis (PCA) of agro-biological traits were calculated using the GenStat software package (GenStat 16th edition SP1, 64 bit developed by VSN Internationa Ltd.) Twelve parameters were selected for PCA: (1) vegetative growth habit before vernalization, (2) leaf width at vegetative state, (3) heading date calculated as number of days after 1st April 2014, (4) vegetative growth after vernalization, (5) plant width after vernalization, (6) dry weight after first cut, (7) plant height, (8) flag leaf length, (9) flag leaf width, (10) head length, (11) number of spikelet per head and (12) number of flowering heads. These parameters were chosen based on previous studies of the genetic diversity of phenotypic traits within *Lolium perenne* (B.P. Loos 1994; Bennet et al. 2000; Statkevičiūtė et al. 2012; Kemešytė et al. 2014).

Results and Discussion

The results obtained for the traits investigated indicate that high variability exists among and within natural perennial ryegrass ecotypes (CV 2.7–61 %) and the cultivars (CV 2.8–93.6 %). The highest variability between ecotypes and cultivars was obtained for dry weight, disease score (mildew, crown rust and stem rust) and phenotypic variability within ecotypes (Table 4.1). This high within-cultivar variation for some traits may be due to selecting cultivars from different geographic origins and breeding programs. PCA was carried out based on morphological characteristics with 12 characters entered to show the differentiation between ecotypes and cultivars. The first three principal components defined 74.29 % of total variation among the 91 studied ecotypes (Table 4.2). PC1 accounted for 35.43 % of total variation, which according to the eigenvector value was best explained by plant height, head length, flag leaf length and width, number of spikelets per head and dry weight. PC2 accounted for 28.45 % of total variation; it was primarily related with heading date. PC3 accounted for 10.41 % of total variation and was explained by vegetative growth habit before and after vernalization. The first three principal components were found to explain 89.73 % of the variation within cultivars (Table 4.2). The most important characters in PC1 (53.42 % of total variation) were heading date, flag leaf length and width, head length and number of spikelets per head but they all showed a negative projection on the first component. PC2 accounted for 26.06 % and was related with plant height and plant width after vernalization. PC3 accounted for 10.25 % was mainly

Table 4.1 Average value and variation of agro-biological traits of *Lolium perenne* natural ecotypes and cultivars

	Ecotypes		Cultivars	
Characters	X±SD	CV %	X±SD	CV %
Winter survival %	97.55±3.78	0	100.00±0.00	0
Heading date	50.92±0.80	2.7	55.73±0.90	2.8
Dry weight	158.30±13.39	14.6	205.73±10.25	8.6
Biomass score	5.57±0.37	11.5	6.33±0.39	10.8
Leaf width at vegetative state	4.10±0.38	15.9	5.19±0.41	13.6
Vegetative growth after vernalization	6.46±0.41	11.1	6.18±0.38	10.6
Plant width after vernalization	5.81±0.42	12.5	6.81±0.46	11.6
Mildew score	3.09±1.19	61.0	3.00±0.88	50.6
Crown rust score	3.70±1.15	53.7	2.61±1.41	93.2
Stem rust score	4.71±0.65	23.8	3.53±0.76	37.1
Phenotypic variability within ecotype	3.19±1.00	54.4	3.22±0.68	36.6
Secondary heading	3.90±0.50	22.1	2.62±0.47	31.2
Regrowth after harvesting	4.61±0.46	17.2	5.42±0.28	9.0
Autumn growth	4.91±0.36	12.6	6.04±0.33	9.6
DDM	59.50±0.54	1.6	62.24±0.59	1.7
WSC	17.95±0.58	5.6	21.49±0.69	5.6

X average, *Sd* Standard deviation, *CV* coefficient of variation

Table 4.2 Basic statistics of Principal Component Analysis (PC1, PC2, PC3) for morphological traits of *Lolium perenne* natural ecotypes and cultivars

	Ecotypes			Cultivars		
Traits	PC1	PC2	PC3	PC1	PC2	PC3
Growth habit before vernalization	0.01	0.29	0.63	0.27	−0.22	0.27
Leaf width at vegetative phase	0.23	−0.39	−0.08	0.11	0.38	−0.48
Heading date	0.13	0.39	−0.43	−0.38	−0.10	0.01
Growth after vernalization	0.05	0.41	0.36	0.14	−0.48	0.08
Plant width after vernalization	0.22	−0.41	0.03	0.02	0.47	0.23
Dry weight	0.39	−0.14	0.03	0.25	0.07	0.65
Plant height	0.41	−0.06	0.09	−0.18	0.49	0.13
Flag leaf length	0.36	0.24	−0.12	−0.38	−0.04	0.16
Flag leaf width	0.33	0.05	−0.0	−0.376	−0.08	0.14
Head length	0.41	0.16	0.01	−0.36	0.02	0.28
Number of spikelets per head	0.30	0.22	−0.06	−0.38	0.03	0.03
Number of flowering heads	0.16	−0.30	0.47	0.25	0.30	0.21
	Ecotype components			Cultivar components		
Eigenvalue	4.25	3.41	1.24	6.41	3.12	1.23
% of variation explained	35.43	28.45	10.41	53.42	26.06	10.25
Cumulative variance	35.43	63.88	74.29	53.42	79.48	89.73

explained by dry weight. Correlations calculated between all the morphological traits were generally low to moderate (0.54–0.72). This is similar to results observed previously in a Dutch *L. perenne* population (Loos 1994). In this study the highest correlation was found between plant width after vernalization and leaf width at vegetative state (0.72), plant width after vernalization and biomass score (0.74), and plant width after vernalization and autumn growth (0.63). Additionally plant height was correlated with flag leaf length (0.54) and head length (0.71). Flag leaf length was correlated with flag leaf width (0.65) and head length (0.74).

Conclusions

A wide phenotypic diversity was observed in the natural *L. perenne* ecotypes and cultivars, but the principal component analysis showed that more cumulative variation was explained for cultivars (89.73 %) than for ecotypes (74.29 %). This may imply that the genetic diversity within cultivars is indeed lower than that found within ecotypes. In this experiment the cultivars used were produced by breeding programs from across Europe, thus is expected that the genetic diversity would be much lower within individual breeding programs. It is possible that the selection for agronomic value during cultivar development may have reduced the potential available genetic diversity. However, these results suggested that the level of diversity in commercial cultivars is still large and needs to be maintained to a certain range for adaptability to both biotic and abiotic stresses (Hu et al. 2011). The analysis showed that a large proportion of the variation was explained by a few characters (plant height, flag leaf length and width, head length, growth habit). and *L. perenne* accessions (despite their widespread distribution) demonstrated surprisingly less phenotypic variation than anticipated. However, the preliminary study used easily measurable traits selected for use only as an initial gauge of the extent of available phenotypic variation present within natural accessions of *L. perenne,* taking little account of potential differences in available adaptations to climatic stresses or diseases. As future safeguards to ensure continued *L. perenne* use within sustainable agricultural systems, crop improvement programs will benefit from further studies on the range of adaptations available within *L. perenne* to stress conditions.

References

Hu T, Huiying L, Deying L, Jianming S, Jinmin F (2011) Assessing genetic diversity of perennial ryegrass (*Lolium perenne* L.) from four continents by inter-simple sequence repeat (ISSR) markers. African Journal of Biotechnology Vol. 10(83): 19365–19374

Humphreys MW, Canter PJ, Thomas HM (2003) Advances in introgression technologies for precision breeding within the Lolium-Festuca complex. Ann. Appl. Biol. 143:1–10

Humphreys M (2010) Ryegrasses. Fodder Crops and Amenity Grasses, Handbook of Plant Breeding, Volume 5:211–260

Thorogood D (2003) Perennial ryegrass (*Lolium perenne* L.). In: Casler MD, Duncan RR (eds.) Turfgrass biology, genetics and breeding. John Wiley & Sons, Inc. Hoboken NJ: 75–105

Statkevičiūtė G, Kemešė V, Lemažienė N, Konik GS, Chomiak M (2012) Genetic diversity of perennial ryiegrass wild ecotypes based on agro-biological traits. Proc Latv Acad Sci Sect B Nat 66(4/5): 20–30

Lemažienė N, Kanapeckas J (2008) Selection efficiency of wild ecotypes for forage and turf grass breeding. Biologija 54(2): 125–128

Loos BP (1994) Morphological variation in Dutch perennial ryegrass (*Lolium perenne* L.) population in relation to environmental factors. Euphytica 74: 97–107

Bennet SJ, Hayward MD, Marshall DF (2000) Morphological differentiation in four species of the genus *Lolium*. Genetic Resources and Crop Evolution 47: 247–255

Kemešytė V, Statkevičiūtė G, Lemežienė N, Brazauskas G, Ivanciv R (2014) Morphological Variation Between Perennial Ryegrass (*Lolium perenne* L.) Natural Ecotypes. Quantitative Traits Breeding for Multifunctional Grasslands and Turf 81–84

5 Comparative Characteristics of Newly Developed Perennial Ryegrass Varieties in Bulgaria

A. Katova, J. Baert, and D. Reheul

Abstract

Three new varieties of perennial ryegrass were developed during the period 1995–2015 in the Institute of Forage Crops in Pleven, Bulgaria. Different breeding methods were applied to develop these varieties. After individual phenotypic selection, a polycross of 91 elite diploid genotypes with local origin was applied for the creation of the variety IFK Harmoniya. The varieties Tetrany and Tetramis were developed by polyploidization of local breeding populations, flowcytometric screening and phenotypic selection, followed by a polycross of 45 and 52 tetraploid elite genotypes. IFK Harmoniya was the first diploid variety registered on the Official Variety List of the Republic of Bulgaria in 2010. The two tetraploid candivars follow the national official variety testing trials for Value of Cultivation and Use (VCU) in three locations and for Distinctness, Uniformity and Stability (DUS) in one location, organized by the Executive Agency for Variety Testing, Field Inspection and Seed Control since 2014. IFK Harmoniya is an early diploid, highly productive, ecologically stable (winter hardy and drought tolerant), persistent, multifunctional perennial ryegrass variety with a dense sward, suitable for pasture, hay-pasture and amenity, in pure stands or in mixtures with white clover for forage or with red fescue for ornamental and sports fields. If Tetrany passes the VCU and DUS trials, it will be the first Bulgarian tetraploid perennial

A. Katova (✉)
Institute of Forage Crops, Pleven, Bulgaria
e-mail: katova@hotmail.com

J. Baert
Plant Science Unit, Institute for Agriculture
and Fisheries Research (ILVO), Melle, Belgium

D. Reheul
Department of Plant Production, Ghent University,
Ghent, Belgium

I. Roldán-Ruiz et al. (eds.), *Breeding in a World of Scarcity: Proceedings of the 2015 Meeting of the Section "Forage Crops and Amenity Grasses" of Eucarpia*,
DOI 10.1007/978-3-319-28932-8_5

ryegrass variety: it is an early to intermediate heading, persistent, winter hardy, drought tolerant, crown rust-resistant, highly productive and leafy candivar with a good feeding value. Tetramis is a tetraploid, very early, persistent, winter hardy, drought tolerant perennial ryegrass candivar that combines a good resistance to crown rust with a high production of forage biomass and seed.

Keywords

Perennial ryegrass • Plant breeding methods • Diploid • Tetraploid • Varieties • Breeder

Introduction

Perennial ryegrass (*Lolium perenne* L.) is one of the most commonly cultivated species in temperate grasslands. It has been subject to substantial breeding activities during the past several decades. In the 1950s and 1960s, breeding companies and public institutes in Europe, Northern America, New Zealand and Australia have performed polycrosses of selected plants of landraces or ecotypes, which are largely responsible for the modern breeding germplasm (Conaghan and Casler 2011). Perennial ryegrass is used for forage (grazing, hay, silage) or for amenity purposes (sports fields, lawns). Perennial ryegrass is preferred by farmers: it has good tillering during sward establishment, quick regrowth, excellent nitrogen assimilation, tolerance to intensive grazing and trampling or frequent cutting and higher nutritive value than other grasses (Frame 1991; Peeters 2004). Worldwide breeding has developed many ryegrass varieties with maximum yield stability and forage quality for given regional ecologies. However, foreign varieties have a low adaptation to Bulgarian agroclimatic conditions. The widespread distribution of perennial ryegrass in natural swards of Bulgaria suggests that local ecotypes are adapted to our environmental conditions. This valuable adaptive potential could be used in the development of Bulgarian varieties of perennial ryegrass (Katova 2005, 2005a). Studies on breeding value of Bulgarian ecotypes started only in 1995. The abovementioned advantages of this species are an argument to include ryegrass in Bulgarian breeding programs. Drought is an important environmental factor that limits the productivity of crops worldwide. Climate change models predict greater variability in rainfall patterns, and increased periods of summer drought will affect many regions, including temperate grasslands. These factors are reflected in the increasing emphasis on design and selection of varieties with better tolerance to prolonged periods of water deficit. Therefore, it is important to develop new forage crop cultivars that can adapt to such climatic changes. The unique agricultural conditions in Bulgaria require grass varieties that are bred to perform and persist in harsh and variable environmental conditions.

Dry matter yield improvement is achievable by developing polyploid cultivars of natural diploid forage grass species. They are characterized by high biomass production with good chemical composition (Sokolovic et al. 2010, 2011). Climatic areas such as Japan that are less favourable for the cultivation of perennial ryegrass have shown good performance of tetraploid varieties (Cai et al. 2014).

Varieties combining improved quality with excellent agronomic performance are our focus. Because the breeding process of perennial ryegrass is significantly longer than other arable plants, work of more than a decade usually precedes cultivar release (Humphreys et al. 2010).

Materials and Methods

Bulgaria is situated in the Mediterranean gene center where the species originated. Genotypes in natural populations combine resistance to unfavourable environmental conditions with high yield potential. A joint expedition between scientists from Bulgaria (IFC – Pleven, Institute of Plant Genetics Resources – Sadovo) and from Japan was held during 1995 to collect seeds of forage species from dry regions with extreme temperatures (i.e., low winter and high summer temperatures). During the period 1995–2010, 3300 genotypes from 14 Bulgarian perennial ryegrass populations and 31 European varieties were phenotypically characterized and several cycles of selection were conducted at IFC-Pleven (Katova et al. 2008; Katova 2005a). The region of the study (Pleven) combines climatic peculiarities (drought and cold) from north-central Bulgaria with a temperate continental climate area. Average annual precipitation was 609 mm during the period 1995–2010 and average annual temperature 12.3°C. During the last decade a maximum temperature for July and August of 46 °C and a minimum temperature for January and February of −28 °C were recorded. The soil type is leached chernozem, poor to medium for humus, with slightly acid to neutral pH. These soil climatic conditions are the background for selection and adaptation of genetic resources to drought and cold.

After 3 years of observation we selected 91 genotypes from 10 Bulgarian populations. A polycross of these elite genotypes is the origin of the diploid variety IFK Harmoniya.

The breeding program for tetraploids began in 2000. Two thousand seedlings of each of two Bulgarian populations were treated with colchicine 3–4 days after germination at ILVO in Melle, Belgium. Seedlings were planted and grown in trays with soil in the greenhouse for 2 months. The surviving plants were screened for tetraploidy using a flow cytometer. In 2001, about 50 tetraploid plants of the two origins were planted in the field in two polycrosses under isolation of winter rye. During 2002–2006 we harvested C1-, C2-, C3-seeds and each generation was checked for ploidy level. Five hundred plants of each C3-populations were scored from 2007 till 2009. Finally, 45 and 52 genotypes were selected in each population. Polycrosses of these selected clones are the origin of the two tetraploid candidate varieties Tetrany and Tetramis.

Variety description was made according CPVO (2011) and UPOV (2006) technical guidelines.

Results and Discussion

IFK Harmoniya is the first Bulgarian diploid variety of perennial ryegrass. IFK Harmoniya's plants are perennial, with a well-developed tuft root system, intensive tillering and semi-erect to erect growth habit (Table 5.1). IFK Harmoniya forms many and very leafy generative and vegetative stems. Generative shoots are fine (1.8 mm stem width) with 4–5 leaves, have a smooth surface and are comparatively resistant to lodging. The plant leafiness at heading is 55–65 %, determined by manual sorting on a dry weight basis. Plants are medium in height at the beginning of heading (30–37 cm) and in full heading (47–57 cm). Leaves are linear, flat or slightly curved, dark green, smooth and bright at the lower surface. The tongue is membranous, medium in height and with a sharp edge. Auricles are short and do not blanket the stem. The length (15–17 cm) and width of the leaves (4–5 mm) are medium in size. The spikes have a length of 17–19 cm and the number of spikelets per spike are 21–23. Seeds are elongate, boat–like. Plants start growing early in spring when soil temperature is 3–4°C and continue vegetation till the first frost. Pasture stage (maturity) is reached around 15–20 April (height of the stand 20–25 cm). The variety persists after many cuttings and grazing. IFK Harmoniya is early heading: between 3 and 10 May. This variety differs from introduced foreign varieties with higher persistency, winter hardiness and tolerance to drought and high summer temperatures. Table 5.2 shows the agricultural characteristics. According to results of competitive testing trials (2001–2003) in Pleven, in very dry summers the variety IFK Harmoniya exceeds the variety Vigor by 18 % for dry matter yield (Katova 2011). The variety forms a dense sward and is multifunctional, suitable for pasture, hay-pasture and amenity, in pure stands or in mixtures with white clover for forage or with red fescue for ornamental and sports fields.

Table 5.1 Comparative results for DUS characters of new perennial ryegrass varieties by the breeder

Character	IFK Harmoniya	Tetrany	Tetramis
Plant: ploidy level	Diploid	Tetraploid	Tetraploid
Plant: growth habit before vernalization	1–3: semi-erect-to erect	3–5: semi-erect-to medium	1: erect
Leaf: length	5: medium	9: very long	7: long
Leaf: width	5: medium	7: wide	7: wide
Leaf: intensity of green colour	7: dark green	9: very dark green	5: green
Plant: width after vernalization	5: medium	5: medium	5: medium
Plant: habit after vernalization	1–3: semi-erect to erect	3–5: semi-erect-to intermediate	1: erect
Plant: height after vernalization	5: medium	7: tall	9: very tall
Plant: tendency for inflorescence formation without vernalization	1: absent	1: absent	1: absent
Plant: time of inflorescence emergence	3: early	3–5: early to medium	1: very early
Plant: natural height during heading	5: medium	7: tall	9: very tall
Plant: width during emergence	5: medium	5: medium	5: medium
Flag leaf: length	5: medium	9: very long	7: long
Flag leaf: width	5: medium	7: broad	7: broad
Flag leaf: length/ width ratio	3: low	7: high	5: medium
Plant: length of longest stem, inflorescence included	5: medium	7: long	9: very long
Plant: length of upper internode	3: short	5 – medium	5 – medium
Inflorescence: length	5- medium	7-long	7 – long
Inflorescence: number of spikelets	7: many	7 – many	7 – many
Inflorescence: density	7: dense	5 – medium	5 – medium
Inflorescence: length of outer glume on basal spikelet	5: medium	5- medium	5- medium
Inflorescence: length of basal spikelet excluding awn	5: medium	7- long	7- long

IFK Harmoniya is the first diploid variety registered on the Official Variety List of the Republic of Bulgaria (OVL) for the years 2010–2015. It is listed on the Organisation for Economic Cooperation and Development (OECD) list for the years 2010–2015 and on the Common EU catalog for agricultural crops for 2010–2015 and it was granted a certificate from the Patent Office of the Republic of Bulgaria in 2010.

Tetrany is the first Bulgarian tetraploid, early to intermediate heading variety of perennial ryegrass. Tetramis is a new tetraploid Bulgarian very early perennial ryegrass variety. Both are very persistent, winter hardy, drought tolerant, tolerant to high summer temperatures and leaf diseases such as crown rust. Herbage and seed yield and feeding value are given in Table 5.2. In trials in Pleven (2007–2009), dry matter yield of Tetrany and Tetramis were 51 and 34 % higher than of the Belgian variety Roy. Their seed yield in 2008 and 2009 were 36 and 35 % higher. Tetrany and Tetramis are suitable for pasture and hay, in pure stands or in mixtures with white clover or bird's foot trefoil or with diploid ryegrass for hay.

The two tetraploid candivars follow the national official variety testing trials for Value of Cultivation and Use (VCU) in three locations (Chepinci, Plovdiv and Selanovci) and for Distinctness, Uniformity and Stability (DUS) in one location (Chepinci) organized by the Executive Agency for Variety Testing, Field Inspection and Seed Control (EAVTFISC) since 2014. The second evaluation year is now ongoing.

Table 5.2 Comparative results for VCU characters of new perennial ryegrass varieties by the breeder

Character	IFK Harmoniya	Tetrany	Tetramis
Crude protein, gkg-1 DM	150–170	170–190	170
Water soluble carbohydrates WSC, %	6–8	6–10	6–8
IVDMD, %	67–72	70–82	70–75
Rust resistance	5	7–9	7–9
Leafiness, %	55–65	53–66	35–43
Group of maturity	Early	Early to intermediate	Very early
Heading date	3–10 May	15–21 May	26–30 April
Stem thickness, mm	1.8	2.8	2.9
Ear length, cm	17–19	19–25	22–29
Number of spikelets in ear	21–23	21–25	21–23
Dry matter yield, kg ha^{-1}	8000–9000	8000–11,000	8000–10,000
Seed yield, kg ha^{-1}	500–600	600–800	700–900
Thousand seed weight, g	1.31–1.51	2.38–3.71	3.71–4.89

The EAVTFISC tests new varieties for DUS with Belgian references Merlinda (early), Roy (intermediate) and Merkem (late).

At this stage, the Institute of Forage Crops is producing high category prebasic and basic seeds with certification from the original breeder's seed stock. Consumer demand is high; implementation and realization is forthcoming.

Conclusions

Perennial ryegrass breeding in Bulgaria has a strong 20 year-long tradition, especially in the Institute of Forage Crops in Pleven. During these 20 years, three cultivars from local origin were created and release has started on the Bulgarian market. IFK Harmoniya is an early diploid variety. Tetrany and Tetramis are tetraploid varieties, early to intermediate and very early, respectively. They are high yielding with excellent quality, very persistent, winter hardy, drought tolerant and tolerant to high summer temperatures. IFK Harmoniya is the first perennial ryegrass variety bred in Bulgaria and the first variety registered on the Official Variety List of the Republic of Bulgaria. The two tetraploid candivars follow National official varieties testing trials for VCU in three locations and for DUS in one location, organized by the Executive Agency for Variety Testing, Field Inspection and Seed Control since 2014.

Acknowledgments The first author is very grateful for the excellent scientific knowledge and practical breeding skills granted by Mr. Joost Baert, Prof. Dirk Reheul and Prof. Erik Van Bockstaele during many stays at ILVO in Melle and visits to Ghent University. Also many thanks to Hilde Carlier for the flow cytometric ploidy screening and to all ILVO staff for their passion and friendly working attitude.

References

Cai, H., Yamada, T. and Kole, C. (2014). Genetics, Genomics and breeding of Forage Crops, Perennial ryegrass – chapter 2, 11–35.

Conaghan, P., and M.D. Casler. (2011). A theoretical and practical analysis of the optimum breeding system for perennial ryegrass. Ir. J. Agric. Food Res. 50:47–63.

Frame J. 1991. Herbage production and quality of a range of secondary grass species et five rates of fertilizer nitrogen application, Grass and Forage Science, 46, 139–151.

CPVO (Community Plant Variety Office) – TP/ 004/ 1/ 23.06.2011 Protocol for distinctness, uniformity and stability test – Ryegrass.

Humphreys, M.W.,U. Feurstein, M. Vandewalle and J. Baert, (2010). Raygrasses, In: B. Boller et al. (eds.) *Fodder Crops and Amenity Grasses*, Handbook of Plant breeding 5, Springer Science Business Media, 211–260.

Katova, A. (2005). Evaluation of Perennial Ryegrass (*Lolium perenne* L.) Local Populations: Forage and Seed Productivity, *Plant Science*, 42, 80–85, Vol. XLII, № 1.

Katova, A. (2005a). Study of morphological traits, biological properties and agricultural value of plant germplasm of perennial ryegrass (*Lolium perenne* L.) with a view to breeding, *Abstract of Thesis for Doctor* (PhD), Pleven, 44p.

Katova, A. (2011). New perennial ryegrass variety (*Lolium perenne* L.) IFK Harmoniya. *Journal of Mountain Agriculture on the Balkans*, 14 (4): 721–739, 2011.

Katova A., Ilieva A., Baert J., Hristov K. and Van Bockstaele E. (2008). Importance of polyploidy for forage and seed productivity and forage quality of perennial ryegrass, In:

Peeters, A. (2004). Wild and Sown Grasses. Publ. FAO, Rome, 2004, pp.311.

Soković, D., J. Radović1 and Z. Tomić (2011). Perennial forage grasses, from breeding to healthy ruminant feed, Biotechnology in Animal Husbandry 27 (3), p 599–614, 2011.

Sokolovic, D., Z. Tomic and S. Babic (2010). Perennial Forage grasses breeding in Serbia: Achievements, limits and prospects, Biotechnology in Animal Husbandry, 25 (spec. issue), Institute for Animal Husbandry, Belgrade – Zemun, 91–106.

UPOV, (2006). TG/4/8 Guidelines for the conduct of tests for distinctness, uniformity and stability (DUS) of Ryegrass, 26 p.

6 Variation for Plant Architecture Traits Within a Perennial Ryegrass Association Mapping Population

V. Kemešytė, G. Statkevičiūtė, A. Aleliūnas, and G. Brazauskas

Abstract

Superior forage quality, rapid establishment rate and good tolerance to grazing or trampling make perennial ryegrass (*Lolium perenne* L.) one of the most important grass species for forage and amenity purposes. The aim of this study was to assess the variation for plant architecture traits within perennial ryegrass association mapping population consisting of single plants from cultivars and natural ecotypes. Phenotypic data were collected for heading date, flag leaf length, flag leaf width, plant height, inflorescence length, spikelet number, seed weight per inflorescence and 1000 seed weight over 2 years. Moderate variation was observed for most of the traits. Higher genetic variation values were determined for cultivars compared with ecotypes.

Keywords

Lolium perenne • Phenotypic traits • Variability

Introduction

Perennial ryegrass (*Lolium perenne* L.) has high forage quality and more tolerance to grazing than most other perennial grasses (Wilkins 1991). However, it is prone to winterkill at northern latitudes.

Biomass and seed yield are important objectives for perennial ryegrass breeding (Quesenberry and Casler 2001). They are related to other plant architecture traits (plant height, leaf size, heading date, inflorescence length etc.) (Studer et al. 2008; Sampoux et al. 2011). The aim of this

V. Kemešytė (✉) • G. Statkevičiūtė • A. Aleliūnas
G. Brazauskas
Institute of Agriculture, Lithuanian Research Centre for Agriculture and Forestry, Akademija, Kedainiai reg., Lithuania
e-mail: vilma@lzi.lt

I. Roldán-Ruiz et al. (eds.), *Breeding in a World of Scarcity: Proceedings of the 2015 Meeting of the Section "Forage Crops and Amenity Grasses" of Eucarpia*,
DOI 10.1007/978-3-319-28932-8_6

study was to assess the phenotypic variability in a perennial ryegrass collection of single plants from cultivars and ecotypes and to evaluate its suitability for association mapping of plant architecture traits.

Material and Methods

One plant each of 208 diploid perennial ryegrass (*Lolium perenne* L.) populations consisting of 91 cultivars (23 of forage, 46 of turf and 22 of turf/forage type), 116 natural ecotypes and one colchicine induced mutant (Pašakinskienė 2005) were used in this work. Plants were established from seeds and clonally propagated in the greenhouse. A set of all genotypes went through a vernalization process for 4 months in the greenhouse, at the constant day/night temperature of 5 °C. A ramet-based field collection was established in April 2013. Another set of un-vernalized genotypes was planted in the field in July 2013 and went through vernalization under natural conditions during winter 2013/2014. In both cases the same panel of genotypes was used, with four clonal plants per genotype planted at 50 × 50 cm distances using a randomized complete block design. Heading date (HD), flag leaf length (FLL), flag leaf width (FLW), plant height (HGT), inflorescence length (IL), spikelet number (SN), seed weight per inflorescence (SWI) and 1000 seed weight (SWT) were assessed in 2013 and 2014. Flag leaf length and width were measured at full emergence of inflorescences. Three leaves of each plant were measured and the stems were marked. The inflorescence of the marked stems were cut at full ripening stage and dried for the IL, SN and SWI determination. Coefficient of variation of phenotypic traits among genotypes and genetic variance ($V_2 = (MS_g - MS_e)/r$, where MS_g, MS_e and r are the mean squares of genotypes, mean squares of experimental error and number of replications, respectively) were estimated with PLABSTAT software package (Utz 2011). NTSYSpc v. 2.2 (*Exeter Software*, Setauket, NY) (Rohlf 2005) was used to perform principal component analysis (PCA).

Results and Discussion

Moderate variation was observed for most of the traits among cultivars and ecotypes over both years (Table 6.1). Lowest polymorphism values between genotypes were identified for heading date as opposed to seed weight per inflorescence. The different growth/vernalization conditions determined that the values of phenotypic traits in 2014 had higher values compared to 2013. May 2014 was cold and dry, delaying the heading date for 1 week as compared to 2013.

For turf grass, short narrow-leaved plants are required, while tall wide-leaved plants are usually required for forage purposes. The plants of the ecotypes were 8–9 cm taller, the flag leaves about 2 cm longer and 3–4 mm wider compared to the cultivars. Since half of the cultivars were turf type, the means of these traits were lower as compared to the ecotypes (Table 6.1).

Significant genetic variance components were obtained for all traits (Table 6.2). Most ecotypes were collected in the similar habitats in Lithuania and Ukraine. Cultivar group consisted of turf and forage genotype of diverse origin, therefore the genetic variation of the traits was higher compared to the ecotype group.

Table 6.1 Variation for phenotypic traits among cultivars and ecotypes in perennial ryegrass association mapping population

	Cultivars				Ecotypes			
	2013		2014		2013		2014	
Trait	Mean	cv%	Mean	cv%	Mean	cv%	Mean	cv%
HD	21 May	1.70	3 June	2.75	21 May	1.89	4 June	2.21
FLL, cm	8.83	16.97	11.91	16.45	10.68	19.25	14.69	15.65
FLW, cm	0.41	13.54	0.45	13.48	0.46	12.85	0.48	13.05
HGT, cm	43.91	11.57	64.39	9.31	50.47	10.51	75.54	8.50
IL, cm	12.93	10.47	18.53	12.57	15.17	10.80	22.79	9.97
SN	17.19	13.00	20.85	11.10	20.01	18.47	24.32	9.87
SWI, g	0.11	67.34	0.17	30.07	0.16	32.59	0.22	28.39
SWT, g	1.91	14.66	2.00	14.65	2.01	14.88	1.86	16.15

HD heading date, *FLL* flag leaf length, *FLW* flag leaf width, *HGT* plant height, *IL* inflorescence length, *SN* spikelet number, *SWI* seed weight per inflorescence, *SWT* 1000 seed weight

Table 6.2 Genetic variance of phenotypic traits in perennial ryegrass cultivars and ecotypes

	HD	FLL	FLW	HGT	IL	SN	SWI	SWT
Cultivars	18.73**	4.80**	0.003**	62.62**	7.80**	5.23**	0.002**	0.094**
Ecotypes	9.42*	1.89**	0.002**	18.63**	2.99**	3.31 **	0.001**	0.072**

HD heading date, *FLL* flag leaf length, *FLW* flag leaf width, *HGT* plant height, *IL* inflorescence length, *SN* spikelet number, *SWI* seed weight per inflorescence, *SWT* 1000 seed weight
* $P<0.05$; ** $P<0.01$

Heading date was mostly influenced by growing condition. Heritability for this trait was 46.44 % for cultivars and 33.26 % for ecotypes. Plant height, inflorescence length and spikelet number are highly heritable traits and could be effectively selected in breeding programs as they are likely to be controlled by additive genes (Yamada et al. 2004; Byrne et al. 2009; Şeker et al. 2014). The results of our study indicate that flag leaf width can also be classified as a highly heritable trait.

Principal component analysis was carried out on phenotypic traits. PC1 accounted 61.06 % of the total variation and was mostly defined by inflorescence length, flag leaf length and width (Fig. 6.1). PC2 accounted for 13.00 % of the observed variation and was explained by 1000 seed weight, heading date and plant height. Ecotypes separated from most of the turf type cultivars. However, no clear separation between ecotypes, forage and turf/forage cultivars was observed in this study. Turf type cultivars characterized by narrow flag leaves and medium late heading formed cluster 1. Genotypes defined by short inflorescences, short and narrow leaves and early heading formed cluster 2. Four ecotypes had long flag leaves, high number of spikelets and late inflorescence emergence

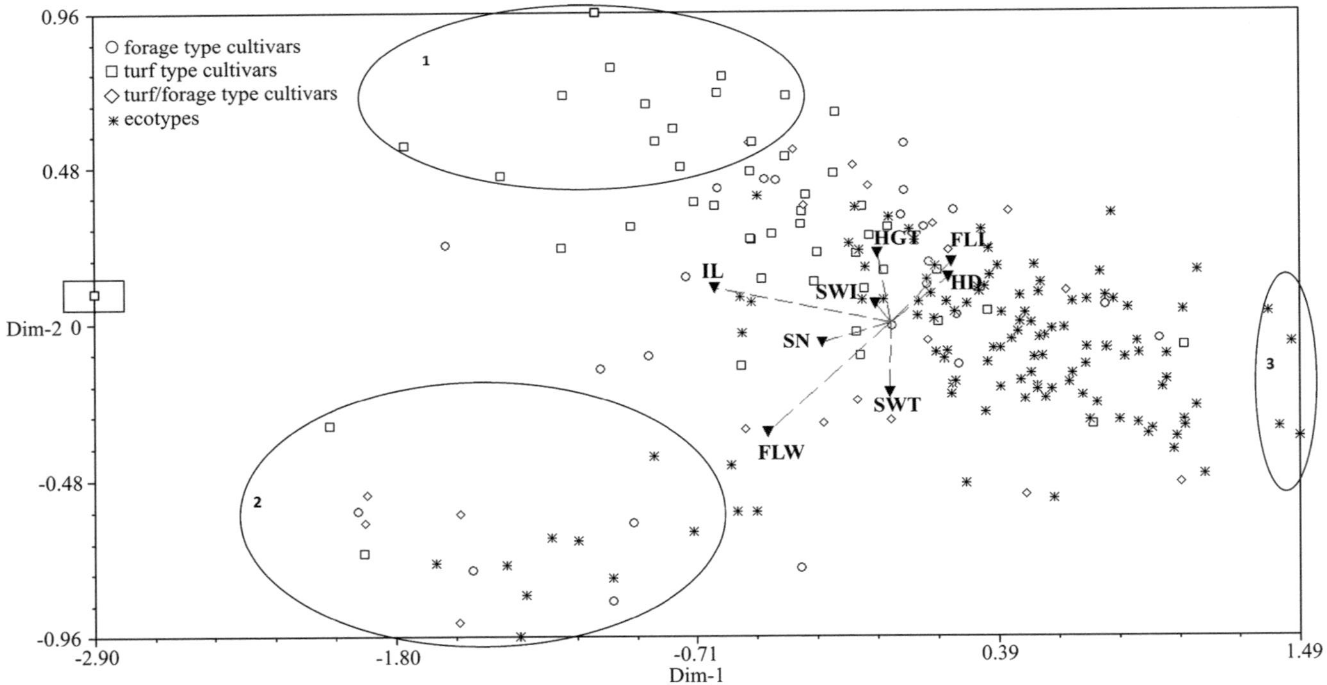

Fig. 6.1 The grouping of perennial ryegrass cultivars and natural ecotypes based on phenotypic traits. *HD* heading date, *FLL* flag leaf length, *FLW* flag leaf width, *HGT* plant height, *IL* inflorescence length, *SN* spikelet number, *SWI* seed weight per inflorescence, *SWT* 1000 seed weight

(cluster 3). A single turf type cultivar characterized by short inflorescences, short and narrow leaves and early heading diverged from the rest (noted by square in Fig. 6.1).

Conclusion

Similar variability for plant architecture traits was determined in both cultivar and ecotype groups. The amount of phenotypic variability present in the collection is sufficient to perform association mapping for plant architecture traits.

Acknowledgments This study was funded through long-term research program "Genetics and purposeful change of genotypes of agricultural and forest plants" implemented by Lithuanian Research Centre for Agriculture and Forestry.

References

Byrne S, Guiney E, Barth S, Donnison I, Mur LAJ, Milbourne D (2009) Identification of coincident QTL for days to heading, spike length and spikelets per spike in *Lolium perenne* L. Euphytica 166:61–70.

Pašakinskienė I (2005) Creeping stem and barren inflorescence: a new mutation in Lolium perenne 'VIROIZ'. Zwierzykowski Z, Kosmala S (eds.), Recent advances in genetics and breeding of the grasses. Institute of Plant Genetics, Poznan. pp 209–213.

Quesenberry KH, Casler MD (2001) Achievements and perspectives in the breeding of temperate grasses and legumes. In: Gomide JA, Mattos WRS, da Silva SC (eds) Proceedings of the International Grasslands Congress, XIX. FEALQ, Piricicaba, Sao Paulo, Brazil. pp 517–524.

Rohlf FJ (2005) NTSYS-pc: numerical taxonomy and multivariate analysis system, version 2.2. Exeter Software, Setauket.

Sampoux JP, Badouin P, Bayle B, Béguier V, Bourdon P, Chosson JF, Deneufbourg F, Galbrun C, Ghesquiére M, Noël D, Pietraszek W, Tharel B, Viguié A (2011) Breeding perennial grasses for forage usage: An experimental assessment of trait changes in diploid perennial ryegrass (*Lolium perenne* L.) cultivar released in the last four decades. Field Crops Research 123: 117–129.

Şeker H, Yazici A, Uysal P (2014) Analysis of variability, heritability, and genetic advance in seed yield and related traits of orchardgrass (*Dactylis glomerata* L.) populations. Turkish Journal of Agriculture and Forestry 38: 633–643.

Studer B, Jensen LB, Hentrup S, Brazauskas G, Kölliker R, Lübberstedt T (2008) Genetic characterisation of seed yield and fertility traits in perennial ryegrass (*Lolium perenne* L.). Theoretical and Applied Genetics 117: 781–791.

Utz HF (2011) PLABSTAT-A computer program for statistical analysis of plant breeding experiments. Version 3A. Germany, Universität Hohenheim.

Wilkins PW (1991) Breeding perennial ryegrass for agriculture. Euphytica 52:201–214.

Yamada T, Jones ES, Cogan NOI, Vecchies AC, Nomura T, Hisano H, Shimamoto Y, Smith KF, Hayward MD, Forster JW (2004) QTL analysis of morphological, developmental, and winter hardiness-associated traits in perennial ryegrass. Crop Science 44(3):925–935.

Extent of Crown Rust Infection in a Perennial Ryegrass (*Lolium perenne* L.) Association Mapping Population

7

S.K. Arojju, D. Milbourne, P. Conaghan, T.R. Hodkinson, and S. Barth

Abstract

One of the most damaging foliar diseases on perennial ryegrass is crown rust, caused by *Puccinia coronata* f. sp. *lolli*. Crown rust causes severe losses to yield and quality of forage and results in decreased palatability and digestibility for ruminants. Crown rust was scored in 1800 individually spaced plants in two replicates on a scale of 1 (no infection) to 9 (severely infected) at three time points (June, August & September 2014) in a block design experiment. These 1800 individuals represent 30 populations with different population structures: ten synthetic varieties, eight full-sib breeding families, eight half-sib breeding families and four Irish ecotypes. The disease progression was observed in all populations, with highest pressure of crown rust seen in September and lowest disease pressure observed in June. Significant differences ($p<0.001$) were recorded among populations, among time points and for the population x time point interaction. The Irish ecotypes were more susceptible to crown rust compared to the other populations. Crown rust phenotypic data will be used to conduct association analysis on these populations to identify significant marker trait associations.

Keywords

Perennial ryegrass • Crown rust • Disease assessment • Association mapping

S.K. Arojju
Crops Environment and Land Use Programme, Crop Science Department, Teagasc, Oak Park Research Centre, Carlow, Ireland

Botany, School of Natural Sciences, Trinity College Dublin, Dublin 2, Ireland

D. Milbourne • S. Barth (✉)
Crops Environment and Land Use Programme, Crop Science Department, Teagasc, Oak Park Research Centre, Carlow, Ireland
e-mail: susanne.barth@teagasc.ie

P. Conaghan
Grassland Science Research Department, Teagasc, Animal and Grassland Research and Innovation Centre, Oak Park, Carlow, Ireland

T.R. Hodkinson
Botany, School of Natural Sciences, Trinity College Dublin, Dublin 2, Ireland

I. Roldán-Ruiz et al. (eds.), *Breeding in a World of Scarcity: Proceedings of the 2015 Meeting of the Section "Forage Crops and Amenity Grasses" of Eucarpia*,
DOI 10.1007/978-3-319-28932-8_7

Introduction

Perennial ryegrass is an important forage grass species in temperate regions. Because of its high palatability and digestibility, this species is highly valued in dairy and sheep forage systems (Hannaway et al. 1999). A total of 76 million hectares are used as permanent pasture in Europe (Fischer et al. 2009), of which 3.36 million hectares (about 80 % of arable land) is used for pasture, hay and silage in Ireland (Central Statistics Office 2012). Improvement of forage quality and increased final biomass are the important traits that need to be addressed.

Crown rust (caused by *Puccinia coronata* f. sp. *lolli*) is a severely damaging foliar disease that causes severe effects to forage quality and dry matter digestibility (DMD) on perennial ryegrass (O'Rourke 1976). This fungus shows a high degree of host adaptation; it is known to infect at least 40 grass genera (Cummins 1971). The disease is prevalent throughout the year but highest pressure is seen in late spring to early autumn (O'Rourke 1976). Severe attack decreases water soluble carbohydrates (WSC) from the leaf blade and also reduces tillering, root growth and plant regrowth after harvesting (O'Rourke 1976; Dracatos et al. 2010). It can lead to yield reduction of up to 30 % with a 56 % reduction of DMD in certain perennial ryegrass cultivars (Lancashire and Latch 1966; O'Rourke 1976; Dracatos et al. 2010). The low biomass and poor forage quality have an impact on milk production and the dead herbage tissue promotes growth of saprophytic fungus *Pithomyces chartarum,* which causes facial eczema in grazing cattle (Lancashire and Latch 1966).

The development of disease resistant cultivars is desirable due to the high economic and environmental costs associated with fungicidal disease control (Dracatos et al. 2010). Studying the genetic architecture of disease resistance using molecular markers contributes to the breeding of resistant varieties. To date, several quantitative trait locus (QTL) mapping studies for rust resistance have been carried out in biparental mapping populations of perennial ryegrass, generally using mixed populations of crown rust, and QTL have been detected on all seven linkage groups (LG). Currently, the nature of the genes underlying these QTL is unknown. Given the use of mixed race populations of crown rust in the experiments, it is possible that QTL effects may be due to classical R genes acting on parts of the pathogen population (creating the impression of quantitative resistance) or due to the action of non-race specific genes of small individual effect (Dracatos et al. 2010). For instance, a QTL identified on LG2 in perennial ryegrass lies in a region of synteny with hexaploid oat in which QTL containing clusters of R genes were previously mapped for crown rust resistance (Jones et al. 2002; Dracatos et al. 2010). Conversely, the wheat peroxidase (class iii) enzyme is important for resistant against leaf rust and was mapped on top of LG2 genetic map where previously QTL was mapped for crown rust (Faris et al. 1999; Muylle et al. 2005; Dracatos et al. 2010) and QTL for herbage quality and crown rust were detected on LG3 and LG7 of *Lolium*, suggesting the association between the lignin synthesis genes and crown rust resistance (Dracatos et al. 2009).

The main problem in linking QTL to candidate genes in biparental mapping experiments is a lack of resolution, since such studies are frequently carried out in relatively small populations. The plant population used in this study are in an on-going project involving genomic selection (GS) and association mapping (AM) at Teagasc. One goal of our study is to test whether it is possible to successfully perform AM for crown rust in this population, and we are considering adopting both genome-wide and targeted mapping approaches. The relatively large, divergent population, which contains both structured and relatively unstructured sub-populations, combined with the previously observed patterns of LD in perennial ryegrass should allow detection of multiple QTLs at a level of resolution allowing better identification of candidate genes underlying the loci. This paper reports the first season of phenotypic data charting the progress of disease in these diploid perennial ryegrass populations under field conditions, and assesses the range of resistance/susceptibility of disease in each family.

Materials and Methods

A field experiment was established in 2013 and first disease scores were conducted during 2014 at Teagasc, Oak Park, Ireland (52°51′34.2″N 6°55′03.0″W) on a grey brown podzolic soil. The rainfall in the year 2014 was 496.2 mm during the growing season (April-October) and was considerably higher than the year before (429.5 mm). However, the dry summer of 2014 was favorable for spread of crown rust via wind-borne urediniospores (Carson 2008). A total of 1800 genotypes were used in this study, which are clonally propagated for replicates. The population consists of 30 distinct subpopulations, each comprising 60 genotypes, which can be broken down into ten released European cultivars (CU) (G01-G10), eight full-sib (FS) breeding families (G11-G18), eight half-sib (HS) breeding families (G19-26) and four Irish ecotypes (EC) (G27-G30). Trials were set up as a balanced incomplete block design in two replicates with four cuts per year. Each block is 7.5 m wide and 4.5 m high surrounded by rust susceptible spreader cultivars. Five check cultivars with one clone each are interspersed in the experimental design.

Disease was scored at three time points (June, August & September) during the growing season in 2014, on a scale of 1–9 on individually spaced plants with 1 = no rust disease, 2 = traces of pustules, 3 = 10 %, 4 = 20 %, 5 = 30 %, 6 = 50 %, 7 = 60 %, 8 = 80 % and 9 = entire plant infected.

The results were subjected to ANOVA (analysis of variance) using GenStat and data presented are based on mean crown rust susceptibility scores of the perennial ryegrass populations.

Results and Discussion

The crown rust susceptibility scores of the perennial ryegrass population at three time points are presented in Fig. 7.1. Considerable and significant ($p<0.001$; Table 7.1) variation in susceptibility was found among populations. As expected, there was a general increase in mean infection score as the growing season progressed, and ANOVA revealed significant differences ($p<0.001$) in crown rust susceptibility among the time points (Table 7.1) and also a significant time point x population interaction. The FS, HS and CU families all followed the same pattern of increasing levels of crown rust infection throughout the season, reaching a peak at time point 3. The ecotypes exhibited a different pattern, with higher mean infection levels than the other families at the first time point, and a peak of infection at time point 2 (Fig. 7.2). This rapid progress of disease in the ecotypes suggests Irish crown rust populations have co-evolved with Irish perennial ryegrass populations and overcome any resistance in these populations, and that modern breeding germplasm (as typified by the CU, FS and HS families) is contributing variation for this trait not available in the ecotype material (Smit et al. 2005).

On an individual family basis, G12 (FS) was the most resistant and families G18 (FS), G30 (EC) were highly susceptible to crown rust (Fig. 7.1). The range of crown rust susceptibility on a per genotype basis varied widely. The lowest variation within families was at time point 1 in families G12 and G24, whereas highest variation was found in the family G30 at time point 2 (Fig. 7.1).

Conclusion

Overall, significant differences in disease susceptibility were found at each scoring time point and a large amount of variation was recorded within the population of 1800 individuals. This good distribution of rust infestation across and within populations is encouraging for further AM analysis since there is sufficient variation in the phenotypic trait for such work. We intend to repeat these measurements for two further years. In addition, the population of 1800 individuals is being subjected to high throughput SNP genotyping using the genotyping by sequencing based approach of Elshire et al. (2011) and we are identifying candidate loci for crown rust resistance from a shotgun sequence of perennial ryegrass we have developed in an inbred line from Teagasc. Combined, these resources should allow detecting significant and stable marker–crown rust trait associations. The resulting information may also provide insights into the race structure of crown rust populations in Ireland.

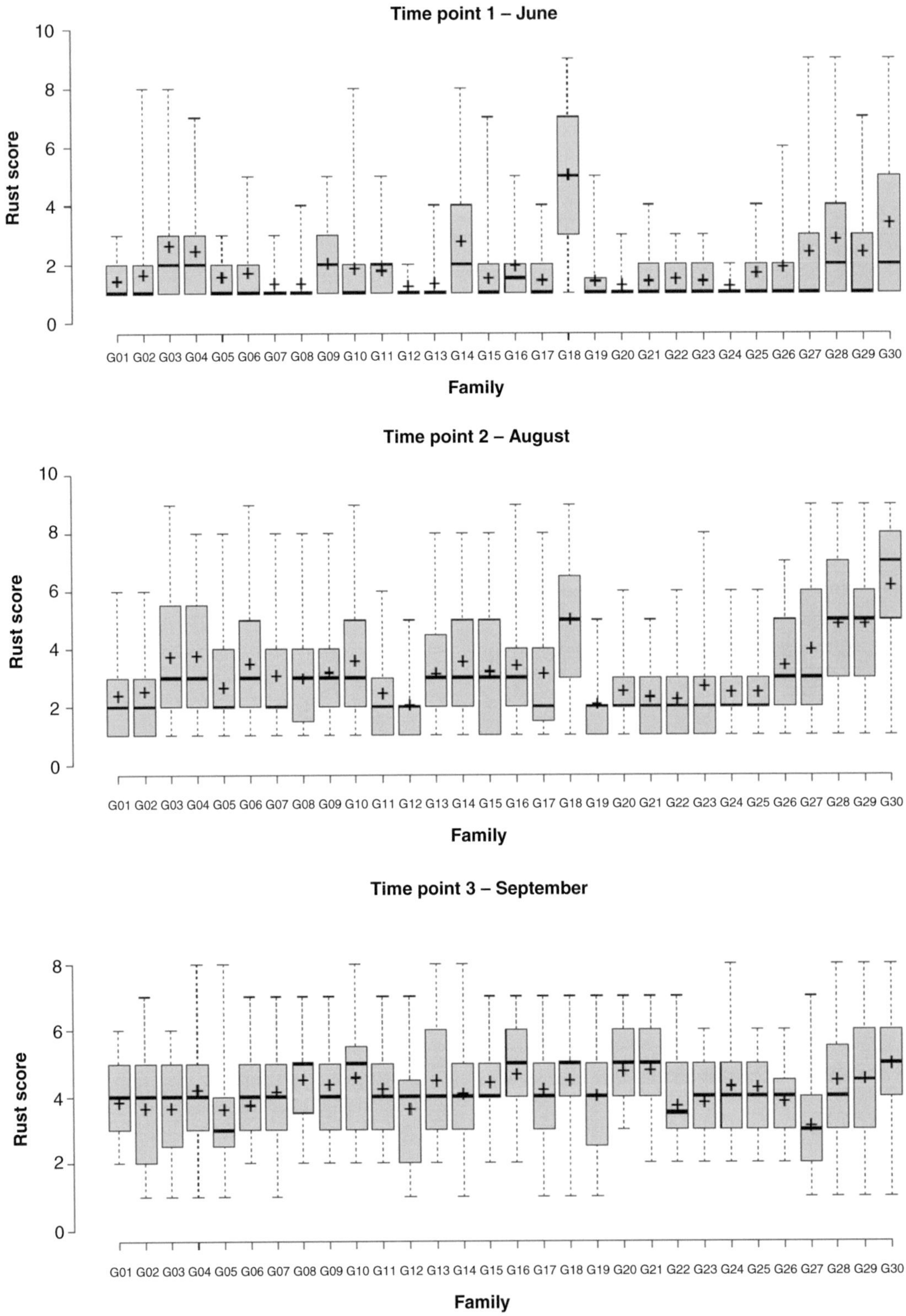

Fig. 7.1 Boxplots representing crown rust susceptibility in the population scored in June (9th to 16th), August (4th to 11th) and September (15th to 19th). Plus (+) represents mean scores of each family and whiskers are the range of scoring within family. Disease was scored on a scale 1–9 with 1=no infection and 9=severely infected. Plants subjected to infrequent cutting (four cuts per year) and disease was scored a month after cutting in time point 2 and 3 to allow plants to grow for scoring

Table 7.1 Analysis of variance of crown rust disease scores for the populations (families) used in this study

Source of variation	Degrees of freedom	F value	P value
Population	29	14.24	<0.001
Time points	2	593.21	<0.001
Time points × population	58	3.56	<0.001

Acknowledgments This work is funded by a Teagasc Walsh PhD fellowship to SK, Teagasc core funding and the Irish Department of Agriculture, Food and The Marine Research Stimulus Fund 11/S/109. We acknowledge the support of the entire Teagasc forage breeding team and Teagasc farm staff.

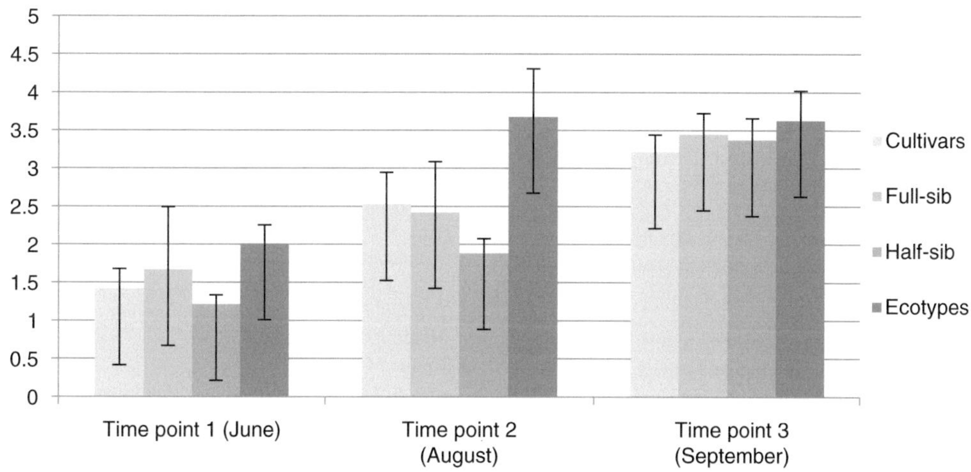

Fig. 7.2 Mean crown rust susceptibility scores of each sub-population ($p < 0.001$) at three time points in the year 2014

References

Carson., M. (2008) Oat crown rust. Retrieved from http://www.ars.usda.gov/Main/docs.htm?docid=9919

Central Statistics Office. (2012). Census of Agriculture 2010 – Final Results. Dublin: The Stationery Office

Cummins, GB. (1971). The rust fungi of cereals, grasses and bamboos. Springer-Verlag, Berlin, Heidelberg, New York.

Dracatos, PM., Cogan, NOI., Sawbridge, TI., Gendall, AR., Smith, KF., Spangenberg, GC and Forster, JW. (2009). Molecular characterisation and genetic mapping of candidate genes for qualitative disease resistance in perennial ryegrass (*Lolium perenne* L.). BMC Plant Biology 9, pp. 62.

Dracatos, PM., Cogan, NOI., Keane, PJ., Smith, KF and Forster, JW. (2010). Biology and Genetics of Crown Rust Disease in Ryegrasses. Crop Science 50, pp.1605–1624.

Elshire, RJ., Glaubitz, JC., Sun, Q., Poland, JA., Kawamoto, K., Buckler, ES and Mitchell, SE. (2011). A Robust, Simple Genotyping-by-Sequencing (GBS) Approach for High Diversity Species. PLoS ONE 6: e19379.

Faris, JD., Li, WL., Liu, DJ., Chen, PD and Gill, BS. (1999). Candidate gene analysis of quantitative disease resistance in wheat. Theoretical and Applied Genetics 98, pp. 219–225.

Fischer, G., Prielera, S., Velthuizena, HV., Berndes, G., Faaijc, A., Londo, M and Wit, MD. (2009). Biofuel production potentials in Europe: Sustainable use of cultivated land and pastures, Part II: Land use scenarios, Biomass and Bioenergy doi:10.1016/j.biombioe.2009.07.009

Hannaway, D., Fransen, S., Cropper, J., Teel, M., Chaney, M., Griggs, T., Halse, R., Hart, J., Cheeke, P., Hansen, D., Klinger, R and Lane, W. (1999). *Lolium perenne*. In: A Pacific Northwest Extension Publication, Vol. PNW 502. Oregon State University, Washington State University, University of Idaho.

Jones, ES., Mahoney, NL., Hayward, MD., Armstead, IP., Jones, JG., Humphreys, MO., King, IP., Kishida, T., Yamada, T., Balfourier, F., Charmet, G and Forster, JW. (2002). An enhanced molecular marker based genetic map of perennial ryegrass (*Lolium perenne*) reveals comparative relationships with other Poaceae genomes. Genome 45, pp. 282–295.

Lancashire, JA and Latch, GCM. (1966). Some effects of crown rust (*Puccinia coronata* Corda) on the growth of two ryegrass varieties in New Zealand. New Zealand Journal of Agricultural Research 9, pp. 628–640.

Muylle, H., Baert, J., Van Bockstaele, E., Pertijs, J and Roldan-Ruiz, I. (2005). Four QTLs determine crown rust (*Puccinia coronata* f. sp. *lolii*) resistance in a perennial ryegrass (*Lolium perenne*) population. Heredity 95, pp. 348–357.

O'Rourke, CJ. (1976). Diseases of grasses and forage legumes in Ireland. The Agricultural Institute, Dublin 4.

Smit, HJ., Tas, BM., Taweel, HZ., Tamminga, S and Elgersma, A. (2005). Effects of perennial ryegrass (*Lolium perenne* L.) cultivars on herbage production, nutritional quality and herbage intake of grazing dairy cows. Grass and Forage Science. 60, pp. 297–309.

The EUCARPIA Multi-site Rust Evaluation – 2013 Results

8

F.X. Schubiger, J. Baert, T. Ball, Z. Both, E. Czembor, U. Feuerstein, S. Hartmann, B. Krautzer, H. Leenheer, C. Persson, W. Pietraszek, L. Poinsard, Y. Quitté, S. Rohde, L. Russi, K. Stixova, S. Schulze, M.C. Tardin, M. Van Nes, E. Willner, L. Wolters, and B. Boller

F.X. Schubiger (✉) • B. Boller
Forage Plant Breeding, Agroscope INH, Zürich CH-8046, Switzerland
e-mail: franz.schubiger@agroscope.admin.ch

J. Baert
Unit Plant Sciences, ILVO, Melle B-9090, Belgium

T. Ball
DSV United Kingdom, Deutsche Saatveredelung AG, Wardington, Banbury, Oxon OX17 1FE, UK

Z. Both
OSEVA PRO Ltd., Grassland Research Station Roznov Zubri, Zubri CZ-75654, Czech Republic

E. Czembor
Plant Breeding and Acclimatization Institute – NRI, Blonie PL-05-870, Poland

U. Feuerstein
Research and Development, Deutsche Saatveredelung AG, Asendorf D-27330, Germany

S. Hartmann
Bayerisches Staatsministerium für Ernährung, Landwirtschaft und Forsten, Institut für Pflanzenbau u. Pflanzenzüchtung der Bayerischen Landesanstalt für Landwirtschaft, Freising D-85354, Germany

B. Krautzer
Department of Ecological Restauration, Agricultural Research and Education centre Raumberg-Gumpenstein, Irdning A-8952, Austria

H. Leenheer
DLF B.V., Moerstraten NL-4727 ST, The Netherlands

C. Persson
Lantmännen Lantbruk, Lantmännen, Svalöv SE-268 81, Sweden

W. Pietraszek
RAGT 2n Centre de Recherche de Montours La Boucaudière, Montours, F – 35460, France

L. Poinsard
Research Department, S.A. Carneau, Orchies F-59310, France

Y. Quitté
Deutsche Saatveredelung AG, La planche, Les Rosiers sur loire F-49350, France

S. Rohde
Saatzucht Steinach, Steinach D-94377, Germany

L. Russi
Dip. Science Agrarie, Alimentari e ambientali, Universita' degli Studi di Perugia, Perugia I-06121, Italy

K. Stixova
Plant Breeding Station, Hladke Zivotice CZ-74247, Czech Republic

S. Schulze
Saatzucht Steinach Bornhof, Bocksee D-17219, Germany

M.C. Tardin
Plant Pathology, RAGT 2n Centre de recherche de Druelle, Druelle F-12510, France

M. Van Nes
Barenbrug, Duitsekampweg 60, Wolfheze NL-6874 BX, The Netherlands

E. Willner
Teilsammlung Nord, IPK Genbank, Malchow/Poel D-23999, Germany

L. Wolters
DSV-Zaden Nederland D.V., Deutsche Saatveredelung AG, Zelder 1, Ven-Zelderheide NL-6599 EG, The Netherlands

I. Roldán-Ruiz et al. (eds.), *Breeding in a World of Scarcity: Proceedings of the 2015 Meeting of the Section "Forage Crops and Amenity Grasses" of Eucarpia*,
DOI 10.1007/978-3-319-28932-8_8

Abstract
In 2013, the fifth EUCARPIA multisite rust evaluation trial took place. A set of 16 Italian (*Lolium multiflorum*), 4 hybrid (*L. boucheanum*) and 34 perennial ryegrass (*L. perenne*) cultivars were sown at 21 sites in 11 European countries. The cultivars were grown as rows in a completely randomized block design with four replicates. Rust incidence was scored in the year of seeding using a scale based on an estimation of the percentage of leaf area affected. Crown rust (*Puccinia coronata* f. sp. *lolii*) was the most frequently observed rust species. Compared to the previous trials, stem rust (*P. graminis* f. sp. *graminicola*) was reported more frequently, i.e., at eight sites for perennial ryegrass but also at five sites for Italian ryegrass. At most sites, there was a highly significant difference in crown and stem rust resistance among Italian and perennial ryegrass cultivars. The ranking of the Italian/hybrid and perennial ryegrass cultivars in terms of crown rust susceptibility was very consistent across the different sites. Moreover, the ranking of the mean crown and stem rust susceptibility of the cultivars was highly correlated with the corresponding ranking of the cultivars in the 2001, 2004, 2007 and 2010 trials, respectively. There is no evidence that crown and stem rust resistance of an individual cultivar was overcome by the rust pathogen over the 12 years of experimental observations.

Keywords
Crown rust • Stem rust • Italian ryegrass • Hybrid ryegrass • Perennial ryegrass

Introduction

The EUCARPIA Multisite Rust Evaluation trial was initiated in 2000 to determine the susceptibility of different Italian (*Lolium multiflorum* Lam.), hybrid (*L. boucheanum* Kunth.) and perennial ryegrass (*L. perenne* L.) cultivars to the most important rust species throughout Europe. Moreover, by repeating the trial every 3rd year, the aim of the study was to find out if rust resistance of individual cultivars breaks down over time and to assess the trends in virulence of different European rust populations. In 2013, the trial was conducted for the fifth time. The results of the four previous trials were published in the proceedings of different meetings of the Eucarpia Fodder Crops and Amenity Grasses section (Boller et al. 2003; Schubiger et al. 2007, 2010a, 2012). In addition, a comprehensive summary of the first three trials was published in 2010 (Schubiger et al. 2010b).

Materials and Methods

In 2013 the trial was sown at 21 sites in 11 European countries. Of these 21 sites, 11 sites have taken part in all five of the trials conducted (since 2001). The same 15 Italian, 3 hybrid and 33 perennial ryegrass cultivars as in the four previous trials were tested. In addition, the Italian ryegrass cultivar Crema, the hybrid ryegrass cultivar Gosia and the perennial ryegrass cultivar Maya were included because they were in the trials of 2007 and 2010.

Up to 24 g of seed (according to the germination rate) of each cultivar were forwarded to each participant in the form of encoded seed lots. At each site the seed was sown in spring in a completely randomized block design with four replicates. The cultivars were grown as rows (3 m long and 0.5 m apart). The Italian and perennial ryegrass cultivars were sown in separate trials. The hybrid ryegrass cultivars were included in the Italian ryegrass trial. The trials were cut and fertilized as was the custom at each site. They were scored for rust incidence between July and October, one to three times during different growth cycles and periods of abundant rust development. Participants were asked to separately score the cultivars for each rust species occurring in the field. A scale from one to nine was used: with 1 = no rust disease, 2 = trace of rust, 3 = 5 %, 4 = 10 %, 5 = 25 %, 6 = 40 %, 7 = 60 %, 8 = 75 % and 9 = more than 75 % of the foliage covered with rust. The rating values represented a relative estimate of leaf area occupied by rust pustules, irrespective of the reaction type. For any particular site, scoring data with an average score of at least two were included in the analysis, provided that there were significant differences among the cultivars at the site. If there were sites with more than one valid scoring moment per year, means of the scorings (per row) were calculated and used in further analysis.

Results

Crown rust (*Puccinia coronata*) was the most frequently evaluated rust species on both ryegrass species. A relevant incidence of crown rust was observed on Italian/hybrid ryegrass at 16 sites and on perennial ryegrass at 14 of 21 sites (Tables 8.1, 8.2 and 8.3). Stem rust (*P. graminis*) occurred on Italian ryegrass at five sites (Table 8.4) and on perennial ryegrass at eight sites (Table 8.5). In Svalöv (Sweden) no rust at all was observed. At all other 20 sites at least one rust species was observed either on Italian or perennial ryegrass or on both species.

Mean crown and stem rust susceptibility scores of Italian, hybrid and perennial ryegrass cultivars from each site where a relevant rust infection was observed are presented in Tables 8.2, 8.3, 8.4, and 8.5.

Across all sites, a highly significant difference ($P < 0.001$) in mean crown and stem rust scores was observed among Italian and perennial ryegrass cultivars (Table 8.1), respectively.

Despite the occurrence of significant interactions of cultivars with sites, the ranking of the Italian and perennial ryegrass cultivars with respect to crown or stem rust susceptibility was very consistent. Spearman rank order correlations between the data of a particular site and the mean of all sites were significant in all but two cases (Tables 8.2, 8.3, 8.4, and 8.5).

The ranking of perennial ryegrass cultivars differed depending on whether crown or stem rust was scored. The Spearman rank order correlation between the mean disease scores of the cultivars for the two pathogens was very low and not significant at $P < 0.05$ ($r_s = 0.28$). Compared to stem rust, cultivars Herault and Kells were relatively more resistant against crown rust. The contrary was true for the cultivars Maja and Kentaur. The two cultivars Gwendal and Bocage were among the most crown and stem rust resistant cultivars.

Discussion

In 2013, the ranking of the cultivars in terms of crown and stem rust resistance was highly consistent over the different sites for both ryegrass species. Moreover, rank order correlations between

Table 8.1 Analysis of variance of mean rust disease scores

Ryegrass species	*Lolium multiflorum*		*Lolium perenne*	
Rust species	*P. coronata*	*P. graminis*	*P. coronata*	*P. graminis*
No. of cultivars	20	20	34	34
No. of sites	16	5	14	8
F-value for cultivars	169.6	24.1	65.3	38.1
F-value for sites	136.5	587.5	246.2	547.1
F-value for cv. x site interaction	4.1	9.5	4.5	5.6

All F-values are significant at the $p < 0.001$ level

Table 8.2 Crown rust (*Puccinia coronata*) disease scores of 16 Italian and 4 hybrid ryegrass cultivars at 16 sites (means of 1–3 scorings per site)

Cultivar	Asendorf D	Aston UK	Bornhof D	Druelle F	Lelystad B. NL	Les Rosiers F	Merelbeke B	Montours F	Ottersum NL	Perugia I	Pulling D	Radzikow PL	Roznov-Zubri CZ	Steinach D	Swifterband NL	Zurich CH	Mean of all sites
Domino (4x)	1.9	2.1	2.8	3.4	2.0	1.8	2.3	2.0	3.3	2.4	3.0	3.3	3.6	4.8	1.8	4.3	2.78
Tarandus (4x)	2.4	2.0	2.5	3.5	2.8	2.5	2.9	3.3	2.5	2.1	3.0	3.3	3.5	4.5	2.0	3.8	2.90
Tonyl (4x)	2.3	2.8	2.5	4.3	3.0	1.6	2.8	2.8	2.4	2.4	3.0	3.3	2.5	5.0	1.9	4.3	2.91
Gosia[a] (4x)	2.1	1.3	6.3	2.6	1.5	1.6	1.9	2.5	2.4	1.9	3.0	5.1	4.6	5.8	2.4	3.1	3.00
Caballo (4x)	1.4	1.9	3.3	4.0	2.8	2.3	2.7	3.5	2.9	2.5	3.0	4.0	3.3	5.0	1.6	4.2	3.01
Zorro (4x)	2.1	2.0	4.3	3.1	2.5	2.5	2.3	3.3	2.0	2.3	2.8	3.9	4.1	5.5	1.6	4.2	3.02
Bolero (4x)	2.1	2.0	2.8	3.4	3.3	1.9	3.3	4.0	2.3	2.5	2.8	4.0	3.6	5.3	2.4	4.0	3.09
Barprisma (2x)	3.1	1.9	2.3	3.1	2.8	2.9	4.9	2.3	3.8	2.3	3.5	4.5	4.4	5.0	1.9	4.0	3.28
Aberexc[a] (4x)	2.6	2.4	2.5	3.9	3.3	4.0	2.8	2.8	3.6	2.8	3.5	3.9	3.0	5.8	2.9	5.8	3.45
Ellire (4x)	2.1	2.1	3.8	5.1	2.3	2.6	3.3	5.3	2.9	3.1	3.5	3.8	4.3	5.0	2.3	6.0	3.58
Fastyl (2x)	3.1	2.4	3.0	3.4	3.0	3.4	3.9	3.3	4.0	3.1	3.5	4.3	5.3	5.8	2.5	4.5	3.64
Pirol[a] (2x)	4.0	2.4	3.5	4.5	4.3	3.8	4.8	4.0	4.6	3.6	3.8	4.6	5.4	6.0	3.5	6.0	4.29
Meryl (2x)	4.0	2.4	3.8	4.9	3.3	4.0	4.3	5.0	5.3	4.0	5.0	4.5	5.1	6.3	4.4	6.5	4.54
Crema (2x)	4.6	5.3	3.3	4.1	5.3	5.9	5.4	5.5	4.1	3.4	4.5	4.9	6.5	6.0	5.1	7.3	5.06
Lolita (4x)	2.8	2.8	5.5	5.6	5.0	5.6	4.6	4.8	4.4	3.9	5.5	5.1	7.8	6.8	3.9	7.6	5.09
Danergo (4x)	2.6	3.4	6.3	5.8	3.8	5.4	4.8	5.0	5.5	3.9	5.8	5.3	7.1	6.8	4.1	7.8	5.19
Ligrande (2x)	4.8	2.9	6.0	5.8	4.8	7.4	5.8	6.0	5.1	4.4	5.8	5.3	7.6	7.0	5.0	8.3	5.73
Gordo (2x)	4.5	2.8	6.5	6.3	3.8	7.4	5.6	6.5	6.0	4.5	5.5	5.8	8.3	7.5	5.4	8.5	5.91
Gumpenst.[a] (2x)	5.3	2.6	6.0	6.3	3.5	7.6	6.3	6.0	6.6	4.8	5.5	6.1	8.4	7.5	5.0	8.4	5.99
Lema (2x)	4.9	2.5	6.8	6.3	4.5	7.8	6.7	7.8	5.8	4.6	5.3	5.5	8.5	7.3	4.4	8.4	6.04
Mean	3.1	2.5	4.2	4.5	3.4	4.1	4.1	4.3	4.0	3.2	4.1	4.5	5.3	5.9	3.2	5.8	4.12
LSD (P<0.05)	0.85	0.71	1.76	1.18	0.92	1.07	0.70	1.26	1.38	1.17	0.74	0.62	1.08	0.89	0.88	0.78	0.47
Correlation[b]	0.84	0.69	0.68	0.81	0.79	0.95	0.88	0.86	0.88	0.93	0.89	0.86	0.90	0.93	0.87	0.91	

Cultivars are ranked according to the mean of all the sites

[a]Hybrid Ryegrass

[b]Spearman rank order correlation with mean of all sites (all values are significant at $P<0.05$)

Table 8.3 Crown rust (*Puccinia coronata*) disease scores of 34 perennial ryegrass cultivars at 14 sites (means of 1–3 scorings per site)

Cultivar	Asendorf D	Aston UK	Bornhof D	Druelle F	Lelystad B. NL	Les Rosiers F	Malchow D	Merelbeke B	Montours F	Ottersum NL	Pulling D	Steinach D	Swifterband NL	Zurich CH	Mean of all sites
Gwendal (4x)	2.5	1.5	4.5	2.3	3.0	1.3	1.9	2.5	2.0	3.8	2.0	2.8	2.9	1.7	2.46
Bocage (4x)	2.5	2.5	6.0	1.9	3.6	2.0	2.9	2.8	2.3	4.6	2.8	3.3	2.8	2.3	3.00
Carrera (2x)	3.9	2.1	4.5	1.9	3.4	2.3	2.9	4.0	4.3	3.9	3.0	2.9	2.1	1.9	3.07
Lacerta (4x)	2.9	3.5	5.5	2.6	3.4	2.5	2.6	2.8	3.0	3.3	3.0	3.4	2.3	3.5	3.16
Option (2x)	3.6	2.5	6.0	2.0	3.0	1.8	2.6	3.7	4.0	4.0	3.5	3.4	2.4	3.4	3.27
Heraut (2x)	3.5	2.3	5.5	3.6	3.3	2.0	2.6	4.9	3.3	4.6	3.3	3.1	2.8	2.8	3.38
Orval (2x)	7.3	1.3	4.5	3.3	2.4	2.3	2.6	3.9	6.0	4.3	2.5	3.9	3.0	1.8	3.49
Barnhem (2x)	4.5	2.3	6.3	3.3	3.1	2.3	3.4	4.6	1.8	5.3	3.0	3.4	3.6	2.7	3.52
Pastoral (4x)	4.1	2.1	7.5	2.5	4.4	3.3	4.1	2.3	2.0	5.2	2.8	3.9	2.8	2.4	3.52
Vincent (2x)	3.4	2.3	5.5	3.1	3.4	2.0	3.3	4.5	5.8	4.3	3.5	3.0	2.5	3.4	3.56
Guru (2x)	4.5	2.5	2.5	2.9	3.0	3.0	1.6	5.3	2.5	4.6	3.8	3.8	4.1	7.0	3.65
Kells (2x)	4.1	2.5	5.8	2.8	3.9	1.8	5.1	5.2	4.5	5.8	3.3	3.9	2.4	2.9	3.84
Aubisque (4x)	4.1	2.6	6.8	3.1	5.3	2.5	4.8	2.9	3.5	5.4	3.3	3.9	3.3	2.9	3.88
Weigra (2x)	4.3	2.9	5.5	3.4	4.5	2.3	3.4	5.6	3.0	4.3	3.8	4.3	3.6	3.7	3.88
Elgon (4x)	3.6	2.5	7.8	3.0	5.3	2.0	4.4	4.2	3.8	4.9	3.3	4.4	3.0	2.5	3.89
Fennema (2x)	4.0	2.5	6.0	3.5	3.9	1.5	4.5	5.8	4.3	5.3	3.8	4.3	3.0	3.6	3.99
Roy (4x)	3.8	2.8	7.5	3.0	4.3	2.8	3.5	4.2	2.8	5.8	3.0	4.1	4.0	4.5	3.99
Aberdart (2x)	4.4	2.5	6.0	3.7	3.8	2.5	3.1	4.5	5.8	5.2	3.0	4.0	4.5	3.9	4.05
Corbet (2x)	5.4	2.3	6.8	2.8	4.5	2.0	5.1	5.8	3.3	6.2	4.0	3.3	3.6	2.3	4.08
Arabella (2x)	4.5	2.6	7.3	2.8	4.5	2.0	4.0	5.8	3.5	5.5	4.0	4.3	3.9	3.5	4.15
Sponsor (2x)	4.1	2.4	6.5	3.4	5.1	2.5	4.8	5.5	4.3	5.8	4.0	3.8	3.9	3.8	4.26
Litempo (4x)	5.0	3.1	7.3	3.6	6.0	4.5	5.9	3.4	2.5	5.8	4.5	4.5	3.4	4.4	4.57
Maja (4x)	5.0	2.8	7.5	2.9	5.6	3.5	6.4	3.3	4.3	6.5	3.8	4.4	3.8	4.5	4.58
Kentaur (4x)	5.4	2.8	8.0	2.8	5.0	4.5	4.6	3.5	5.0	5.7	4.3	4.5	4.3	4.7	4.63
Tivoli (4x)	3.6	3.5	7.8	3.3	5.9	2.5	5.9	3.9	5.0	6.8	4.5	4.5	4.1	4.4	4.68
Gladio (2x)	5.4	2.5	7.0	2.4	5.3	3.3	5.1	5.9	4.8	6.1	4.3	4.9	4.3	4.6	4.68
Terry (4x)	5.0	2.9	7.8	2.5	5.8	4.0	5.8	4.1	4.5	6.0	4.0	4.6	4.1	4.6	4.68
Lipresso (2x)	5.1	2.8	6.3	4.0	5.5	2.8	3.3	5.8	5.0	6.2	4.8	4.6	3.8	7.2	4.77

(continued)

Table 8.3 (continued)

Cultivar	Asendorf D	Aston UK	Bornhof D	Druelle F	Lelystad B. NL	Les Rosiers F	Malchow D	Merelbeke B	Montours F	Ottersum NL	Pulling D	Steinach D	Swifterband NL	Zurich CH	Mean of all sites
Aristo (2x)	4.5	3.1	7.0	2.8	6.8	2.8	5.8	5.3	4.5	7.5	4.0	4.3	5.4	3.4	4.79
Foxtrot (2x)	5.1	2.5	6.5	4.6	5.5	3.3	6.9	5.9	4.3	6.7	4.3	4.3	4.8	3.9	4.88
Sirocco (4x)	5.9	2.9	8.5	3.0	4.9	4.5	5.5	4.7	5.5	6.4	4.5	4.9	4.3	6.0	5.09
Helmer (4x)	6.0	3.3	8.0	3.5	6.4	4.8	6.0	4.6	5.8	6.8	4.5	5.0	4.8	6.2	5.39
Condesa (4x)	5.6	2.9	8.8	3.6	8.3	5.8	6.5	5.4	6.8	7.6	4.8	5.0	5.8	5.5	5.86
Aurora (2x)	7.5	4.4	8.0	5.6	7.5	2.8	5.9	7.3	8.0	5.8	5.5	5.8	5.7	8.0	6.26
Mean	4.5	2.6	6.5	3.1	4.7	2.8	4.3	4.5	4.1	5.5	3.7	4.1	3.7	3.9	4.15
LSD (P=0.05)	1.12	0.54	1.30	0.99	1.43	1.32	1.30	1.02	1.15	0.95	0.89	0.54	1.17	0.82	0.44
Correlation[a]	0.72	0.65	0.77	0.48	0.88	0.70	0.83	0.52	0.61	0.88	0.88	0.88	0.87	0.78	

Cultivars are ranked according to the mean of all the sites

[a]Spearman rank order correlation with mean of all sites (all values are significant at $P<0.05$)

Table 8.4 Stem rust (*Puccinia graminis*) disease scores of 16 Italian and 4 hybrid ryegrass cultivars at 5 sites (means of one scoring per site)

Cultivar	Gunpenstein A	Hladke Zivotice CZ	Montours F	Orchies F	Perugia I	Mean of all sites
Zorro (4x)	2.0	2.5	2.3	3.3	2.0	2.40
Domino (4x)	2.3	2.5	1.0	4.3	2.0	2.40
Caballo (4x)	2.3	2.5	2.3	3.5	2.3	2.55
Bolero (4x)	1.8	2.5	2.3	4.5	2.0	2.60
Ellire (4x)	2.3	2.5	3.0	4.8	1.8	2.85
Tonyl (4x)	2.3	2.8	3.0	4.8	2.0	2.95
Tarandus (4x)	2.3	4.0	2.3	5.8	2.5	3.35
Fastyl (2x)	2.8	5.0	2.0	4.5	2.5	3.35
Lolita (4x)	2.0	3.8	1.0	7.5	2.8	3.40
Danergo (4x)	2.5	4.5	1.0	7.5	2.3	3.55
Barprisma (2x)	2.8	4.8	3.0	5.5	2.3	3.65
Aberexcel (4x)[a]	2.0	4.5	3.8	6.0	2.3	3.70
Gosia (4x)[a]	2.3	4.8	3.3	6.5	1.8	3.70
Meryl (2x)	2.8	4.3	1.3	8.8	2.5	3.90
Pirol (2x)[a]	3.0	4.8	2.8	8.0	2.8	4.25
Gumpensteiner (2x)[a]	2.8	5.8	1.5	9.0	2.3	4.25
Gordo (2x)	3.5	6.5	1.0	8.8	2.3	4.40
Crema (2x)	3.5	5.5	2.5	8.3	2.8	4.50
Ligrande (2x)	3.8	6.3	1.5	8.5	2.5	4.50
Lema (2x)	4.0	6.3	1.3	9.0	2.8	4.65
Mean	2.6	4.3	2.1	6.4	2.3	3.55
LSD (p=0.05)	0.77	1.00	0.81	1.37	0.66	1.25
Correlation[b]	0.79	0.90	n.s.	0.92	0.60	

Cultivars are ranked according to the mean of all the sites
[a]Hybrid Ryegrass
[b]Spearman rank order correlation with mean of all sites (all values are significant at $P<0.05$ except n.s. = not significant)

the trial of 2013 and the four previous trials showed a consistent ranking of Italian and perennial ryegrass cultivars over the 12-year test period. In terms of crown rust resistance, the six top ranking Italian ryegrass cultivars in 2001 were still the top six in the 2013 trial and the same four cultivars were the most susceptible in 2001 and in 2013. In addition, Gwendal and Bocage were the two most crown rust resistant perennial ryegrass cultivars in all of the five experimentation trials. Condesa and Aurora were always the most crown rust susceptible cultivars. Not only the ranking but also the overall level of scoring stayed very consistent over the 12 years of observation.

Perennial ryegrass cultivars with resistance to crown as well as to stem rust are available. However, the low correlation between crown and stem rust resistance of the cultivars suggests different unlinked resistance genes to be effective against the two rust species. Breeding rust-resistant ryegrass cultivars in an environment where crown and stem rust naturally occur may result in genetic material with resistance to both species.

Table 8.5 Stem rust (*Puccinia graminis*) disease scores of 34 perennial ryegrass cultivars at 8 sites (means of 1–2 scorings per site)

Cultivar	Gunpenstein A	Hladke Zivotice CZ	Les Rosiers F	Malchow D	Montours F	Orchies F	Radzikow P	Roznov Zubri CZ	Mean of all sites
Gwendal (4x)	2.4	4.3	1.8	2.3	2.0	6.8	3.3	2.1	3.09
Pastoral (4x)	1.8	3.8	1.9	1.8	2.3	7.8	4.0	2.5	3.20
Maja (4x)	2.3	4.0	1.6	1.5	3.5	6.0	4.8	2.1	3.22
Kentaur (4x)	2.1	4.8	1.6	2.0	3.0	6.3	4.3	2.4	3.30
Bocage (4x)	2.0	4.8	2.0	2.3	1.8	8.0	3.1	2.5	3.30
Aubisque (4x)	2.3	4.5	2.1	2.5	2.0	6.5	3.8	2.9	3.31
Aberdart (2x)	3.0	6.5	2.8	1.3	1.5	5.5	4.0	2.5	3.38
Carrera (2x)	2.5	5.8	2.0	2.3	3.8	5.8	3.3	2.0	3.41
Tivoli (4x)	1.8	4.8	2.3	1.5	3.0	7.0	4.5	2.8	3.44
Lacerta (4x)	2.1	5.5	2.3	2.5	3.3	6.3	3.5	2.4	3.47
Orval (2x)	2.3	6.0	2.5	2.3	1.3	7.5	3.0	3.5	3.53
Litempo (4x)	2.1	5.3	2.3	1.8	2.8	7.3	4.5	2.5	3.55
Roy (4x)	2.4	4.3	2.1	2.3	3.3	5.0	3.8	5.5	3.56
Terry (4x)	2.0	4.5	2.3	1.8	3.0	6.5	4.5	4.0	3.56
Condesa (4x)	2.3	5.0	2.3	1.8	1.8	7.0	4.9	4.0	3.61
Helmer (4x)	2.3	4.5	2.4	1.8	3.0	6.5	5.0	4.0	3.67
Elgon (4x)	2.3	4.5	2.0	2.5	4.3	6.8	4.1	4.1	3.81
Foxtrot (2x)	3.1	5.8	2.3	1.8	3.0	7.0	5.0	3.5	3.92
Sirocco (4x)	2.3	6.0	2.4	1.8	3.0	7.8	5.4	3.8	4.03
Gladio (2x)	2.5	6.3	3.3	3.3	2.5	6.3	5.3	3.1	4.05
Aristo (2x)	2.8	6.0	2.8	2.0	5.0	7.5	3.8	3.6	4.17
Option (2x)	3.1	6.5	3.6	3.8	3.8	7.0	3.5	5.8	4.63
Guru (2x)	4.1	6.8	4.1	3.0	4.8	6.8	3.3	4.9	4.70
Barnhem (2x)	3.4	7.0	3.8	3.0	4.8	7.0	4.9	4.1	4.73
Vincent (2x)	3.6	6.3	4.6	2.8	3.8	7.0	3.9	6.5	4.80
Arabella (2x)	3.1	6.5	4.8	3.8	4.5	5.8	5.1	5.1	4.83
Weigra (2x)	3.1	6.8	4.6	2.8	4.5	5.8	4.8	6.8	4.88
Sponsor (2x)	3.4	7.0	4.6	4.3	4.8	5.8	4.0	5.3	4.88
Heraut (2x)	3.4	6.3	3.6	3.8	5.0	7.8	3.9	5.8	4.92

Cultivar	Gunpenstein A	Hladke Zivotice CZ	Les Rosiers F	Malchow D	Montours F	Orchies F	Radzikow P	Roznov Zubri CZ	Mean of all sites
Kells (2x)	3.3	6.0	4.4	2.8	4.3	8.0	4.1	6.9	4.95
Corbet (2x)	3.3	6.8	6.0	3.0	5.3	7.8	4.4	4.3	5.08
Fennema (2x)	3.4	5.8	5.8	3.3	4.5	8.0	4.9	7.5	5.38
Lipresso (2x)	3.9	7.0	5.1	5.0	4.0	6.3	5.8	6.8	5.47
Aurora (2x)	4.6	7.0	7.3	6.8	1.5	8.0	5.8	7.4	6.03
Mean	2.8	5.6	3.2	2.7	3.4	6.8	4.3	4.2	4.11
LSD ($P=0.05$)	0.51	0.65	0.96	1.20	0.77	1.79	0.92	0.85	0.85
Correlation[a]	0.82	0.77	0.92	0.74	0.61	n.s.	0.43	0.90	

Cultivars are ranked according to the mean of all the sites

[a]Spearman rank order correlation with mean of all sites (all values are significant at $P<0.05$ except n.s. = not significant)

Conclusion

Resistant cultivars showed a rather stable resistance in different environments as well as over time. There is no evidence for a breakdown of resistance of an individual cultivar during the 12-year testing period.

References

Boller B, Schubiger FX, Streckeisen P (2003) The Eucarpia multisite rust evaluation – results 2001. Vortr. Pflanzenzüchtung 59:198–207.

Schubiger FX, Streckeisen P, Boller B (2007) The EUCARPIA Multisite Rust Evaluation – Results of the trials 2004. D. Rosellini and F. Veronesi (eds.) Breeding and seed production for conventional and organic agriculture. Proceedings of the Eucarpia meeting Perugia:154–158.

Schubiger FX, Boller B et al. (2010a) The EUCARPIA Multi-site rust evaluation – Results of the trials 2007. C. Huyghe (ed) Sustainable use of genetic diversity in forage and turf breeding. Springer Science+Business Media Dortrecht. pp 331–340.

Schubiger FX et al. (2010b) Susceptibility of European cultivars of Italian and perennial ryegrass to crown and stem rust. Euphytica 176: 167–181.

Schubiger FX et al. (2012) The EUCARPIA Multi-site rust evaluation – Results 2010. S. Barth and D. Milbourne (eds.), Breeding Strategies for Sustainable Forage and Turf Grass Improvement. Springer Science+Business Media Dortrecht. pp 209–217.

Variation in Drought Tolerance of Perennial Ryegrass (*Lolium perenne* L.)

9

P. Westermeier, A. Wosnitza, E. Willner, U. Feuerstein, W. Luesink, S. Schulze, A. Schum, and S. Hartmann

Abstract

Global climate change is one of the biggest challenges for research in agricultural science. In Europe, expected temporary periods of reduced precipitation during the growing season will affect crop yields. Perennial ryegrass (*Lolium perenne* L.) is one of the most important forage grasses in Europe due to high yields and nutritional composition. Because perennial ryegrass has no distinct tolerance to drought, it is likely to be particularly affected by global climate change. Perennial ryegrass is found in many geographical regions across Europe, thus genetic variation for drought tolerance is likely. We have evaluated persistence under temporary drought conditions of 200 accessions, representing gene bank material from several countries with differing amounts of precipitation as well as breeding material. The evaluation was based on field trials at drought-prone locations. A contrasting drought response could be detected within the accessions and a representative subsample of 54 accessions could be identified for use in a two location rain-out shelter trial. Different methods for phenotyping recovery after drought stress were compared and traits were scored on a single plant basis in the semi-controlled environment rain-out shelter. A huge variation was found not only between, but also within accessions. It became obvious that the most important mechanism under Central European drought conditions is not maintenance of biomass

P. Westermeier (✉) • A. Wosnitza • S. Hartmann
Institute for Crop Science and Plant Breeding; Forage Plant Research Breeding and Grassland Systems, Bavarian State Research Center for Agriculture, Freising, Germany
e-mail: peter.westermeier@LfL.bayern.de

E. Willner
IPK Gatersleben Gene Bank, Satellite Collection North, Malchow/Poel, Germany

U. Feuerstein
Deutsche Saatveredelung AG, Lippstadt, Germany

W. Luesink
Norddeutsche Pflanzenzucht Hans-Georg Lembke KG, Holtsee, Germany

S. Schulze
Saatzucht Steinach GmbH & Co KG, Steinach, Germany

A. Schum
Institute for Resistance Research and Stress Tolerance, Julius Kühn-Institut, Groß Lüsewitz, Germany

I. Roldán-Ruiz et al. (eds.), *Breeding in a World of Scarcity: Proceedings of the 2015 Meeting of the Section "Forage Crops and Amenity Grasses" of Eucarpia*,
DOI 10.1007/978-3-319-28932-8_9

production under severe drought stress but rather fast recovery after a limited period of water shortage. Single clones with contrasting recovery performance could be identified and will be used to investigate drought tolerance mechanisms and breed new varieties in future projects.

Keywords
Perennial ryegrass • Drought tolerance • Recovery • Genetic diversity • Rain-out shelter • Phenotyping

Introduction

The impact of global climate change and its consequences for plant production is one of the major challenges for current and future agricultural research. Depending on the assumed increase of the CO_2 concentration in the atmosphere, mean summer temperatures are expected to increase between 2 and 3 °C and mean precipitation rate is expected to decline between 20 and 40 % in Central Europe, together with increased incidence of extreme weather situations (IPCC 2007). The drought type expected to occur more often is called the Mediterranean drought type, characterized by rain-fed winters and irregular rainfall and periods of drought in summer. This change will have a negative influence on the plant-available soil water content, leading to drought-mediated crop damage and yield losses (Lobell and Gourdji 2012). Especially perennial ryegrass (*Lolium perenne* L.), one of the most important forage grasses in Europe (Sampoux et al. 2013) is known to have no distinct drought tolerance (Sheffer et al. 1987; Thomas and Evans 1990). Perennial ryegrass occurs naturally in a wide range of habitats in Central Europe, Asia and North Africa. Therefore genetic variation for drought tolerance was expected to be found in accessions of diverse geographic origin.

Drought stress response of plants can be divided into two categories: stress adaptation and stress tolerance (Ludlow and Muchow 1990). Stress adaptation is often associated with a faster growth cycle to escape drought stress and thus implies a lower yield potential (Kumar 2005). Stress tolerance can be divided into the subcategories stress avoidance and real stress tolerance. Stress avoidance mechanisms activate mechanisms to reduce transpiration rate or to explore unused soil water, which seems promising to bridge the time-limited drought periods expected to occur more often in the future. Real stress tolerance addresses mechanisms to maintain physiological functions under water shortage, which is often correlated with slow growth rates and low yield under an optimal water supply. For perennial ryegrass grown under fluctuating water conditions, the most promising strategy to secure yield is not the maintenance of biomass production during a time-limited drought period (short term response), but the ability to survive and recover after rewatering to secure persistence and yield over an entire year (long-term response) (Kemp and Culvenor 1994; Lelièvre and Volaire 2009).

Screening germplasm collections and using them to breed new varieties with enhanced drought tolerance is one of the most promising approaches to overcome the problems rising with climate change (Challinor et al. 2014). Sanna et al. (2012) found a wide variability for persistence in natural populations of perennial ryegrass originating from Sardinia and Corsica under semi-arid rain fed conditions. Cheplik et al. (2000) showed genetic variation for recovery after drought stress in perennial ryegrass independent from endophyte infection. Genetic progress for the target trait "*yield under drought stress*" is slow, due to low heritability and unsteady selection environments for drought tolerance over the years. Therefore secondary selection traits are used with higher heritabilities as well as good correlation with yield under temporary drought conditions. Rain-out shelter trials in a semi-controlled environment can complement the selection for drought tolerant individuals and to increase selection gain.

Materials and Methods

Plant Material

A total of 186 accessions of perennial ryegrass were tested for drought tolerance. Included in the set were wild collections (73) originating from Europe and the Near East (from rain-fed Ireland to the drought-prone countries Turkey and Iran) as well as historical European varieties (2) provided by IPK-Gatersleben gene bank. 111 current varieties and candidates from commercial breeding programs were provided by breeding companies. The set was supplemented by the potentially drought tolerant species *Festulolium* (*Festuca* spec. x *Lolium* spec.; ten varieties), meadow fescue (*Festuca pratensis* Huds.; two varieties) and tall fescue (*Festuca arundinacea* Schreb.; two varieties) as checks. Field trials were sown as microplots in autumn 2011 at five locations (Bornhof, Kaltenhof, Malchow, and Triesdorf in Germany; Les Rosiers sur Loire in France) in a 20×10 alpha-lattice design with four replicates and were phenotyped in 2012 (first harvest year). Based on 2012 data for visual drought response scoring, 54 accessions with diverse drought response were selected for testing in rain-out shelters (locations Freising and Kaltenhof; Germany, RCB-design, two replicates) supplemented by two additional diploid *Festulolium* (F_1 hybrid of meadow fescue × perennial ryegrass). For the rain-out shelter trials, 40 individuals of each accession were cloned (20 plants per replication) for testing identical genotypes at both rain-out shelter sites. Observation of plants in the rain-out shelters was conducted in 2013 and 2014.

Phenotyping

In the field trials under natural drought conditions, the amount of biomass directly before each harvest time point was scored visually as a plot mean on a 1–9 scale with 9 as maximum biomass growth. Additionally, drought response was scored visually on a 1–9 scale with 9 indicating strong symptoms of drought stress including leaf rolling, wilting and leaf senescence. In rain-out shelter trials, two drought phases per vegetation period were induced. The first drought period started at the end of April, simulating a spring drought, and ended after volumetrically determined soil water content dropped below field capacity (approx. 35–40 days). After rewatering and recovery of the plants, a second drought phase started at the end of June, simulating an early summer drought. Visually scored traits (biomass production, drought response) were recorded on a single plant basis to enable the selection of the most tolerant individuals and were calculated as plot mean for statistics. Yield parameters (fresh weight, dry matter content and dry weight) were determined manually on a plot basis comprising all 20 plants of a replicate.

Statistical Analysis

The software package PLABSTAT version 3A (Utz 2005) was used for ANOVA analysis, calculation of heritabilities and adjusted means based on the following statistical models. The model $y_{ijk}=\mu+\alpha_i+l_j+\beta_{kj}+(al)_{ij}+e_{ijk}$ was used for the RCB design in the rain-out shelters with y_{ijk} *being the* trait observation, μ as overall mean, α_i *as* fixed effect of genotype *i*, l_j *as* random effect of location *j*, $(al)_{ij}$ *as* random interaction effect of genotype *i* with location *j*, β_{kj} *as* fixed effect of replication *k* nested within location *j*, *and* e_{ijk} *as* random residual error. The model $y_{ijkm}=\mu_j+\alpha_{ij}+r_{kj}+\beta_{mkj}+e_{ijkm}$ was used for the single location alpha lattice analysis with y_{ijkm} being the trait observation of location *j*, μ_j as location-specific mean, α_{ij} as fixed effect of genotype *i*, r_{kj} as random effect of replication *k*, β_{mkj} as fixed effect of block *m* nested within replication *k*, and e_{ijkm} as random residual error. Adjusted means were committed to the analyses across locations using the model $y_{ij}=\mu+\alpha_i+l_j+(al)_{ij}+e_{ij}$ with y_{ij} being the adjusted mean of genotype i from location j, μ as overall mean, α_i as fixed effect of genotype, l_j as random effect of location, $(\alpha l)_{ij}$ as random genotype x environment interaction, and e_{ij} as residual error with $\mathbf{e} \sim N(0, \sigma^2_e)$, where σ^2_e is pooled residual error of all single location analyses. The software package R 3.1.2 (R Development Core Team 2014) was used for mean comparison analysis.

Results and Discussion

Genotypic Variation for Persistence and Recovery Under Drought Tolerance

Phenotyping a total of 200 accessions across three harvest years at drought-prone locations enabled an evaluation of the persistence under drought conditions. Visual scoring of biomass growth showed significant genotypic variance across locations in all 3 years. Heritabilities for single cutting time points ranged from 18.1 to 80.1 % in 2012, 54.6–70.5 % in 2013 and 63.1–74.7 % in 2014. This indicates existing genetic variance in the set of accessions, as every single time point (including those after a drought stress phase) showed significant genotypic variance for biomass production. To estimate drought response of the plants, visual impact of drought on the plant phenotype was scored. The visual drought stress response only showed acceptable repeatabilities when severe drought conditions occurred. This was not the case in 2013 and was not given in every environment in the years 2012 and 2014. Heritability reached 40.8 % in 2012 and 42.0 % in 2014 with single repeatabilities of up to 72.5 % (location Kaltenhof in 2012), indicating the difficulties of direct selection for drought tolerance.

For this reason semi-controlled phenotyping platforms like rain-out shelters provide a more stable and reliable selection environment under otherwise nearly practical growing conditions. Visual scoring of drought response showed increased repeatability for single locations under semi-controlled conditions and reached a maximum of 84.7 % in 2013 for location Freising with a significant correlation to biomass scoring at cutting time point 4 (after second drought stress period) of 0.38. Heritability for visual drought stress response was also low and confirmed the difficulties of multi-location comparison for this trait even under semi-controlled conditions. Heritabilities for visual biomass scoring ranged between 63.5–79.5 % in 2013 and 79.9–91.1 % in 2014 with a highly significant correlation to yield measurement (between 0.83 and 0.92) in both years. That type of scoring seems to have much more potential for assessing drought tolerance.

Figure 9.1 shows the time course of the visually scored biomass production of the most susceptible and tolerant perennial ryegrass accessions (diploid and tetraploid) spanning two harvest years compared to *Festulolium* and *Festuca* species. Perennial ryegrass showed a high variation in the potential of biomass production as well as in the ability to recover after drought periods (time points 2/1, 4/1 and 2/2, 4/2). While the best rated accessions could recover nearly completely after the second drought phase in 2013, the worst rated remained at a low level of biomass scoring, which also continued in the second year. The importance of the ability to recover after a drought period was already described as one possible mechanism for drought tolerance by Kemp and Culvenor (1994). The overall level of biomass rating was higher in the tetraploid material group than in the diploid. Furthermore tetraploids showed a higher ability to recover after drought. One reason is that tetraploid plant material has enhanced resistance against abiotic stress compared to their diploid counterparts (Liu et al. 2011) because of increased hydraulic conductivity in the xylem (Maherali et al. 2009). Variation in biomass rating was higher within the diploid accessions than in the tetraploid group. Besides breeding material the diploid group contained all gene bank accessions, which were not selected for yield, as well as historical varieties. This could be one explanation for the higher variability and the lower mean level of biomass rating, whereas the tetraploid accessions exclusively originate from current

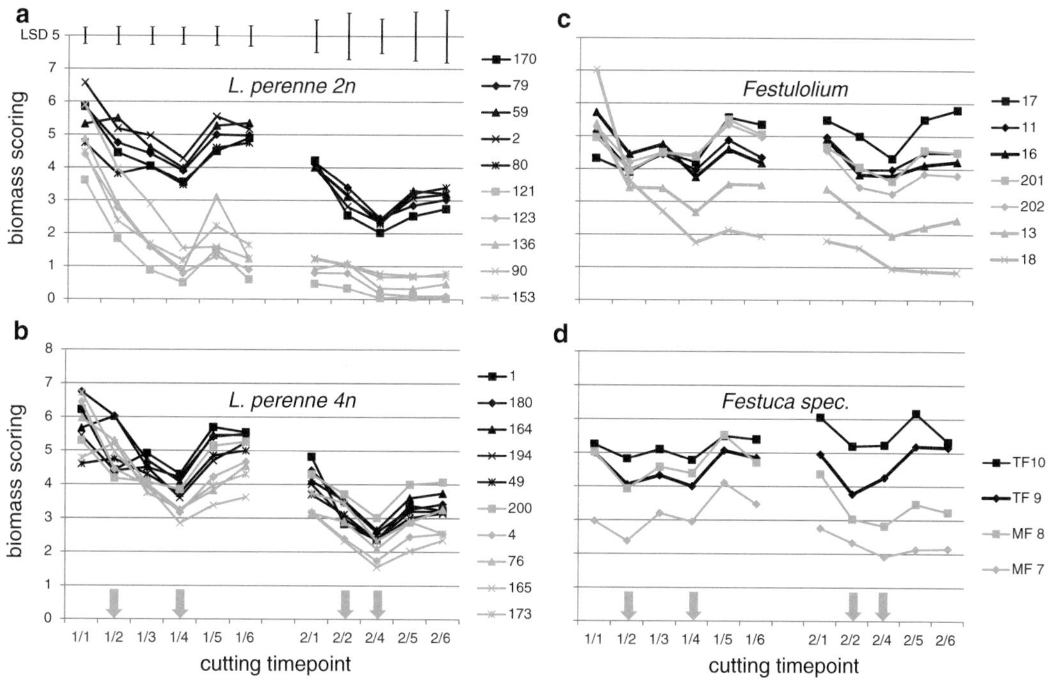

Fig. 9.1 Visual scoring of biomass growth before harvest on a 1–9 scale in a two-location rain out shelter trial in year 2013 (timepoints x/1) and 2014 (timepoints x/2). Red arrows indicate the induction of drought stress. (**a**): extremes of diploid *ryegrass* accessions; (**b**): extremes of tetraploid *ryegrass* accessions; (**c**): 7 *Festulolium* accessions; (**d**): Each two accessions of *tall fescue* (TF) and *meadow fescue* (MF). LSD5: least significant difference at p ≤ 5 % level

breeding material. *Festuca* species and *Festulolium* are supposed to be more drought tolerant than perennial ryegrass. Interestingly, the seven *Festulolium* and four *Festuca* species checks in the rain-out shelter trial showed a similar variation in biomass production compared to the *Lolium* accessions. In the first harvest year, both tall and meadow fescue showed a relatively high level of recovery after drought stress, while in the second harvest year meadow fescue showed a severe decline. One reason for the differences in the persistence could be deeper rooting of tall fescue in the second year. Especially in the second harvest year, also *Festulolium* displayed a large variation in recovery with phenotypic scores between 1 and 6. Hexaploid *Festulolium*, similar to tall fescue (accessions 11, 16 and 17), showed the best persistence to drought analogous to pure tall fescue. The tetraploid *Festulolium* 13 (a cross between *L. perenne* L. and *F. pratensis* Huds.) showed persistence and recovery similar to diploid perennial ryegrass and the Italian ryegrass type 18 (cross between *L. multiflorum* Lam. and *F. arundinacea* Schreb.) was the most susceptible accession in this group. The diploid *Festulolium* accessions 201 and 202 (F_1 of the cross between *L. perenne* L. and *F. pratensis* Huds.) showed similar biomass production than the tall fescue type *Festulolium*.

Conclusion and Outlook

A large variation could be detected for drought response and biomass production in perennial ryegrass as well as in *Festulolium*, meadow and tall fescue. Selecting genotypes with the ability to recover fast after a temporary limited drought period seems promising to enhance drought tolerance for the Mediteranean type of drought stress.

With single clones identified within the accessions showing contrasting recovery it will be possible to build segregating populations. This genetic resources will be used to answer questions concerning the morphological (e.g. root depth) and physiological (e.g. ingredients, water use efficiency) background as well as the genetic mechanisms of persistence ability under drought stress.

References

Challinor AJ, Watson J, Lobell DB, Howden SM, Smith DR, Chhetri N (2014) A meta-analysis of crop yield under climate change and adaptation. Nature Climate Change 4: 287–291.

Cheplick GP, Perera A, Koulouris K (2000) Effect of drought on the growth of *Lolium perenne* genotypes with and without fungal endophytes. Functional Ecology 14: 657–667.

IPCC (2007) Climate Change 2007 – Impacts, adaptation and vulnerability. Contribution of working group II to the fourth assessment report of the Intergovernmental Panel on Climate Change. Cambridge University Press, Cambridge/NY.

Kemp DR, Culvenor RA (1994) Improving the grazing and drought tolerance of temperate perennial grasses. New Zealand Journal of Agricultural Research 37: 365–378.

Kumar D (2005) Breeding for drought resistance. In: Ashraf M, Harris PJC (Hg.) Abiotic stresses. Plant resistance through breeding and molecular approaches. Food Products Press, New York. S. 145–176.

Lelièvre F, Volaire F (2009) Current and potential development of perennial grasses in rainfed mediterranean farming systems. Crop Science 49: 2371–2378.

Liu S, Chen S, Chen Y, Guan Z, Yin D, Chen F (2011) In vitro induced tetraploid of *Dendrathema nankingense* (Nakai) Tzvel. shows an improved level of abiotic stress tolerance. Scientia Horticulturae 127: 411–419.

Lobell DB, Gourdji SM (2012) The influence of climate change on global crop productivity. Plant Physiology 160: 1686–1697.

Ludlow MM, Muchow RC (1990) A critical evaluation of traits for improving crop yields in water-limited environments. Advances in Agronomy 43: 107–153.

Maherali H, Walden AE, Husband BC (2009) Genome duplication and the evolution of physiological responses to water stress. New Phytologist 184: 721–731.

Sampoux JP, Baudouin P, Bayle B, Béguier V, Bourdon P, Chosson JF, De Bruijn K, Deneufbourg F, Galbrun C, Ghesquière M, Noël D, Tharel B, Viguié A (2013) Breeding perennial ryegrass (*Lolium perenne* L.) for turf usage: an assessment of genetic improvements in cultivars released in Europe, 1974–2004. Grass and Forage Science 68: 33–48.

Sanna F, Saba P, Sassu M, Franca A (2012) Preliminary selection of *Lolium perenne* L. natural populations for pasture improvement purpose in rainfed Mediterranean conditions. In : Acar Z, López-Francos A, Porqueddu C (eds.). New approaches for grassland research in a context of climate and socio-economic changes. Zaragoza : CIHEAM, 2012: 213–217 (Options Méditerranéennes : Série A. Séminaires Méditerranéens; n. 102)

Sheffer KM, Dunn JH, Minner DD (1987) Summer drought responses and rooting depth of three cool-season turfgrasses. HortScience 22: 296–297.

R Development Core Team (2014) R: A language and environment for statistical computing. R Foundation for Statistical Computing, Vienna, Austria.

Thomas H, Evans C (1990) Influence of drought and flowering on growth and water relations of perennial ryegrass populations. Annals of Applied Biology 116: 371–382.

Utz HF (2005) PLABSTAT – plant breeding statistical program. Version 3A. Institut für Pflanzenzüchtung, Saatgutforschung und Populationsgenetik, Universität Hohenheim, Stuttgart.

Drought Stress-Induced Changes in Gas Exchange, Chl *a* Fluorescence Parameters and Yield in *Lolium*, *Festuca* and *Festulolium*

10

A. Fariaszewska, J. Aper, M. Staniak, J. Baert, J. Van Huylenbroeck, and J. De Riek

Abstract

Changes in yield and physiological processes in drought stress (40 % field water capacity) and well-watered conditions (70 % field water capacity) in forage grasses were investigated. Nine varieties from five species (*Lolium perenne*, *Lolium multiflorum*, *Festuca pratensis*, *Festuca arundinacea* and *Festulolium braunii*) were examined in a pot experiment in a greenhouse in 2013–2014 in Poland. The aim of the study was to assess the effects of water deficit on different forage grasses. All measured physiological parameters were affected by drought stress in the different cuts. Intensity of photosynthesis, transpiration rate and dry matter yield were significantly lower in drought stress than under well-watered conditions in all varieties. The differences in chlorophyll fluorescence (the *Fv/Fm* ratio, an estimation of the PSII phytochemical efficiency) were less pronounced, although most of varieties showed a minor but significant decrease of the *Fv/Fm* ratio under drought stress.

Keywords

Photosynthesis • Transpiration • Fluorescence • Drought stress • Grasses

Introduction

Forage grasses occupy a very important place in the group of forage crops. In view of the expected climate change that will likely include longer periods of drought stress, researchers and grass breeders are interested in grasses performing better during periods of mild drought. *Lolium perenne* and *L. multiflorum* are considered to be optimal species for northwestern European

A. Fariaszewska (✉) • M. Staniak
Department of Forage Crop Production, Institute of Soil Science and Plant Cultivation, State Research Institute, Puławy, Poland
e-mail: afariaszewska@iung.pulawy.pl

J. Aper • J. Baert • J. Van Huylenbroeck • J. De Riek
Plant Sciences Unit, Institute for Agricultural and Fisheries Research (ILVO), Melle, Belgium

I. Roldán-Ruiz et al. (eds.), *Breeding in a World of Scarcity: Proceedings of the 2015 Meeting of the Section "Forage Crops and Amenity Grasses" of Eucarpia*,
DOI 10.1007/978-3-319-28932-8_10

grasslands as they combine high yields with nutritious forage. However, they lack resilience against drought. *Festuca* species are more drought tolerant but they present problems with digestibility and palatability. *Festulolium* hybrids are expected to combine the advantages from both the *Festuca* and *Lolium* complexes. Obviously, longer drought stress interferes with plant metabolism and limits productivity of crop plants (Bohnert et al. 1994). A drought period limits the rate of photosynthesis as a first reaction to stress conditions (Kalaji and Łoboda 2010). Another plant mechanism to protect against water deficit is reduced transpiration through stomata closure. Nevertheless, the reaction to drought stress differs between grass species and even varieties due to specific morphological-biological characteristics (Goliński and Xi 1998). In addition, chlorophyll fluorescence can be used as indicator of photosynthetic changes in PSII under different abiotic stresses (Rahbarian et al. 2011). The aim of this study was to determine the physiological reactions of *Festuca*, *Lolium* and *Festulolium* species under mild drought stress. More specifically, we investigated differences at the level of forage yield, photosynthesis, transpiration rate and chlorophyll fluorescence.

Materials and Methods

Plant Material

Nine varieties of forage grasses were evaluated: *L. perenne* var. 'Melluck' (2n), 'Meltador' (4n), *L. multiflorum* var. 'Meldiva' (2n), 'Melmia' (4n), *F. pratensis* var. 'Merifest' (2n), 'MerifestTp' (4n), *F. arundinacea* var. 'Barolex' (6n), 'Callina' (6n) and *Festulolium braunii* var. 'Felopa' (4n).

Structure of Experiment

A completely randomized block experiment was carried out in a greenhouse in 2013–2014 in the Institute of Soil Science and Plant Cultivation, State Research Institute (IUNG-PIB) in Poland. On 16 April 2013, 16 Mitscherlich pots for each variety were sown with 29 seeds and later thinned to 15 plants per pot (144 pots total). When the height of the plants reached approximately 50 cm, they were cut: three times in 2013 and one time in 2014. All pots were divided into four groups of 36 pots: one control group was kept at 70 % field water capacity (FWC, well-watered conditions) during all cuts, one was kept at 70 % FWC and reduced to 40 % FWC (drought stress) during 3 weeks before the first cut (around 20 June–3 July 2013), one was kept at 70 % FWC during the first cut but reduced to 40 % FWC during 3 weeks before the second cut (around 24 July–8 August 2013) and one group was kept at 70 % FWC during the first and second cut and reduced to 40 % FWC during 3 weeks before the third cut (around 2–15 September 2013). In spring 2014, eight pots were randomly selected for each variety (in total 72 pots) and one control group of 36 pots was kept at 70 % FWC while the other group of 36 pots was kept at 70 % FWC but reduced to 40 % FWC during 3 weeks before the 4th cut (around 15–20 May 2014).

Physiological Measurements

All physiological measurements were made after 2 weeks of drought treatment. The intensity of photosynthesis and transpiration measurements were performed on the second fully developed leaf in three replications per pot using a Ciras-2 Portable Photosynthesis System (PP Systems, Amesbury, MA, USA): temp 22–27°C, PAR 1000–1200 $\mu mol \cdot m^{-2} \cdot s^{-1}$, CO_2 – 380 μmol $CO_2 \cdot mol^{-1}$ air.

Chl *a* fluorescence (the *Fv/Fm* ratio, created by dividing variable fluorescence *Fv* by maximum fluorescence *Fm*) which estimates of the PSII phytochemical efficiency was measured on the leaf after 25 min of adaptation to darkness in five replications/pot using a Pocket PEA fluorimeter (Hansatech Instruments ltd, Norfolk, UK).

Statistical Analysis

All data from both experiments were subjected to analysis of variance (ANOVA) and means were analyzed using the LSD (Least significant difference) test. All calculations were performed using Statgraphic Centurion v. 16 (Statpoint Technolgies, Inc).

Results and Discussion

Drought stress caused a significant reduction in dry matter yield in all varieties during all four investigated cuts (Table 10.1). The reduction was greatest at the third cut (on average by 33.7 % compared to the plants grown under optimal moisture conditions). 'Merifest Tp' had on average the lowest relative reduction in dry matter yield for all cuts (15.6 %). The highest reduction was found for 'Meltador' (on average 32.7 % for the four cuts). In absolute terms, 'Merifest Tp' had on average the highest total production in drought conditions for the four cuts (28.7 g pot^{-1}).

Under conditions of reduced soil moisture to 40 % FWC, the intensity of photosynthesis and transpiration was significantly lower in all grass varieties. CO_2 assimilation decreased in all varieties at 40 % FWC compared to 70 % FWC (on average 25.6 % for the four cuts) (Table 10.2). The highest reduction was observed before the second cut (on average 31.2 %). The reduction was greatest for 'Merifest' (on average 31.9 %) and the lowest for Felopa (on average 18.8 %). The transpiration process was significantly lower in drought stress than optimal soil moisture level in all the studied species and varieties of grasses. The highest reduction of this process was found before the second cut: on average 37.8 %. 'Meldiva' showed the greatest reduction of transpiration: on average by 39.2 % for all cuts. In absolute terms, 'Merifest' had on average the lowest value of transpiration in water deficit for the four cuts (1.57 mmol $H_2O{\cdot}m^{-2}{\cdot}s^{-1}$). On the other hand, the decrease in transpiration under drought stress was lowest for 'Meltador': an average of 15.1 % (Table 10.3). Across the species, the transpiration process was strongly reduced in *L. multiflorum* and *F. pratensis,* indicating a better protection mechanism against drought stress than *L. perenne* and *Festulolium braunii.*

The *Fv/Fm* ratio was decreased under drought stress in all four cuts in most varieties. Drought stress had minor impact on efficiency of PSII, but differences between optimally watered plants and drought stress were significant (Table 10.4).

Table 10.1 Dry matter yield of nine forage grass varieties after 3 weeks of drought stress

	Dry matter yield (g pot^{-1})							
	2013						2014	
	Cut I		Cut II[b]		Cut III[b]		Cut IV	
Varieties	C[a]	S[a]	C	S	C	S	C	S
L. perenne var. Melluck	32.2	23.8	33.9	18.3	22.7	12.9	26.5	24.5
L. perenne var. Meltador	43.9	36.3	39.9	23.1	21.1	10.3	32.9	26.2
L. multiflorum var. Meldiva	41.9	36.6	33.4	27.0	19.3	13.9	38.3	29.3
L. multiflorum var. Melmia	46.7	38.9	32.8	28.2	20.9	14.2	39.1	32.2
F. pratensis var. Merifest	34.2	27.7	36.3	27.9	27.5	20.1	30.0	27.5
F. pratensis var. Merifest Tp.	27.8	26.6	37.9	31.4	34.2	29.1	37.4	27.6
F. arundinacea var. Barolex	34.0	22.6	34.2	27.6	29.6	25.5	40.0	29.8
F. arundinacea var. Callina	30.8	23.4	33.3	26.1	36.4	21.0	38.7	26.1
F. braunii var. Felopa	34.3	33.3	23.2	21.9	34.1	16.0	35.4	31.0
LSD varieties/treatment	2.94		1.95		1.57		2.22	

LSD least significant difference ($\alpha=0.05$)

[a]C-70 % FWC (well-watered), S-40 % FWC (water stress)

[b]Pots in cut II were not subjected to drought stress in cut I. Pots in cut III were not subjected to drought stress in cut II and cut I

Table 10.2 Intensity of photosynthesis of nine forage grass varieties after 2 weeks of drought stress

	Intensity of photosynthesis (µmol $CO_2 \cdot m^{-2} \cdot s^{-1}$)							
	2013						2014	
	Cut I		Cut II[b]		Cut III[b]		Cut IV	
Variety	C[a]	S[a]	C	S	C	S	C	S
L. perenne var. Melluck	9.6	7.4	7.0	3.5	7.5	6.9	8.2	7.3
L. perenne var. Meltador	12.0	7.3	8.4	5.7	7.8	7.1	10.6	9.3
L. multiflorum var. Meldiva	11.9	9.5	9.9	7.2	7.3	5.1	9.6	7.4
L. multiflorum var. Melmia	13.0	8.5	10.6	8.1	8.4	6.1	11.4	8.9
F. pratensis var. Merifest	10.9	6.5	6.9	4.5	8.3	6.7	10.2	6.8
F.pratensis var. Merifest Tp.	10.0	6.7	8.1	4.8	10.2	8.9	11.4	8.3
F. arundinacea var. Barolex	11.1	7.4	8.3	6.0	10.8	8.4	10.6	8.3
F. arundinacea var. Callina	8.1	6.5	8.3	6.0	9.5	7.1	11.4	7.5
F. braunii var. Felopa	11.7	9.6	10.6	8.8	9.2	6.6	10.9	9.6
LSD varieties/treatment	0.66		0.43		0.38		0.30	

LSD least significant difference ($\alpha = 0.05$)
[a]C-70 % FWC (well-watered), S-40 % FWC (water stress)
[b]Pots in cut II were not subjected to drought stress in cut I. Pots in cut III were not subjected to drought stress in cut II and cut I

Table 10.3 Transpiration of nine forage grass varieties after 2 weeks of drought stress

	Transpiration (mmol $H_2O \cdot m^{-2} \cdot s^{-1}$)							
			2013		2014			
	Cut I		Cut II[b]		Cut III[b]		Cut IV	
Varieties	C[a]	S[a]	C	S	C	S	C	S
L. perenne var. Melluck	3.5	2.3	2.8	1.8	2.5	2.4	1.5	1.2
L. perenne var. Meltador	3.2	2.2	2.6	2.3	2.6	2.6	1.7	1.4
L. multiflorum var. Meldiva	4.1	2.1	2.8	2.1	2.3	1.4	1.6	0.9
L. multiflorum var. Melmia	3.9	2.2	2.7	2.6	2.0	1.3	1.6	1.1
F. pratensis var. Merifest	4.7	2.9	1.3	1.0	1.7	1.3	2.2	1.1
F. pratensis var. Merifest Tp.	3.9	2.3	2.0	0.9	2.3	1.8	1.7	1.3
F. arundinacea var. Barolex	5.2	2.7	2.8	2.3	2.3	1.8	1.8	1.5
F. arundinacea var. Callina	4.3	2.7	2.7	2.1	2.4	1.6	2.0	1.1
F. braunii var. Felopa	4.6	3.8	3.0	2.6	2.7	2.2	2.4	1.7
LSD varieties/treatment	0.48		0.29		0.27		0.10	

LSD least significant difference ($\alpha = 0.05$)
[a]C-70 % FWC (well-watered), S-40 % FWC (water stress)
[b]Pots in cut II were not subjected to drought stress in cut I. Pots in cut III were not subjected to drought stress in cut II and cut I

Table 10.4 *Fv/Fm* ratio of nine forage grass varieties after 2 weeks of drought stress

	Fv/Fm							
	2013						2014	
	Cut I		Cut II[b]		Cut III[b]		Cut IV	
Varieties	C[a]	S[a]	C	S	C	S	C	S
L. perenne var. Melluck	0.80	0.78	0.80	0.78	0.80	0.79	0.78	0.78
L. perenne var. Meltador	0.80	0.79	0.80	0.79	0.80	0.80	0.79	0.79
L. multiflorum var. Meldiva	0.80	0.79	0.80	0.76	0.81	0.80	0.80	0.76
L. multiflorum var. Melmia	0.80	0.79	0.80	0.78	0.81	0.81	0.80	0.78
F. pratensis var. Merifest	0.80	0.79	0.80	0.75	0.82	0.80	0.79	0.77
F. pratensis var. MerifestTp.	0.80	0.80	0.80	0.76	0.80	0.81	0.78	0.77
F. arundinacea var. Barolex	0.79	0.78	0.78	0.77	0.81	0.80	0.78	0.76
F. arundinacea var. Callina	0.79	0.78	0.79	0.77	0.80	0.80	0.76	0.75
F. braunii var. Felopa	0.80	0.79	0.81	0.81	0.82	0.81	0.80	0.79
LSD varieties/treatment	0.005		0.007		0.004		0.090	

LSD least significant difference ($\alpha = 0.05$)
[a]C-70 % FWC (well-watered), S-40 % FWC (water stress)
[b]Pots in cut II were not subjected to drought stress in cut I. Pots in cut III were not subjected to drought stress in cut II and cut I

Conclusions

All varieties of forage grasses in the present study showed significantly lower values for yield, intensity of photosynthesis, transpiration and *Fv/Fm* ratio under drought stress versus optimal water conditions. In general, the reduction in dry matter yield per pot was greatest at the third cut. The smallest relative reduction in yield was found in *F. pratensis* and this species had on average the highest total production in drought conditions for the four cuts (28.7 g pot^{-1}). *L. perenne* was highly sensitive as compared with the other grasses. The transpiration process was strongly reduced in *L. multiflorum* and *F. pratensis,* indicating a better protection mechanism against drought stress than *L. perenne* and *Festulolium braunii*. Intensity of photosynthesis was the least reduced for *Festulolium braunii* compared with the other grasses. Water deficit also caused reduction of chlorophyll fluorescence *a* (*Fv/Fm* ratio). This study can help breeders to explain the differences between forage grasses in their reaction to mild drought stress.

References

Bohnert HJ, Nelson DE, Jensen RG. (1994) Adaptations to environmental stresses. Plant Cell, 7: 1099–1111.
Goliński P, Xi Q. (1998) Biological and chemical properties of creeping red fescue from point of view of its utilization in difficult site conditions. Grass. Sci. Eu., 2: 699–702.
Kalaji MH, Łoboda T. (2010) Chlorophyll fluorescence in the studies of the physiological condition of plants. SGGW Warsaw, p. 116–118.
Rahbarian R, Khavari-Nejad R, Ganjeali A, Bagheri A, Najafi A, (2011) Drought stresses effects on photosynthesis, chlorophyll fluorescence and water relations in tolerant and susceptible chickpea *(Cicer Arietinum* L.) genotypes. Acta Biol. Cracovensia, 1: 47–53.

Toward a Standard Test for Summer Dormancy in Tall Fescue

11

E.C. Brummer and M. Trammell

Abstract

Tall fescue (*Lolium arundinaceum* (Schreb.) Darbysh.) is a broadly adapted forage grass that tolerates diverse biotic stresses. Summer dormant tall fescue persists in hotter and drier environments than summer active tall fescue, but productivity of summer dormant cultivars lags behind both summer active cultivars and alternative crops for dryland regions. Therefore, yield improvement of summer dormant cultivars is a breeding target. We have made hybrids between summer dormant and summer active genotypes in an effort to transfer the summer dormant trait into a summer active genetic background. Limiting our progress is the lack of a standard test to assess summer dormancy. In this paper, we propose a standard test and highlight the need for assessing the depth and length of the dormancy period.

Keywords

Summer dormancy • Phenotyping • Breeding

Introduction

Tall fescue can be broadly classified into two major ecotypic groups – summer active "Continental" germplasm and summer dormant "Mediterranean" germplasm. Summer dormancy enables plants to survive hot, dry summers typical of the Mediterranean regions of southern Europe, the Middle East, and North Africa. Summer active ecotypes are not summer dormant and tend to die or be severely injured by hot, dry summers. Thus, summer dormant tall fescue may be a cool-season grass of choice in areas such as

E.C. Brummer (✉)
Department of Plant Sciences, Plant Breeding Center, University of California, Davis, USA
e-mail: ecbrummer@ucdavis.edu

M. Trammell
Samuel Roberts Noble Foundation, Ardmore, OK, USA
e-mail: matrammell@noble.org

I. Roldán-Ruiz et al. (eds.), *Breeding in a World of Scarcity: Proceedings of the 2015 Meeting of the Section "Forage Crops and Amenity Grasses" of Eucarpia*,
DOI 10.1007/978-3-319-28932-8_11

northern California and western Oklahoma and Texas. Although summer active grasses can produce acceptable yields in northern California under irrigation, the increasingly limited amount of water available for pasture irrigation suggests that summer dormant grasses, un-irrigated over summer, may be a viable alternative.

Over the past several years, we have conducted a number of experiments to assess the potential value of summer dormant tall fescue cultivars, especially in regions where summer active cultivars tend to perform poorly. In this paper, we review the main results from these trials. Recently, we have developed hybrids between ecotypes and backcross progenies to summer active genotypes in an attempt to transfer the summer dormant trait to summer active germplasm. As a consequence of our project, we have realized that a clear phenotypic assay of dormancy is necessary in order to fully characterize the trait and to ensure its introgression.

Material and Methods

In the fall of 2011, trials were established across three states in the USA in order to evaluate persistence and yield of summer active and summer dormant tall fescue cultivars in multiple environments. At each location, paired trials were established, with one trial harvested for yield and the other grazed to evaluate persistence. Locations selected for testing were Vashti, TX (33.55N; 98.04W), Ardmore, OK (34.10N; 97.05W), Woodward, OK (36.27N; 90.24W) and Tifton, GA (31.27N; 83.30W). Experimental design was a randomized complete block (RCB) with four to five replications per trial. Plot sizes were 1.5 by 6.1 m sward plots sown using a Hege (Colwich, KS) small plot cone-drill into clean, tilled seed beds with a seeding rate of 22 kg ha^{-1} at all locations.

All persistence trials were grazed heavily with beef cattle (*Bos taurus*) to maintain a stubble height of approximately 5 cm or less. Grazing began in the spring following establishment and continued through 2014, except at the Tifton, GA location where grazing was terminated in 2013. Data from the forage yield trials were taken by cutting plots with a Hege (Colwich, KS) sickle bar plot harvester at a height of approximately 7 cm. Sub-samples were collected from each plot during the time of harvest and dried in a forced air oven at approximately 60 °C to determine dry weight. All plot yields were adjusted to a dry weight basis and include the sample weights. Stand ratings for all trials were collected using the frequency grid method described by Hopkins (2005).

Results

Yield and Persistence of Summer Dormant and Summer Active Cultivars

In four trials in the southern USA, we observed that summer dormant cultivars had superior survival in the three southern Great Plains locations while summer active cultivars, particularly those with a fungal endophyte, performed better in the southeastern USA at Tifton, GA (Fig. 11.1). These results suggest that summer dormant cultivars are superior to summer active cultivars, regardless of endophyte status, in the western region of the southern Great Plains.

In our trial at Ardmore, OK, we found that summer dormant cultivars had higher yield in early spring (March) and similar yield in May to summer active cultivars (Table 11.1). However, the summer active cultivars had only ~10 % stands. Therefore, the summer dormant cultivar yields were actually quite low when viewed in terms of stand percentages. We should caution against over-interpreation of these data: these plots were grazed heavily during the establishment year, which is not recommended management (but which is ideal pressure to select persistent genotypes). In general, low yield of summer dormant relative to summer active cultivars indicates that improving the yield could make summer dormant cultivars more profitable and therefore more widely adopted in regions where they are best suited.

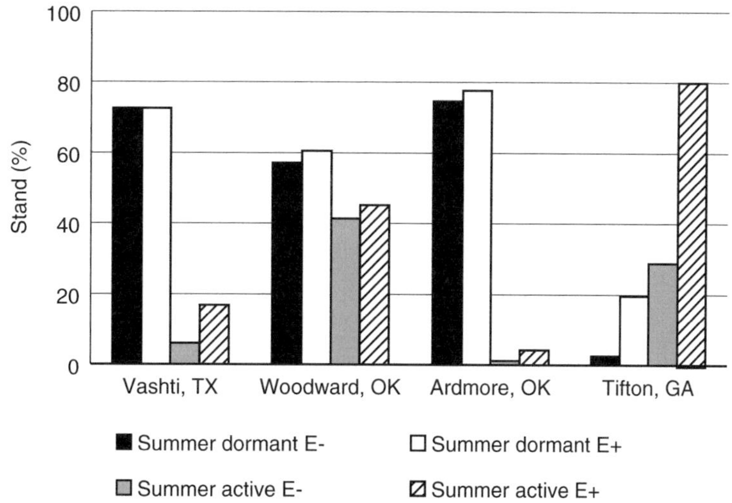

Fig. 11.1 Persistence of summer active and summer dormant tall fescue germplasm with (E+) or without (E−) infection with a fungal endophyte at four locations in the southern USA. Stands were established in 2011 and uniformly excellent. Ratings shown in this figure were taken in 2015

Table 11.1 Yield and stands of summer active and summer dormant tall fescue germplasm with (E+) and without (E−) infection with a fungal endophyte in a trial sown in October 2011 at Ardmore, OK

	Summer active		Summer dormant	
	E+	E−	E+	E−
No. of entries	6	3	6	12
Yield – march 2013 (kg/ha)	170	200	646	808
Yield – may 2013 (kg/ha)	1496	1589	1381	1453
Stand % (March 2013)	9	10	94	91

Discussion

Approaches to Yield Improvement in Summer Dormant Cultivars

We have undertaken several approaches to improve summer dormant yield, including direct selection within summer dormant populations, strain crosses between divergent summer dormant germplasm, and screening a broad collection of summer dormant germplasm for high yield potential. Though we expect that all of these could result in yield improvement, none appear capable at this point to contribute the substantial increases needed.

An alternative approach is to attempt to transfer summer dormancy into summer active germplasm. Hybrids between the two ecotypes express partial sterility, but fertility can be restored through backcrossing. The primary difficulty of this introgression, however, is an inability to reliably phenotype summer dormancy. Consequently, we are attempting to develop a standard test, including a series of standard, check cultivars or germplasms, to assess summer dormancy in a repeatable and simple manner. Once developed, this test would serve not only to assist in selecting and evaluating summer dormant germplasm and breeding populations; it would also enable the genetic mapping of the trait, potentially leading the way to facile marker-based selection for summer dormancy.

Toward a Phenotypic Standard Test of Summer Dormancy

A number of possible methods to assess summer dormancy have been developed over the years, as summarized by Norton et al. (2009). A successful assay will differentiate summer dormancy from senescence induced by drought avoidance, a strategy that may be employed by both summer dormant and summer active genotypes. In addition to differentiating dormancy from senescence in the summer, we also would like to be able to differentiate among dormant germplasms in terms of the amount of dormancy expressed during the summer (whether plants are fully dormant or only partially dormant) and the length of time plants are dormant – that is, the time of recovery in autumn. Ideally, we would want to develop a cultivar that is reliably dormant through the summer, but which comes out of dormancy as soon as possible in autumn to provide adequate forage production prior to winter.

Beyond simply documenting dormancy, an ideal method would be one that could reliably categorize dormancy into clearly defined and repeatable identified groups, as is done with alfalfa autumn dormancy (Teuber et al. 1998). In the case of alfalfa, a series of eleven standard check cultivars have been identified that span the range of dormancy reaction. The autumn plant height of these cultivars is regressed on their dormancy scores, and the regression equation can then be used to predict dormancy of other germplasm based on its autumn growth.

We are attempting to adapt the phenotypic evaluations described by Norton et al. (2009) into a standardized test. The two primary methodologies promoted by Norton et al. (2009) included (1) a ratio of dry matter production of a candidate cultivar with productivity of a known non-dormant control under fully irrigated conditions, and (2) a visual estimation of growth following a midsummer irrigation that had been preceded by an extended drought. One additional benefit of this so-called "midsummer storm" methodology is that it was the only method tested that was able to identify dormancy induction even in spring-sown plants (i.e., plants that had not experienced low temperatures and short photoperiods). This could be very useful in a breeding and introgression program.

We have initiated a project to assess dormancy prediction building on the outline provided by Norton et al. (2009). The project is being conducted in two locations, Ardmore, OK (34.17N; 97.14W) and Davis, CA (38.54N; 121.74W). Davis experiences a true Mediterranean climate, with virtually no rainfall in the summer and high daytime temperatures being common (~40 °C). Ardmore, by contrast, can experience hot, dry summers, but typically has periods of cooler temperatures and rainfall throughout summer, and thus is not a true Mediterranean climate in most years.

We are evaluating a series of cultivars and germplasms of summer dormant and summer active tall fescue under three irrigation treatments: (1) full irrigation throughout summer, (2) no irrigation throughout summer, and (3) no irrigation, except for a significant event in mid-July and early-September. Our intent is to conduct visual evaluations following the irrigation events in treatment (3) to identify dormant vs. active plants in July, and to differentiate dormancy levels in September. In July, only non-dormant plants would regrow following watering. In September, plants that were dormant in July but which regrow in September would have a more desirable length of dormancy than dormant plants that do not regrow until later in autumn. Similar data will be taken in the other irrigation treatments.

In addition to visual evaluations, we plan to measure plant height and plant biomass, and also to use proximal sensors to develop an equation to categorize dormancy levels, based on the methods of Pittman et al. (2015). Summer dormant orchardgrass (*Dactylis glomerata* L.) and hardinggrass (*Phalaris aquatic* L.) are also being evaluated.

Conclusion

We have shown that summer dormancy provides a means for superior persistence in the hot, dry environments of the southern Great Plains when compared to summer active types. However, in order for producers within these regions to fully realize the crop's potential, the limitation of yield must be addressed. A critical factor in improvement would be limiting the length of dormancy so that plants will regrow as soon as moisture is available in the autumn. The hybridization of summer dormant and summer active genotypes in order to transfer the dormancy trait into a higher yielding "Continental" background is one approach to this limitation. A method for assessing summer dormancy is also needed. Existing tests can assess a germplasm's ability to go dormant, but the depth and length of the dormancy is not adequately addressed and can be highly influenced by environmental conditions. Understanding the dormancy trait is important to the successful adoption of summer dormant tall fescue cultivars in the southern Great Plains of the USA.

References

Hopkins A (2005) Grazing tolerance of cool-season grasses planted as seeded sward plots and spaced plants. Crop Science 49:1559–1564.

Norton MR, Volaire F, Lelièvre, F, Fukai S (2009) Identification and measurement of summer dormancy in temperate perennial grasses. Crop Science 49: 2347–2352.

Pittman JJ, Arnall DB, Interrante SM, Moffet CA, Butler TJ (2015) Estimation of biomass and canopy height in bermudagrass, alfalfa, and wheat using ultrasonic, laser, and spectral sensors. Sensors 15: 2920–2943.

Teuber LR, Taggard KL, Gibbs LK, McCaslin MH, Peterson MA, Barnes DK (1998) Fall dormancy. North American Alfalfa Improvement Conference. Available at http://www.naaic.org/stdtests/Dormancy2.html. Verified 6 March 2015.

White Clover (*Trifolium repens* L.) Germplasm Selection for Abiotic Stress Tolerance from Naturalized Populations Collected in the Southern Regions of Chile

12

H. Acuña, L. Inostroza, D. Pinochet, O. Balocchi, F. Ortega, O. Romero, I. López, I. Seguel, G. Tapia, A. Torres, N. Teuber, R. Galdames, and M. Rebuffo

Abstract

Nine white clover populations (WCP) naturalized in Chile were evaluated under controlled (greenhouse pot experiments) and field conditions for water and P stress. The objective of this work was to determine dry matter production and water use and P efficiencies to identify which naturalized white clover populations would be most useful for breeding programs. In the field experiments the clover was grown mixed with perennial ryegrass, but in pots the clover was grown alone. The treatments were the factorial combinations of WCP by water and P levels; completely randomized and complete randomized block experimental designs were used for pot and field experiments, respectively. The population WCP 9-1-X was the most efficient for using water under controlled conditions and was among the three with the highest density of stolons under non-irrigated field conditions. The naturalized WCP 8-1-X produced more clover DM at the low P level than at the high P level and reached the same DM yield as cultivar Huia, which surpassed all the naturalized populations. These genetic materials have high potential to be included in white clover breeding programs whose objective is to generate cultivars for water-restricted and/or P-deficient marginal soils.

H. Acuña (✉)
Facultad de Agronomía, Universidad de Concepción, Chillán, Chile
e-mail: gacunap@udec.cl

L. Inostroza • G. Tapia
CRI Quilamapu, Instituto de Investigaciones Agropecuarias INIA, CRI Quilamapu, Chillán, Chile

D. Pinochet • O. Balocchi • I. López
Facultad de Ciencias Agrarias, Universidad Austral de Chile, Valdivia, Chile

F. Ortega • O. Romero • I. Seguel • R. Galdames
CRI Carillanca, INIA, CRI Carillanca, Temuco, Chile

A. Torres
CRI Remehue, INIA, CRI Remehue, Osorno, Chile

N. Teuber
Departamento Técnico, COVEPA SpA, Osorno, Chile

M. Rebuffo
EEA La Estanzuela, Instituto Nacional de Investigación Agropecuaria INIA, EE La Estanzuela, Colonia, Uruguay

I. Roldán-Ruiz et al. (eds.), *Breeding in a World of Scarcity: Proceedings of the 2015 Meeting of the Section "Forage Crops and Amenity Grasses" of Eucarpia*,
DOI 10.1007/978-3-319-28932-8_12

Keywords
Water stress • Water use efficiency • Soil P stress • P efficiency • Pot and field experiments

Introduction

White clover is the most important forage legume in grazing areas of the central-south, south, and austral regions of Chile, located between 30 and 55° S. The predominant soil types in this large area are of volcanic ash origin, but they vary widely in pH, Al saturation, and P availability (Besoain 1985). All soils are P deficient mainly because of their high P fixation ability; they can fix a high proportion of P applied as inorganic fertilizer (Escudey et al. 2001). The northern part of this area has a Mediterranean climate where white clover is used in irrigated fields. Natural soil moisture availability improves southward due to increased rainfall. Between 38 and 39° S the Mediterranean climate transitions to a humid zone with exceptionally favorable water conditions for growing white clover. The last 30–40 years have been characterized by summer drought periods, which depress clover growth. Further south, the species is distributed through the humid sectors of the austral zone of Chile (Aysén and Magallanes regions). The existing development of naturalized populations in these specific environments suggests that it is possible to select white clover genetic material tolerant to extreme conditions. Germplasm collections of forage species, including white clover, were carried out between 1990 and 2000. One of these collections was used to select the naturalized populations evaluated in this study (Ortega et al. 1994).

Drought is one of the abiotic stresses that most limits plant growth and crop yield. Identifying genotypes that are tolerant to drought and/or have greater water use efficiency is currently a global challenge because of the increasing world population and reduced water resources available for agriculture (Ruane et al. 2008). White clover is negatively affected by soil P deficiency because of its high P requirements (Singh et al. 1997). In mixed grass-clover swards, the companion grass is a very strong competitor for P uptake in situations where its root system allows it to absorb P from greater soil depths than white clover can (Goodman and Collison 1982). Four experiments were therefore planned, under controlled and field conditions, as part of the activities of a project financed by the Regional Fund for Agricultural Technology (FONTAGRO – 787/2005) and carried out by INIA-Uruguay and INIA-Chile. Two of these four experiments have been published (Inostroza and Acuña 2010; Acuña and Inostroza 2012), while full-length results of the other two have not yet been published. The first objective of this study was to evaluate nine white clover naturalized populations and two introduced cultivars under controlled (greenhouse pot experiments) and field conditions (mixed with perennial ryegrass, under grazing), in terms of their tolerance to moisture restriction and soil P deficiency, mainly using dry matter production, water use efficiency and P use efficiency as indicators. The second objective was to identify the naturalized white clover populations with the highest water use and P use efficiencies that would be useful in breeding programs aimed at obtaining cultivars that are better adapted to marginal soils with water and soil P availability restrictions.

Materials and Methods

The water stress pot experiment (WSPE) and the P stress field experiment (PSFE) were carried out at INIA-Quilamapu research center (36° 36′ S, 72° 02′ W), while the P stress pot experiment (PSPE) and a water stress field experiment (WSFE) were carried out at the Universidad Austral de Chile (39° 48′ S, 73° 15′ W) and INIA-Carillanca research center (38° 41′ S, 72° 25′ W), respectively, during the years 2007–2010. The pot experiments lasted 9 months; the field experiments ran for three complete growing seasons. The latter experiments were grazed, after DM production evaluations, at a high stocking density for a short time.

Plant Materials

The nine studied WCP were: 2-3-X, 7-1-X, 9-1-X, 5-2-X, 8-2-X, 12-2-X, 8-1-X, 9-2-X, and 6-1-X (all small-leaved). Details of population collection sites and agronomic characterization are found in the study by Ortega et al. (1994). The white clover cultivars Huia and Will were used as controls. Sowing for pot and field experiments was done in 27 cm^3 (3×3×3 cm) capacity seedbeds with peat as substrate (Biolan B3L, pH 5.5–6.0) in autumn 2007. Seedlings were inoculated 1 week after emergence with a *Rhizobium trifolii* solution. They were then transferred to pots (30 cm and 100 cm depth, in WSPE and PSPE, respectively) or the field (1.5×3.0 m plots). In the latter case, seedlings were planted in spring (September), 20 cm apart, between ryegrass (cv. Nui) rows sown directly (20 kg ha^{-1} of seed) in autumn also 20 cm apart.

Abiotic Stress Treatments

The WSPE soil moisture treatments were: no water stress (NWS) with 54 % soil moisture (dry basis) and −0.01 MPa water potential, and with water stress (WWS) with 25 % soil moisture (dry basis) and −0.5 MPa water potential. Non-irrigated and irrigated treatments were included in WSFE. The irrigated treatment received 30–50 mm of water in eight and three events between November and March 2007–2008 and 2008–2009, and between January and March 2010, respectively. The non-irrigated treatment (rain-fed) received 120, 135 and 309 mm in spring-summer (November to March) of the first, second and third growing seasons, respectively. The PSPE considered five levels of soil P availability: 4.0, 8.7, 15.2, 20.4, and 27.3 mg kg^{-1} of Olsen-P, while treatments in the PSFE were 8.0 and 16.0 mg kg^{-1} of Olsen-P.

Measurements

Shoot DM production and partitioning were the main variables measured in both pot and field experiments. Plant morphological traits such as leaf, leaflet, and stolon dimensions and numbers of clover growing points were measured in all the experiments. In the WSPE the variables needed to estimate water use efficiency (WUE) and other physiological traits were also measured, as described in Inostroza and Acuña (2010). The PSPE and PSFE included the necessary determinations to calculate P absorption efficiency (PAE) and P utilization efficiency (PUE): soil Olsen-P and herbage P concentration, as described in Acuña and Inostroza (2012); Pinochet et al. (2009).

Experimental Design and Statistical Analyses

Treatments in all the experiments corresponded to all the factorial combinations of the 11 WCP (nine naturalized plus two cultivars) by the number of levels of the respective stress factor. An exception was the WSFE where two experiments were conducted in two adjacent fields, one for each moisture condition. The ex perimental designs of complete randomized block and completely randomized were used for the field and pot experiments, respectively, with four replicates (except for PSPE, in which three replicates were considered). Data were analyzed with ANOVA and/or regression analysis, and the relationships between the variables were analyzed by principal component analysis (PCA) with the XLSTAT software (Addinsoft, USA).

Results and Discussion

Water Stress and Use Efficiency

The PCA of seven traits evaluated under controlled conditions (WSPE) showed that 94.3 % of the variability of the data was explained by the first two principal components. Principal component 1 (PC1) and PC2 explained 77.3 % and 17.0 % of the variability, respectively (Fig. 12.1). Variables contributing more to PC1 were stolon DM (17.7 %), LAR (leaf area ratio, 17.3 %), WUE (16.4 %), LWR (leaf weight ratio, 16.3 %), and shoot DM (13.0 %). Populations investing a high proportion of their biomass in the transpiring area (high LAR and LWR) achieve lower shoot DM production and low WUE (Huia, 8-1-X). However, populations investing a higher proportion of biomass in non-transpiring organs (stolons) achieve higher shoot DM and WUE (9-1-X, Fig. 12.1).

The results of the WSFEs showed that DM yield of WCP did not significantly vary (P >0.05) in the non-irrigated experiment. However, cultivar Will under irrigation was significantly higher (P <0.05) than some of the naturalized WCP. Percentages of WC (clover contribution to clover/ryegrass mixture) were higher under non-irrigated conditions as compared with those under irrigation (Table 12.1).

Means of stolon length per unit area (m m^{-2}) were 26.5, 72.6, and 27.4 with irrigation and 48.9, 35.8, and 51.8 without irrigation, at 14, 30, and 33 months after transplanting the white clover, respectively. The stolon internodes were shorter under non-irrigated conditions. During most of the experimental period, the higher stolon density in the non-irrigated environment could be explained by lower ryegrass growth under these conditions. Fothergill et al. (1997) report this type of relationship. The 8-1-X, 12-2-X, and 9-1-X WCP were noted for this trait (stolon density) in the non-irrigated experiment, while 12-2-X, 5-2-X, and 7-1-X were highlighted in the irrigated experiment. In the non-irrigated WSFE, white clover shoot DM increased from the first to the third growing season (172, 606 and 1794 kg ha^{-1} of DM, respectively). In the irrigated WSFE, dry matter production means of WCP reached a maximum in the second season and then fell in the third (357, 1562, and 903 kg ha^{-1}, respectively).

The analysis of WSPE and WSFEs confirmed the importance of developing water-stress-tolerant germplasm. Water stress decreased DM production and altered biomass partitioning. The high potential yield cultivar Will had DM production similar to some of the naturalized populations when under water stress. Biomass partitioning is perceived as being important in both cases, since the higher stolon production under water-stressed conditions helps clover survival and productivity.

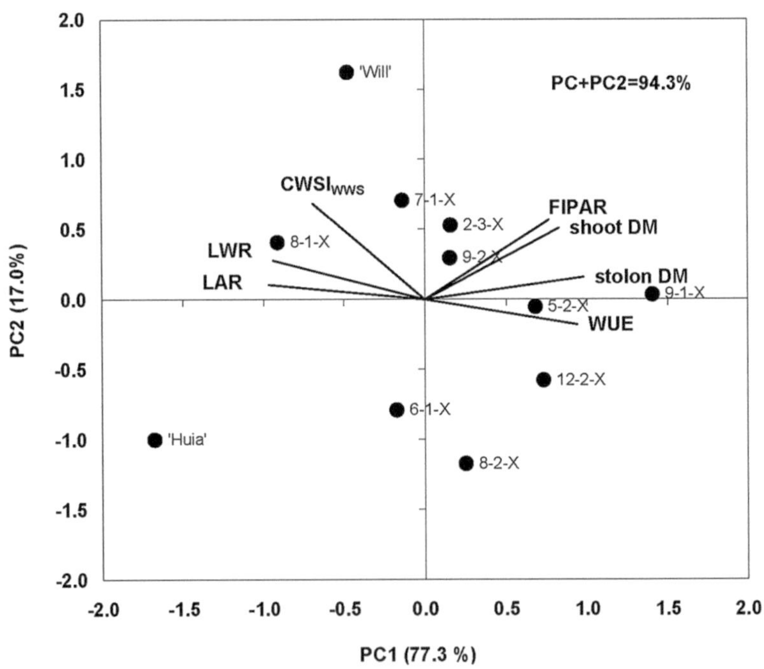

Fig. 12.1 Biplot of the first two principal components (PC1 and PC2) for the PC analysis of seven traits evaluated in 11 white clover populations, under two water treatments. The traits are: water use efficiency (*WUE*), stolon dry matter (*stolon DM*), shoot dry matter (*shoot DM*), leaf weight ratio (*LWR*), leaf area ratio (*LAR*), fraction of intercepted photosynthetically active radiation (*FIPAR*) and crop water stress index under non watered stress treatment ($CWSI_{NWS}$) (Inostroza and Acuña 2010)

Table 12.1 Shoot dry matter (DM) production of white clover populations (WCP) (g pot^{-1}) and water use efficiency (WUE, g kg-1) in the water stress pot experiment, and shoot DM production (total of the three growing seasons) of WCP (t ha^{-1}) and percentage of legume in total herbage (clover + ryegrass) in the water stress field experiments

			Field experiments			
	Pot experiment		Non-irrigated		Irrigated	
WCP	DMshoot	WUE	DMshoot	WC(%)	DMshoot	WC(%)
2-3-X	30.9	4.8	2.7	20.0	2.8	11.0
7-1-X	27.2	4.6	2.5	21.0	3.0	11.0
9-1-X	28.9	6.2	1.9	16.0	1.6	6.0
5-2-X	26.6	5.0	2.2	15.0	2.5	10.0
8-2-X	20.1	5.2	2.8	21.0	2.6	11.0
12-2-X	28.6	5.5	3.4	26.0	2.1	8.0
8-1-X	22.5	4.2	2.5	18.0	2.0	10.0
9-2-X	27.5	5.3	2.1	18.0	2.9	11.0
6-1-X	24.6	4.8	2.5	17.0	2.8	12.0
Huia	14.7	3.4	2.1	16.0	3.1	12.0
Will	26.9	3.9	3.5	27.0	4.7	18.0
Significance	***	***	ns	*	*	*
LSD	3.64	1.03		6.69	1.36	5.20
WS means						
NWS	33.6	4.2				
WWS	17.6	5.4				
Significance	***	***				
LSD	1.55	0.43				

Phosphorus Absorption and Utilization Efficiency

In the PSPE the DM production showed a linear increase as available soil P increased. Significant differences between WCP were found at all P levels, but the rankings of WCP were not consistent between different P levels. Notably, WCP 8-1-X reached its highest yield at the lower P levels (up to 15 mg kg^{-1}), but at the higher P levels yielded equal to or less than the rest of the WCP. In the PSFE in the second growing season, the total WCP dry matter yields of the high P level were significantly higher ($P < 0.05$) than those of the low P level for the populations 7-1-X, 5-2-X, 8-2-X, 9-2-X, and 6-1-X. The rest of the populations, including the control cultivars, did not differ significantly ($P > 0.05$), with the exception of WCP 8-1-X where production from the low P treatment was 25 % higher than the high P treatment ($P < 0.01$). The PAE (Table 12.2) was clearly higher when soil P was low. The high absorption efficiency (0.47) attained by WCP 8-1-X significantly surpassed the control cultivars and all the other naturalized populations, with the exception of WCP 6-1-X. The PUE means of WCP did not differ significantly ($P > 0.05$) but 9-2-X showed the highest value; the mean PUE values for P treatments were higher for low than high soil P ($P < 0.001$) (Table 12.2), (Fig. 12.2).

The WCP 8-1-X is of interest because of its P absorption ability when soil P is scarce, and WCP 9-2-X is noted for its ability to use absorbed P to produce DM. Both are similar to Huia for DM production under soil P shortage and surpass the Huia absorption capacity in the field experiment. The WCP 9-2-X surpasses the P utilization ability of Huia for producing herbage DM (Table 12.2). It can be inferred that WCP dry matter production response is similar under controlled and field conditions, so that populations such as WCP 8-1-X express their highest P absorption ability at low soil P levels under both conditions. Pot PUE had similar values to those found in the field under the low P treatment, but

Table 12.2 Phosphorus absorption efficiency (PAE, P abs (kg ha^{-1})/Soil P (mg kg^{-1})) and phosphorus utilization efficiency (PUE, DM yield (kg ha^{-1})/ P abs (kg ha^{-1})) of the eleven white clover populations (WCP), calculated as average of five levels of soil Olsen–P (4.0, 8.7, 15.2, 20.4 and 27.3 mg kg^{-1}) of the pot experiment, and for two contrasting levels of available P (low and high, 8 and 16 mg kg^{-1}, respectively) of the field experiment

	Pot experiment		Field experiment	
WCP	PAE	PUE	PAE	PUE
2-3-X	0.63	491	0.28	389
7-1-X	0.61	452	0.43	368
9-1-X	0.57	431	0.40	371
5-2-X	0.52	429	0.37	382
8-2-X	0.59	427	0.41	375
12-2-X	0.62	434	0.37	373
8-1-X	0.51	449	0.47	371
9-2-X	0.56	469	0.42	410
6-1-X	0.70	463	0.48	371
Huia	0.54	432	0.32	382
Will	0.71	612	0.43	389
Significance	*	*	***	NS
LSD	0.07	35.0	0.069	–
		Means (field experiment)		
		Low P	0.49	423
		High P	0.30	337
		Significance	***	***
		LSD	0.030	15.1

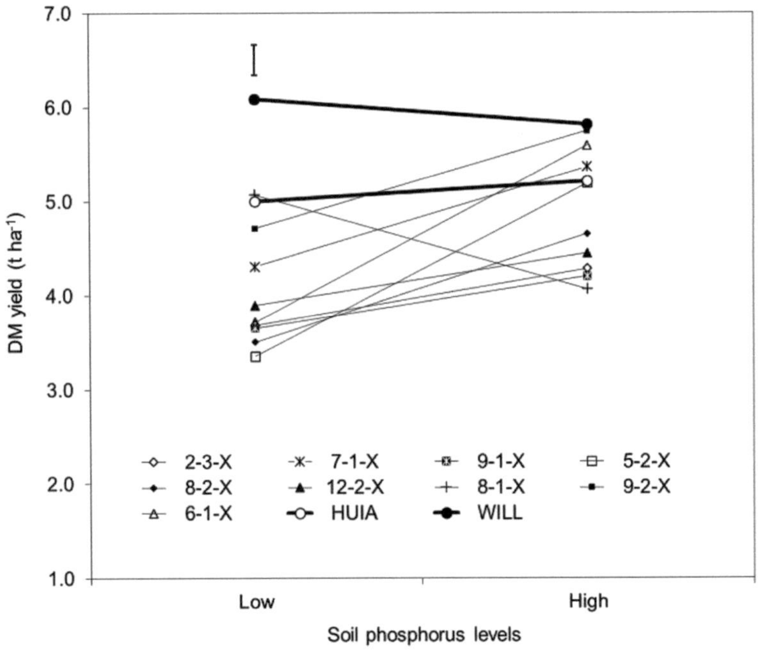

Fig 12.2 Effect of contrasting soil phosphorus levels (Low and High) on DM yield (t ha^{-1}) of white clover populations (WCP), in mixed grass/clover sward, in the second growing season (2008–2009) of the phosphorus stress field experiment. *Vertical bar*: Standard error of mean (SEM) to compare white clover populations x phosphorus means

they were clearly lower for the high P treatment. Pot and field results are not consistent when comparing PAE and PUE among WCP.

It is found that pot results can be reliable and useful when field data are not available, given that the latter are comparatively more difficult to obtain because of time and resource requirements. Traditional pots were not used in these evaluations, but rather 1-m-deep cylinders, to allow unlimited deep root development, including those with shallow rooting such as white clover, because plants grown from seed develop a much longer taproot than those multiplied from stolon.

Conclusions

The WCP 9-1-X was the most efficient for using water under controlled conditions and was among the three with the highest density of stolons under non-irrigated field conditions. At low soil P level, the WCP 8-1-X reached the same DM yield as Huia, surpassing all the naturalized populations, and had the highest PAE. High variability was found in the characterized WCP in terms of response to water stress, WUE, P use efficiency, and tolerance to P deficient soils. The selected naturalized white clover germplasm from southern Chile has a high potential utility in white clover breeding programs whose objective is to generate cultivars for water-restricted and/or P-deficient marginal soils.

Acknowledgments The authors thank FONTAGRO, the institutions involved in executing the project, INIA-Uruguay, INIA-Chile, and the Universidad Austral the Chile, UACH, and "Proyecto FONDECYT 1130340" to support the paper presentation. They also thank the technical staff from INIA-Chile and UACH and the students who helped with project activities.

References

Acuña H. and Inostroza L. (2012) Phosphorus efficiency of naturalized Chilean white clover in a grazed field trial. Grass and Forage Science 68: 125–137

Besoain E. (1985) Los suelos. En: Tosso J. (ed.) Suelos Volcánicos de Chile, Capítulo 1 [The soils. In: Tosso J. (ed.) Volcanic Soils of Chile, Chapter 1], pp. 25–106. Santiago, Chile: INIA, Ministerio de Agricultura.

Escudey M., Galindo G., Forster J., Briceño J., Díaz P. and Chang A. (2001) Chemical forms of phosphorus of volcanic-ash derived soils in Chile. Communications in Soils Science and Plant Analysis 32, 601–616.

Fothergill, M., Davies, D.A., Morgan, C.T. and Jones, J.R. (1997) White Clover Crashes. In: Younie. D. (Ed). Legumes in Sustainable Farming Systems, Occasional Symposium of the British Grassland Society, No. 30, pp 172–176.

Goodman P.J. and Collison M. (1982) Varietal differences in uptake of ^{32}P labelled phosphate in clover plus ryegrass swards and monoculture. Annals of Applied Biology 100, 559–565.

Inostroza L. and Acuña H. 2010. Water use efficiency and associated physiological traits of nine naturalized white clover populations in Chile. Plant Breeding 129: 700–706

Ortega F., Demanet R., Paladines O. y Medel M. (1994) Colecta y caracterización de poblaciones de trébol blanco (*Trifolium repens* L.) en la zona sur de Chile (Collection and characterization of white clover populations in the south zone of Chile). Agricultura Técnica (Chile) 54(1): 30–38.

Pinochet D., Balocchi O., López I., Pereira M. and Carrasco G. (2009) Eficiencia de uso de fósforo de accesiones de trébol blanco naturalizado en Chile [Phosphorus use efficiency of white clover accessions naturalized in Chile]. Libro de Resúmenes, XXXV Congreso Sociedad Chilena de Producción Animal. Coyhaique, Chile, 27–29 October. pp. 59–60.

Ruane J., Sonnino A., Steduto P. and Deane C. (2008) Coping with Water Scarcity: What Role for Biotechnologies? FAO, Rome.

Singh D., Sale P. and McKenzie B. (1997) Water relations of white clover (*Trifolium repens* L.) in a drying soil, as a function of phosphorus supply and defoliation frequency. Australian Journal of Agricultural Research 48, 675–681.

13 Evaluation of Orchardgrass (*Dactylis glomerata* L.) Autochthonous Serbian Germplasm in Pre-breeding

D. Sokolović, S. Babić, J. Radović, M. Petrović,
G. Jevtić, Z. Lugić, and A. Simić

Abstract

Forage breeding of cocksfoot (*Dactylis glomerata* L.) has been very intense in recent decades as it is one of the most important cool season forage grasses in temperate Europe. It is naturally present in Serbia from lowland to mountains over 1000 m above sea level. It shows great adaptability, a high yield of quality forage and is a regular component of several grass-legume mixtures. The aim of this investigation was to evaluate cocksfoot germplasm previously formed from autochthonous populations for breeding of late maturity cultivars and to define genetic diversity. Investigation was conducted on 24 accessions in a 2-year period (2011–2012). Special attention was directed to maturity (heading date) and dry matter yield. Twelve other important traits, components of forage production and quality were measured. All data were processed by ANOVA, multivariate principal component and cluster analyses. The difference in heading averages was more than 20 days. Some populations showed highest DMY and lateness at the same time, which qualifies them for breeding for a new medium to late cultivar. The most important traits for grouping of populations and variability in germplasm were plant height, tiller number and DMY. The first two principal components which showed statistically significant correlations with these traits describe more than 56 % of variability in germplasm. Very few matches of geographic origin and grouping of populations in cluster analysis were obtained.

D. Sokolović (✉) • S. Babić • J. Radović
M. Petrović • G. Jevtić • Z. Lugić
Department of Breeding , Institute for Forage Crops, Globoder, Kruševac 37251, Serbia
e-mail: dejan.sokolovic@ikbks.com

A. Simić
Faculty of Agriculture, Belgrade University, Zemun-Belgrade 11070, Serbia

I. Roldán-Ruiz et al. (eds.), *Breeding in a World of Scarcity: Proceedings of the 2015 Meeting of the Section "Forage Crops and Amenity Grasses" of Eucarpia*,
DOI 10.1007/978-3-319-28932-8_13

Keywords
Cocksfoot • Breeding • Late maturity • DMY • Germplasm • Multivariate analysis

Introduction

Cocksfoot (*Dactylis glomerata* L.) is one of the most important perennial fodder grasses of temperate climate. It has a high and stable quality biomass yield with a large range of climate tolerance. It is frost-, shade-, heat- and drought-tolerant with good regrowth ability. Among 18 subspecies with three ploidy levels (Lumaret 1988), tetraploid *D. glomerata* ssp. *glomerata* is the most common. It has high morphological variability (Lindner et al. 2004) and represents the major cocksfoot subspecies for forage production in Europe (Peeters 2004). More than 200 cultivars of cocksfoot have been bred since 1950s in the world; 133 of them are now maintained in Europe (Sanada et al. 2010). Within these cultivars the differences in maturity (early-late) range from 15 to 20 days depending on the environment. Common breeding objectives were to increase forage yield, disease resistance, and tolerance to drought and cold. Improvement of resistance to stem rust (*Puccinia graminis*) has been one of the major objectives in many cocksfoot breeding programs on all continents (Alderson and Sharp 1995). The most important cocksfoot breeding criteria in Serbia, besides DMY and quality, persistency and rust resistance, is earliness or lateness, because it is a regular component of grass-legume mixtures. The optimal timing of cutting or ensiling (i.e., optimal balance between DMY and quality) for cocksfoot does not coincide with most other species. Cocksfoot can be classified as early to intermediate in maturity in comparison with other grasses and legumes (Dent and Aldrich 1963); in Serbia it flowers very early, at the end of April or beginning of May. Most of autochthonous material and released cultivars in Serbia are early to medium-early in maturity.

Therefore our breeding objective is to develop cocksfoot populations that mature several days later than the existing cultivars on the Serbian market. At the same time, new material should be productive and adapted to local agro-ecological conditions. Very often required characteristics are available in natural populations, what refers them to direct utilization in breeding (Posselt and Willner 2007). Their value has long been recognized by plant breeders (Boller and Green 2010; Soković et al. 2011). In grasses, there are two possible breeding pathways to develop late cultivars. First is population breeding (phenotypic recurrent selection within population) if the starting material is characterized with late maturity and is already adapted and productive (Hurst and Hall 2007). The second is introgressing of late maturity in existing breeding germplasm by incorporating some promising plants in the breeding scheme. In both cases, however, starting germplasm must be either created or defined.

The main goal of this trial was to evaluate starting germplasm in pre-breeding. During seed collection, the population was formed based on the criteria of late maturity and the plant's robust habitus. An additional investigation task was to estimate genetic variability, distances and relations of populations and traits in the collection using multivariate methods.

Material and Methods

Plant Material

In this study a collection of cocksfoot (*Dactylis glomerata* L.) autochthonous populations was evaluated. Twenty-two accessions were collected during the period 2006–2009, from various places across Serbia (Fig. 13.1) at altitudes between 600 and 1600 m and little or no anthropogenic influence. Seeds were gathered from a minimum of 50 plants from each place, according to Marshall and Brown (1983). The basic

Fig. 13.1 Collecting localities

criteria for collecting the populations were late maturity and the robust habitus of plants. Two cultivars, K40 (Serbia) and Hallmark (USA) were also included in trial as standards.

Experimental Design

Two years of investigation (2011–2012) were conducted on experimental fields at the Institute for Forage Crops in Kruševac in Globoder, Serbia. The nursery was a spaced plant (60×60 cm) design with 30 plants per population. The trial was a completely randomised block system with three blocks, each with ten plants of each population. Traits related to forage production (Table 13.1) were measured on plants at the same phenological phase of heading (80 % of tillers with emerged panicles). Two cuts were done every year, in first 10 days of May and in first 10 days of September. Basic parameters of biomass chemical composition were also determined in the second year: crude protein content (CPC) according to the Weende system (AOAC 1990), content of neutral detergent fiber (NDF) (Van

Table 13.1 Measured traits on spaced plants during 2 years

Trait	Unit	Abbreviation
Heading date	days from April 1st	HD
Plant height	cm	PH
Leaf length	cm	LL
Leaf width	mm	LW
Leaf number	–	LN
Generative tillers number	–	GT
Vegetative tillers number	–	VT
Total tillers number	–	TN
Dry matter yield in first cut	g per plant	DMYI
Dry matter yield in total	g per plant	DMY
Leaf share	% of leaf DM in total DM	LS

Soest et al. 1991), acid detergent fiber (ADF) and lignin content (ADL), both according to AOAC 973.18 (1997).

Data Analysis

All results are presented as 2-year average value of the selected traits. Basic parameters of variability (interval of variation (IV) and coefficient of variation (CV)) were calculated and ANOVA was performed. PCA was carried out on population averages. The dimension of factor area was determined based on principal components with eigenvalues above one. Hierarchical cluster analysis was performed on raw data using the Ward method with Euclidean distances. Regression analyses of yield and heading date was also performed. All data were processed by multivariate principal component (PCA) and cluster analyses using Statistica 8.0 (Stat Soft, inc.).

Results and Discussion

Analysis of maturity, morphological traits influencing dry mater yield and quality, and the biomass production and chemical composition, all revealed very high within population variability (Table 13.2). This was expected since populations were collected from locations without intensive agricultural management. Many papers have demonstrated the high diversity available in cocksfoot germplasm (Jafary 2004; Zeng et al. 2008) and shown that populations from extensively managed locations harbor a high variability for heading and flowering date (Last et al. 2014). Coefficients of variation were especially high in production traits such as TN and DMY. The level of inter-population variability was also very high and there were statistically significant differences between population averages for all traits ($Lsd_{0.01}$ values in Table 13.2). The difference in heading date between the latest populations and the Serbian cultivar K-40 was more than 20 days on average and about 12 days with cv. Hallmark, respectively. Several populations had a later HD than the average of the collection (37.47).

Analysis of maturity showed success in assembling a collection with lateness as target trait. Ghesquiere et al. (2014) showed that dry matter yield was positively correlated with lateness in cocksfoot. Multiple regression analysis of HD and productivity of the investigated collection in general refuted this, since yield dropped with lateness (Fig. 13.2). Notably, the population collected from higher altitudes showed later heading in comparison with other populations.

Nevertheless, some populations showed the highest DMY and lateness at the same time. Maljen 1, Golija 1, Suvobor 1 and even Kopaonik can be used directly for population improvement through phenotypic recurrent selection toward new medium-late cocksfoot forage cultivars. Other populations (Maljen 2, 3, Povlen 1 and especially Radočelo) are promising for choosing late genotypes. Their averages for yield were failing but genetic coefficients of variation and interval of variation within populations confirm the existence of genotypes with promising traits.

The first four principal components (PC) described almost 80 % of variability available in the dataset.

The first principal component (PC1) defined a large amount of variability among populations

Table 13.2 Average values and variability for investigated traits of cocksfoot populations

	HD	PH	LL	LW	LN	TN	GT	VT	DMYI	DMY	LS	CPC[a]	NDF[a]	ADF[a]	ADL[a]
J. Kučaj 1	39.4	**59.1**	27.0	8.8	2.9	53.9	43.3	10.6	**70.5**	89.2	37.3	119.4	704.9	392.7	49.9
J. Kučaj 2	36.7	67.0	31.2	**11.9**	3.4	53.3	**22.6**	30.7	70.7	111.8	49.6	130.4	722.7	388.3	57.5
Tupižnica	36.4	79.5	30.3	11.0	3.5	87.5	70.3	17.2	183.1	223.0	38.0	98.7	683.2	397.8	59.3
Rtanj 1	32.9	77.7	29.3	11.0	3.5	81.9	54.0	27.9	149.7	193.4	32.3	125.0	712.7	379.3	37.7
Rtanj 2	32.2	78.5	30.7	10.1	3.5	88.4	65.3	23.1	135.2	159.8	46.1	123.7	683.4	378.3	31.5
Homolje	34.9	63.4	29.5	9.6	3.8	92.2	52.2	39.9	127.6	168.4	40.6	125.5	665.2	394.2	39.9
Rudnik	32.0	76.3	28.4	10.3	3.7	93.2	54.6	38.6	148.0	184.8	40.8	131.3	667.6	366.2	34.4
Suvobor 1	37.3	83.2	**33.6**	11.2	3.5	80.9	65.8	15.2	173.5	239.8	41.9	86.5	702.8	376.3	**28.3**
Suvobor 2	36.6	77.1	30.3	10.6	3.5	90.8	75.8	15.0	208.2	228.1	32.2	88.3	695.9	387.7	38.6
Maljen 1	41.2	92.2	31.5	11.2	3.4	101.6	**84.4**	17.2	**229.7**	**260.2**	36.4	92.4	744.1	397.6	38.5
Maljen 2	43.5	77.5	25.1	8.8	3.4	74.0	59.6	14.4	132.3	150.6	31.1	**71.7**	732.8	428.4	49.4
Maljen 3	41.9	86.6	29.2	10.2	3.7	59.6	38.7	20.9	118.6	144.2	34.7	74.7	682.9	399.3	53.1
Povlen 1	42.1	64.1	26.8	9.4	**2.9**	72.8	60.1	12.6	109.9	138.8	47.1	103.2	738.2	391.5	37.0
Povlen 2	35.9	81.3	29.1	11.2	3.8	73.7	38.0	35.7	116.3	159.4	40.0	104.1	**586.7**	**332.3**	29.1
Ćelije	31.4	66.0	24.7	8.9	3.7	82.1	62.1	20.0	113.1	143.7	42.3	119.1	**791.1**	401.8	**66.7**
Kopaonik	41.3	86.9	28.7	9.6	3.7	72.1	63.7	8.4	160.6	178.0	**22.3**	78.0	753.7	**432.2**	42.4
Radočelo	**44.8**	66.9	28.6	9.2	3.0	43.6	35.2	8.4	89.6	104.5	54.1	101.9	672.3	421.5	52.9
Golija 1	39.0	83.5	32.9	10.5	3.6	84.1	68.5	15.6	181.0	211.2	68.5	90.7	764.2	428.8	51.0
Golija 2	39.9	60.9	**23.7**	**7.4**	3.5	**42.1**	33.9	**8.3**	77.0	**84.8**	44.5	94.5	672.1	357.1	33.2
Javor	33.2	72.3	26.7	9.6	3.7	92.3	72.1	20.3	145.2	165.0	47.5	118.0	718.5	397.7	43.4
Sjenica	39.1	70.8	28.5	9.5	3.5	83.8	67.8	16.0	150.5	174.1	41.6	108.1	602.6	374.8	36.7
Vlasina	32.9	77.6	29.6	10.8	3.6	97.4	56.7	**40.6**	162.4	205.8	**68.7**	**138.9**	651.5	373.0	40.9
K-40	**24.6**	**101.3**	22.1	9.7	**3.8**	**102.4**	83.9	18.4	196.3	229.4	32.4	99.8	757.3	429.8	34.5
Halmark	32.7	90.6	27.9	10.9	3.8	102.3	56.2	46.1	172.9	216.7	53.5	106.3	668.4	364.9	28.8
IV	7–24	26–95	12–27.4	3.5–8.5	1–2.5	88–176	77–173	16–92	125–449	160–606	3.1–7.2	2.7–4.9	2.3–6.3	0.6–4.4	1.3–4.8
CV (%)	12.6	12.9	8.19	9.6	6.2	21.4	68.1	49.5	29.5	26.1	25.5	17.8	6.9	6.0	26.2
Lsd 0.01	3.99	16.8	5.8	1.6	0.5	21.1	20.0	10.7	51.7	57.6	3.7	5.2	18.0	31.1	5.2

Minimal and maximal values in bold

[a] $mg\ kg^{-1}$

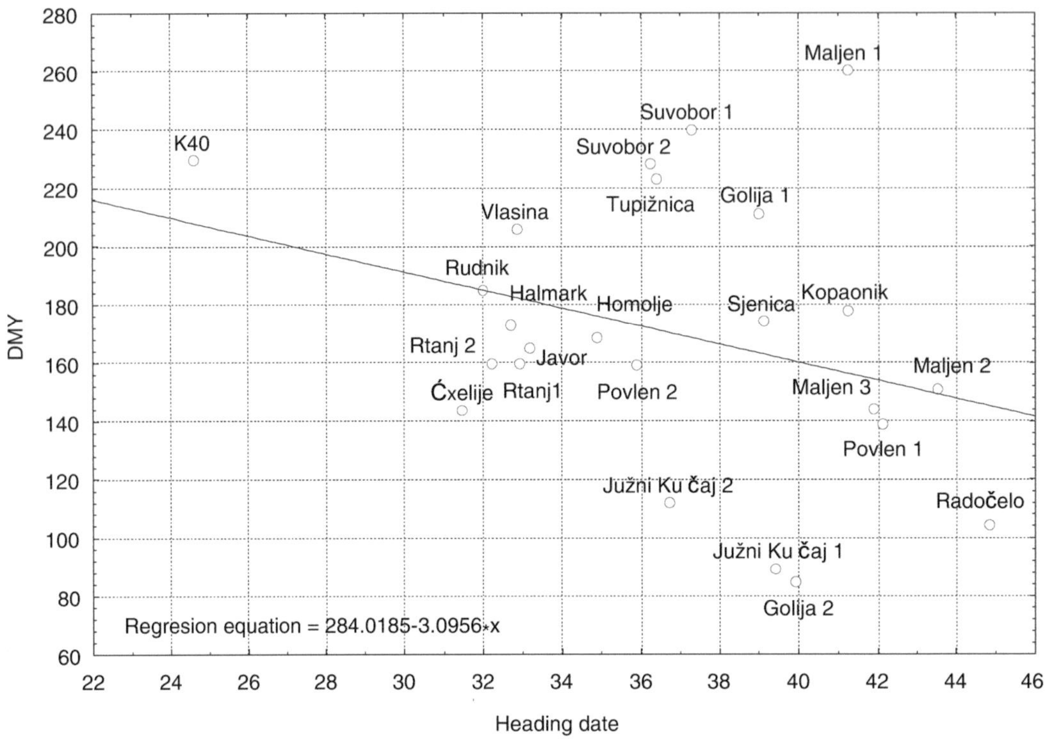

Fig. 13.2 Regression analysis of DMY and heading date

(34.63 %), and showed significant correlation with PH, GT, TN, DMYI and DMY. Figure 13.3 shows the directions of trait vectors between first and second PC axes, and Fig. 13.4 the distribution of populations. Notably, populations with pronounced lateness gathered in the lower left corner, while in the opposite corner populations with good yield abilities were grouped. This confirms that populations Vlasina, Rudnik and Povlen 2 are characterized by a high number of vegetative tillers and DM quality, but they are the earliest populations. The most important traits for PC2 axis are properties related to the chemical composition of DM.

Populations Kopaonik, Maljen 2 and Golija 1 are promising for both maturity and productivity, but since they are located in direction of trait vectors for ADF and NDF (Fig. 13.3) and opposite of CPC vector, their DM quality needs improvement.

Hierarchical cluster analysis (not shown) grouped the populations in three main clusters with minor matches with geographic origin. Therefore, in general, the phenotypic similarity among Serbian populations was not related to geographical proximity. This is probably due to the small collecting area (300 km in diameter). In such small distances and areas, grass pollen flow and exchange of genes are probably significant. Also, there was no clear differentiation between natural populations and commercial varieties.

Correlation coefficients confirm that HD is negatively correlated with productivity in this cocksfoot germplasm, but this was not statistically significant (Table 13.3).

In addition, lateness influenced the reduction of VT and LN. Since these traits are carriers of forage quality, late maturity is negatively correlated with CPC. Correlations among forage quality traits are always affected by the degree of heading and number of generative tillers (Saiga 1981). CPC was also positively correlated with VT. Most of traits which are components of DMY were mutually positively correlated, as well as both type of fiber NDF and ADF. A posi-

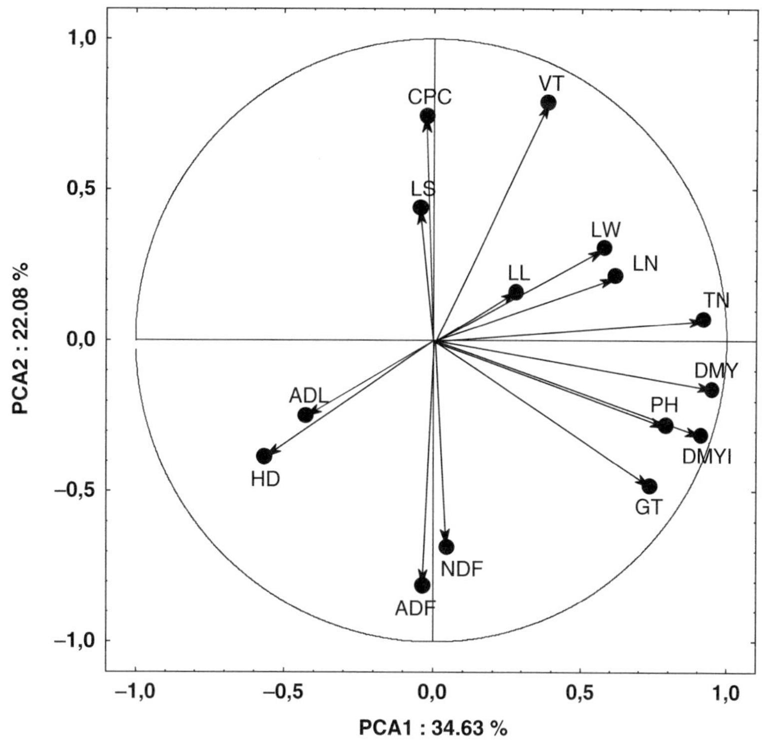

Fig. 13.3 PCA biplot for traits between first two principal components

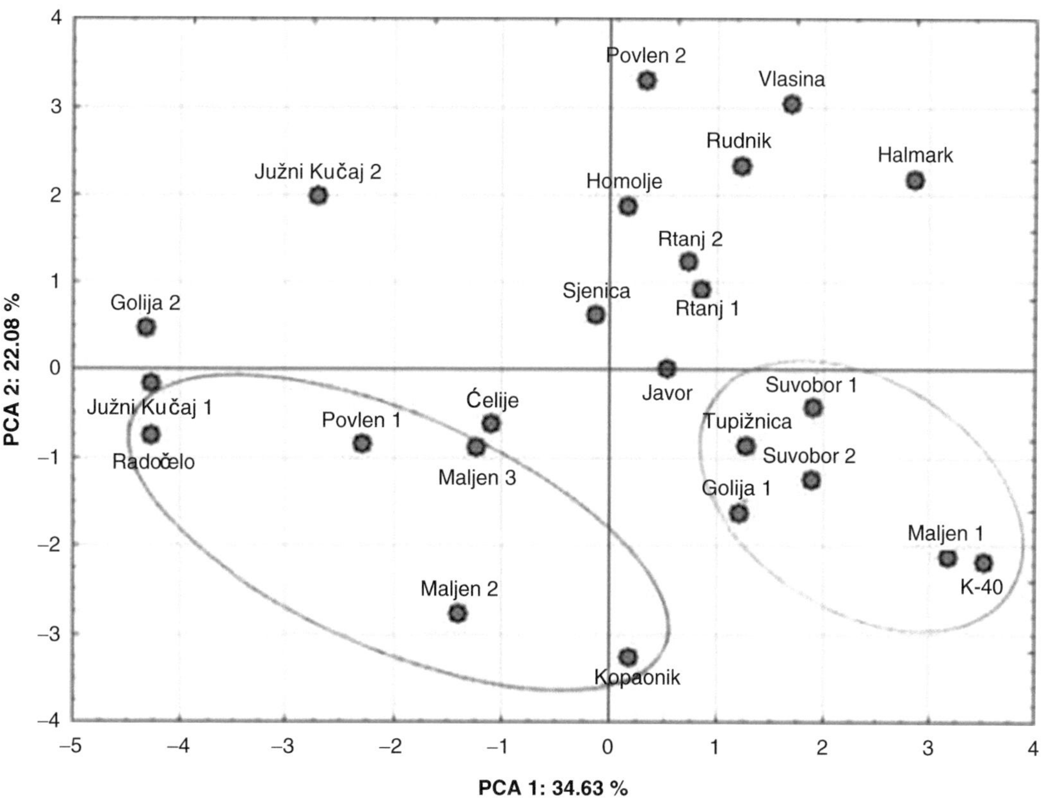

Fig. 13.4 PCA biplot for populations between first two principal components

Table 13.3 Correlation coefficients between analyzed traits

	HD	PH	LL	LW	LN	TN	DMY	GT	VT	DMYI	LS	CPC	NDF	ADF	ADL
HD	1	−0.3	0.22	−0.20	**−0.59**[a]	**−0.62**	−0.36	−0.31	**−0.54**	−0.28	−0.06	**−0.55**	−0.03	0.22	0.22
PH		1	0.14	0.47	**0.52**	**0.57**	**0.75**	**0.53**	0.12	**0.76**	−0.18	−0.40	0.15	0.20	−0.31
LL			1	**0.73**	−0.07	0.12	0.39	0.02	0.13	0.3	0.26	0.01	−0.16	−0.14	−0.01
LW				1	0.20	0.41	**0.6**	0.1	0.47	0.44	0.15	0.17	−0.13	−0.23	−0.03
LN					1	**0.53**	0.44	0.23	**0.51**	0.40	−0.07	0.01	−0.11	−0.15	−0.24
TN						1	**0.84**	**0.79**	0.47	**0.82**	0.03	0.18	0.07	−0.02	−0.37
DMY							1	**0.77**	0.20	**0.97**	−0.05	−0.17	0.11	0.06	−0.35
GT								1	−0.18	**0.86**	−0.18	−0.21	0.36	0.34	−0.26
VT									1	0.05	0.3	**0.6**	−0.4	−0.51	−0.19
DMYI										1	−0.11	−0.27	0.16	0.18	−0.34
LS											1	0.4	−0.14	−0.15	0.08
CPC												1	−0.22	−0.41	0.03
NDF													1	**0.72**	0.42
ADF														1	0.49
ADL															1

[a]Bold: correlation is significant at $p < .01$

tive correlation between forage yield and ADF and NDF was found also by Sanada et al. (2007).

Conclusions

The differences in heading date were more than 20 days among populations. Some populations showed high DMY and lateness at the same time, which qualifies them for population breeding of new medium-to-late cultivars. Traits which are most important for such variability and characteristics of populations were plant height, tillers number and DMY. No clear differentiation from standard varieties was observed. Very few matches of geographic origin and grouping of populations in cluster analysis were obtained. This study has resulted in the selection of genotypes for further breeding of a late forage cocksfoot cultivar.

Acknowledgments This research was funded by Ministry of Education, Science and Technological Development of the Republic of Serbia (Project TR 31057).

References

Alderson J, Sharp WC (1995) Grass varieties in the United States. CRC Press, Florida. USA.

AOAC (1990) Official Method 984.13. Crude protein in animal feed, forage, grain and oil seeds. Official Methods of Analysis of AOAC International. 15th edition.

AOAC (1997) Official Method 973.18. Fiber (acid detergent) and lignin in animal feed. Official Methods of Analysis of AOAC International. 16th edition. Arlington, VA, USA, 28–29.

Boller B, Green S (2010) Genetic resources, pp. 13–37. In: Boller B. et al (eds.) Fodder crops and amenity grasses. Handbook of plant breeding, Springer Science+Business Media, New York.

Dent JW, Aldrich DTA (1963) The interrelationships between heading date, yield, chemical composition and digestibility in varieties of perennial ryegrass, timothy, cocksfoot and meadow fescue. J. Natl. Inst. Agric. Bot. 9:261–281.

Ghesquiere M, Barre P, Durand J-L, Litrico I, Sampoux J-P, Mournet P, Volaire F (2014) Genetic resources to climate scenarios in *Dactylis* and *Festuca* of temperate versus Mediterranean origin, pp 41–48. In: Soković et al. (eds.) Quantitative traits breeding for multifunctional grassland and turf, Springer, the Netherlands.

Hurst A, Hall E (2007) Plant variety descriptions submitted for registration of plant breeders rights in Australia up to 3 August 2007. Plant Var. J. 20:146–148.

Jafary AA (2004) Evaluation and seed yield characteristics in 29 accessions of cocksfoot (*Dactylis glomerata*) through a multivariate analysis. Ira. J. Agric. Sci. 35:817–825.

Last L, Boller B, Widmer F, Kolliker R (2014) Influence of management and habitat on genotypic and phenotypic diversity of *Dactylis glomerata*, pp. 91–95. In: Soković et al. (eds.) Quantitative traits breeding for multifunctional grassland and turf, Springer, the Netherlands.

Lindner R. Lema M, Garcia A (2004) Extended genetic resources of *Dactylis glomerata* subsp. *izcoi* in Galicia (northwest Spain). Genet. Resour. Crop Evol. 51:437–442.

Lumaret R (1988) Cytology, genetics and evolution in the genus *Dactylis*. CRC Crit. Rev. Plant Sci. 7:55–91.

Marshall DR, Brown AHD (1983) Theory of forage plants collecting, p.135–148. In: McIvor JC, Bray RA (eds.) Genetic resources of forage plants. CSIRO, East Melbourne.

Peeters A (2004) Wild and sown grasses. FAO and Blackwell publishing. 127–134.

Posselt UK, Willner E (2007) Performance of *Lolium perenne* ecotypes in comparison with cultivars. Proceedings of EUCARPIA fodder crops and amenity grasses section, Perugia, Italy, 56–59.

Saiga S (1981) Studies on breeding for improvement of forage quality of cocksfoot (*Dactylis glomerata* L.). Res. Bull. Hokkaido Natl. Agric. Exp. Stn. 129:25–92.

Sanada Y, Takai T, Yamada T (2007) Inheritance of the concentration of ware-soluble carbohydrates and its relationship with the concentrations of fibre and crude protein in herbage of cocksfoot (*Dactylis glomerata* L.). Grass Forage Sci. 62:322–331.

Sanada Y, Gras M, van Santen E (2010) Cocksfoot. pp. 317–328. In: Boller B et al.(eds.) Fodder crops and amenity grasses, Handbook of plant breeding, Springer Science+Business, New York.

Soković D, Lugić Z, Radović J, Živanović T, Babić S, Simić A, Maletić R (2011) Evaluation of morphological traits.dry matter yield and quality of *Lolium perenne* L. autochthonous populations from Serbia through multivariate analysis. Genetika, 43, 1:129–140.

Van Soest PJ, Robertson JB, Lewis BA (1991) Methods for dietary fibre, neutral detergent fibre and non-starch polysaccharides in relation to animal nutrition. J. Dairy Sci., 74:3583–3597.

Zeng B, Zhang XQ, Lan Y, Yang WY (2008) Evaluation of genetic diversity and relationship in orchardgrass (*Dactylis glomerata* L.) germplasm based on SRAP markers. Can. J. Plant Sci. 88:53–60.

Part II

Scarcity of Breeders

Towards Model-Assisted Evaluation of Perennial Ryegrass Varieties

14

T. De Swaef, A. Ghesquiere, P. Lootens, and I. Roldán-Ruiz

Abstract

As in other crops, evaluation of grass genotypes requires objective measures of the performance of individual plants or swards. However, because of genotype-by-environment-by-management interactions, the genetic value of a genotype cannot always be easily assessed. Detailed ecophysiological models that describe physiological processes determining phenotypic traits consist of parameters that can be considered as genetic coefficients that better represent the genetic value of the genotype. We used the LINGRA grass model to estimate model parameters for a perennial ryegrass variety based on harvest data from seven growing seasons. Despite its relative simplicity, the LINGRA model succeeded in simulating the dry matter harvest of different years, although it substantially underestimated the yield in the driest year. The grass-specific LINGRA features should therefore be combined with a more extensive ecophysiological model to enable reliable estimation of genotype specific parameters.

Keywords

Ecophysiological modeling • Perennial ryegrass • Genotype-by-environment

Introduction

As in other crops, evaluation of grass genotypes requires objective measures of the performance of individual plants or swards. Quantitative traits (like yield) are the result of genotype by environment by management interactions (G×E×M), and can consequently not be attributed solely to the genotype. Therefore, it is very difficult to

T. De Swaef (✉) • A. Ghesquiere • P. Lootens
I. Roldán-Ruiz
Plant Sciences Unit, Institute for Agricultural and Fisheries Research (ILVO), Melle, Belgium
e-mail: tom.deswaef@ilvo.vlaanderen.be

I. Roldán-Ruiz et al. (eds.), *Breeding in a World of Scarcity: Proceedings of the 2015 Meeting of the Section "Forage Crops and Amenity Grasses" of Eucarpia*,
DOI 10.1007/978-3-319-28932-8_14

assess whether a genotype that is performing well in a specific situation will perform similarly in other conditions.

Recently, interest in ecophysiological modelling in breeding has been steadily increasing (Tardieu and Tuberosa 2010). Such models use the environmental conditions as input variables for a set of mechanistic equations that describe specific physiological processes such as nitrogen uptake, transpiration, photosynthesis, organ development, etc., and ultimately the growth of the crop. If such a model is sufficiently complex, the model parameters can be regarded as genetic coefficients (Yin and Struik 2010). These parameters are less (or not) variable with environmental conditions and therefore better represent the plant's genetic value compared to directly measured phenotypic traits. These parameter values can be estimated by calibration procedures using different phenotypic measurements. This approach additionally enables virtual investigation of the performance of a variety/genotype in many different environments.

Here we present the application of the LINGRA grass model (Schapendonk et al. 1998) to data of one perennial ryegrass variety (*Lolium perenne* 'Cancan'). Specific goals were: (1) to estimate model parameters for this variety and (2) to evaluate whether this simple model has enough complexity to describe the dry matter yield across a range of growing seasons.

Material and Methods

Plant Material and Measurements

Plots of the perennial ryegrass variety 'Cancan' were sown at the field trial site (50°59′02″N 3°46′17″E) in a sandy loam soil at the Institute for Agricultural and Fisheries Research (ILVO) in spring. Yield data from the different cuttings were collected in the next growing season. Data used to optimize the parameter values for this study originate from seven growing seasons (harvests in 2003, 2005, 2007, 2008, 2009, 2010, 2011). Swards were harvested using the Haldrup Plot Harvester at a mowing level of 5 cm, and were dried for at least 48 h at 70°C. Meteorological data was collected from a nearby weather station.

Modeling

We implemented the LINGRA code (Schapendonk et al. 1998) in R (R Core Team 2014), and ran simulations of the set of algebraic and differential equations for 365 days per year with a time step of 1 day, using the Euler integration method of the *deSolve* package (Soetaert et al. 2010). The simplex algorithm was used to optimize parameter values using the *neldermead* package (Bihorel and Baudin 2015; Nelder and Mead 1965), by minimizing the sum of squared errors between simulated and measured dry matter harvests.

Results and Discussion

Of the complete set of model parameters (>40), a subset of the most sensitive parameters, defined via a sensitivity analysis, was optimized for the 'Cancan' variety using the data of all seven growing seasons (Table 14.1). These parameters could be expected to be genotype-dependent and therefore to provide information on the genotype's behavior in different environments. Further details of the LINGRA model can be found in Schapendonk et al. (1998).

Figure 14.1 presents simulation data for 3 years, differing in annual precipitation (Fig. 14.1a), where 2003 was the driest year, 2007 the wettest and 2011 intermediate.

Table 14.1 Set of parameters optimized for cultivar 'Cancan'

Parameter	Description	Unit	Value
NITMAX	Optimal organic leaf nitrogen content	%	3.64
RDRD	Base relative death rate of leaves	d^{-1}	0.0033
RDM	Maximum root depth	cm	150
LUEMAX	Maximum light use efficiency	$g\ MJ^{-1}$	3.2814

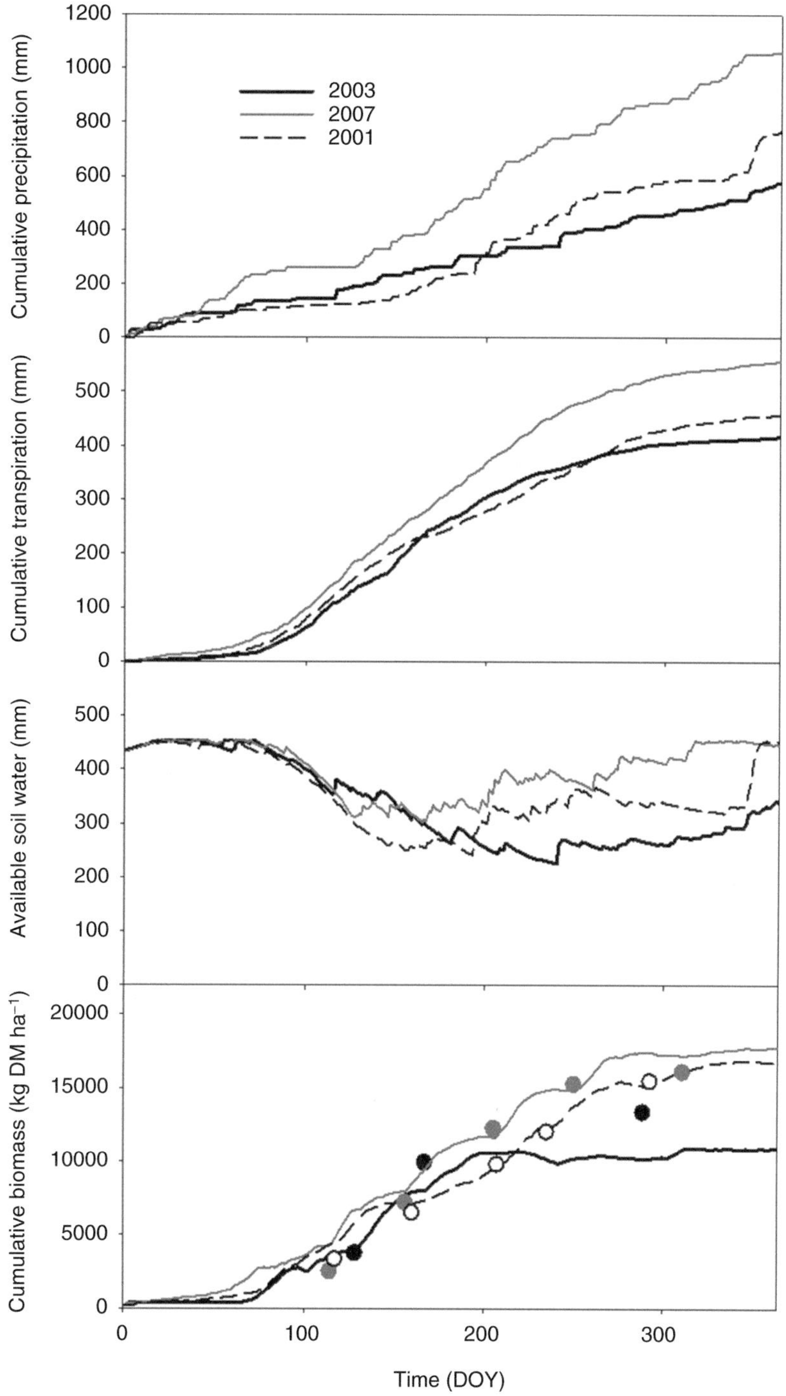

Fig. 14.1 Example of simulated cumulative transpiration (**b**), available soil water (**c**) and cumulative dry matter harvest (**d**) for 3 years (2003, 2007, 2011) differing in annual precipitation (**a**). *DOY* denotes Day Of the Year

Simulated cumulative transpiration was highest for the wettest year (2007). Transpiration in the driest year (2003) was limited, especially near the end of the growing season (Fig. 14.1b), because of the combined effect of stomatal closure and limited biomass development due to reduced soil water availability (Fig. 14.1c). The simulated dry matter yield corresponded well with yield data in 2007 and 2011, but substantially underestimated yield in 2003 (Fig. 14.1d).

The effects of limited water availability on biomass growth are thus potentially inadequately represented in the version of the LINGRA model used here.

Furthermore, the present work reports on the estimation of genetic coefficients based on single measurements taken over many years. In breeding practice, such an approach might not be ideal because of time constraints.

Therefore, future work should focus on the merging of specific grass features included in the LINGRA model with a more detailed ecophysiological crop model (e.g. GECROS from Yin and van Laar 2005). Parameter estimation could then be done in a shorter timeframe, using measurements of different physiological processes quasi-simultaneously within a growing season.

Conclusion

Despite its relative simplicity, the LINGRA model successfully simulated dry matter harvest of different years quite well, except for a substantial underestimation of yield in the driest year. The LINGRA features that are specific for grass should therefore be combined with a more extensive ecophysiological model to enable reliable estimation of genotype-specific parameters (Fig. 14.1).

References

Bihorel S., Baudin M. (2015). neldermead: R port of the Scilabneldermead module. R package version 1.0-10. http://CRAN.R-project.org/package=neldermead

Nelder J.A., Mead R. (1965). A simplex method for function minimization. Computer Journal 7, 308–313.

R Core Team (2014). R: A language and environment for statistical computing. R Foundation for Statistical Computing, Vienna, Austria. URL http://www.R-project.org/.

Schapendonk A.H.C.M., Stol S., van Kraalingen D.W.G., Bouman B.A.M. (1998). LINGRA, a sink/source model to simulate grassland productivity in Europe. European Journal of Agronomy 9, 87–100.

Soetaert K., Petzoldt T., Setzer R.W. (2010). Solving Differential Equations in R: Package deSolve Journal of Statistical Software, 33(9), 1–25.

Tardieu F., Tuberosa R. (2010). Dissection and modelling of abiotic stress tolerance in plants. Current Opinion in Plant Biology 13, 206–212.

Yin X., Struik P.C. (2010). Modelling the crop: from system dynamics to systems biology. Journal of Experimental Botany 61, 2171–2183.

Yin X., van Laar H.H. (2005). Crop systems dynamics, an ecophysiological simulation model for genotype-by-environment interactions. Wageningen Academic Publishers.

Real-Time Growth Analysis of Perennial Ryegrass Under Water Limiting Conditions

15

K. Jonavičienė, S. Yates, S. Nagelmüller, F. Liebisch, N. Kirchgessner, A. Walter, G. Brazauskas, and B. Studer

Abstract

Understanding plant growth under abiotic stress conditions is very important for the development of tolerant crops. Here, we present a novel integrated phenotyping platform for investigating real-time growth of perennial ryegrass under abiotic stress using new technologies. In this work we demonstrate its use in studying leaf elongation rate with respect to temperature and soil moisture deficit. The method used is non-destructive, low labor intensive and applicable to other grass species.

Keywords

Leaf elongation rate (LER) • Perennial ryegrass (*Lolium perenne* L.) • Real-time growth • Water deficit • Phenotyping

Introduction

High forage yield and quality makes perennial ryegrass (*Lolium perenne* L.) the predominant forage grass species used in both grazing and hay production systems. Perennial ryegrass is known for its large variation in morphological and growth characteristics as well as its rapid response to drought because it requires a relatively large amount of water to sustain growth (Norris 1985).

In agriculture, drought is causing major yield losses. Even temperate environments are subjected to mild summer droughts reducing yield. Despite some success, direct selection for high yield under water stress has proven to be difficult (Nelissen et al. 2014). A very important aspect to consider when breeding for drought tolerant crops is the

K. Jonavičienė (✉)
Institute of Agricultural Sciences, ETH Zürich, Universitätstrasse 2, Zürich 8092, Switzerland

Institute of Agriculture, Lithuanian Research Centre for Agriculture and Forestry, Instituto a. 1, Akademija, Kėdainiai distr 58344, Lithuania
e-mail: kristinaja@lzi.lt

S. Yates • S. Nagelmüller • F. Liebisch
N. Kirchgessner • A. Walter • B. Studer
Institute of Agricultural Sciences, ETH Zürich, Universitätstrasse 2, Zürich 8092, Switzerland

G. Brazauskas
Institute of Agricultural Sciences, Lithuanian Research Centre for Agriculture and Forestry, Instituto a. 1, Akademija, Kėdainiai distr 58344, Lithuania

I. Roldán-Ruiz et al. (eds.), *Breeding in a World of Scarcity: Proceedings of the 2015 Meeting of the Section "Forage Crops and Amenity Grasses" of Eucarpia*,
DOI 10.1007/978-3-319-28932-8_15

severity and duration of the stress, as increased survival under severe drought does not equate to improved performance under mild drought (Skirycz and Inzé 2011). Plant growth slows and arrests before the leaf water content decreases (Parent et al. 2010), which is advantageous for survival under extreme stress. But when the stress does not threaten the survival, growth arrest leads to reduced biomass which is the main yield target in forage crops. Therefore, understanding plant growth by characterizing leaf elongation rates (LER) under favorable and unfavorable conditions is of high value for yield improvement in perennial ryegrass.

Most water stress studies focus on plant growth and soil water availability, while other environmental factors are unaccounted variables (Tardieu 2011). However, plant growth is dependent upon soil water availability, water loss (evaporative demand) and temperature. These environmental variables can be integrated into models describing growth, referred to as ecophysiological models, to determine the genotypic response of a plant to its environment (Reymond et al. 2003). But to precisely describe the genotypic response, the environmental parameters along with plant growth need to be monitored frequently. Here, we present a novel, high throughput and non-destructive phenotyping platform, to monitor the leaf growth under adverse conditions limiting growth.

Material and Methods

Plant Material and Growth Conditions

Two natural ecotypes of perennial ryegrass, designated 3380 and 3809, along with a selected genotype of the cultivar 'Vigor' (1299) were vegetatively propagated into four replicates consisting of 20 tillers each. Plants were grown in 370 g commercial potting mix substrate ('Spezialmischung 209', RICOTER Erdaufbereitung AG, Aarberg, Switzerland) under regular irrigation and fertilization in a greenhouse for 3 weeks. Leaf growth measurements were performed in a climate chamber (Conviron, Winnipeg, Canada) with a light/dark photoperiod of 16/8 h and a light intensity of 275 μmol photosynthetically active radiation (PAR) m^{-2} s^{-1}. The average day/night temperature was 25/15 °C and the relative humidity was kept at 50 ± 15 %.

Fixation of the Leaves and Image Acquisition

For each clonal replicate (four per genotype), a young leaf (<5 cm length outside of stem) was selected for leaf elongation analysis. The leaf was attached to a string by a hair clip at the tip and kept taut using weights of 20 g. White plastic beads (ø 20 mm) were threaded onto the strings and placed on the growth tracking array to provide artificial landmarks that allowed registration of marker movements (Nagelmüller et al. 2016) (Fig. 15.1a). In front of this system, a LupusNET HD camera (LUPUS-Electronics® Gmbh, Landau, Germany) was installed. Digital images (20MP) were taken every 120 s and analysed with the software 'Martrack Leaf' to monitor LER (Mielewczik et al. 2013).

Drought Treatment

The soil water potential (hPa) was determined in three out of four replicates every four hours with a Koubachi Wi-Fi Plant Sensor (Koubachi AG, Zürich, Switzerland) at a depth of 7 cm (Liebisch et al. 2014). The temperature of the meristem of six plants was also measured with a type K thermocouple (ø 0.2 mm) inserted into the tiller and recorded using a CR10X Campbell Scientific Datalogger (Campbell Scientific Inc, Logan, USA). Fully watered plants with all sensors attached (0 h) were subjected to a 5 days water deprivation and re-watered after 5 days.

Data Processing

The data of room temperature (RT (C^0)), relative humidity (RH (%)) and meristem temperature (MT (C^0)) were summarized hourly for the first five days

Fig. 15.1 Leaf elongation measurement platform for time series measurements of leaf elongation rates, soil water potential and meristem temperature, where (**a**) shows the growth tracking array with white plastic beads on black board, (**b**) the thermocouple type K, (**c**) the Koubachi plant sensor and (**d**) a perennial ryegrass plant with (**a–c**) attached

of the experiment (120 h total). LER ($mm.h^{-1}$) was expressed on an hourly basis, using Eq. 15.1

$$\mathrm{LER} = \frac{D_1 - D_0}{t_1 - t_0} \times 60 \qquad (15.1)$$

where D_0 is the displacement (in mm) at the start of the hour, D_1 the displacement at the end of the hour, t_0 the time at the start (min) and t_1 the time at the end (min).

The soil water potential (ψ) was calculated as hygrometric water potential (hPa), which was $\log_{10}$ transformed. A Loess fit was used to calculate and impute the hourly hPa per genotype.

For data analysis, linear models were created in the R statistical environment, using the 'lm' function. LER was fitted against soil moisture or meristem temperature with 'genotype' as a factor. The summary function was used to extract adjusted R^2 (hereafter reported as R^2) and P-value (P). To characterize the relationship between meristem temperature and LER, only data from the first 24 h were used in order to remove the effect of soil moisture on LER.

Results and Discussion

The LER of three perennial ryegrass genotypes was evaluated under water limiting conditions using a novel integrated phenotyping platform (Fig. 15.2a–d). After 5 days of water deprivation, the plants were re-watered and resumed growth within a few hours (data not shown). Figure 15.2b shows the linear relationship between meristem temperature and LER within the first 24 h of the experiment (without water stress, $R^2=0.42$, all genotypes). The genotypes 1299 and 3380 showed a similar response whereas an increased LER was found for the genotype 3809 with respect to meristem temperature ($P < 0.05$). Figure 15.2c shows the relationship between LER and soil moisture ($\log_{10}$(hPa)) with linear regression fits ($R^2=0.57$, all genotypes). Here, the genotypes 1299 and 3809 showed similar trends, in contrast to genotype 3380 which exhibited a decreased response to soil moisture ($P < 0.05$). These results are consistent with results observed by Laidlow (2009) showing the linear correlation between soil moisture and LER. Figure 15.2d shows the relationship between soil moisture and time, demonstrating that soil water loss for the three different genotypes was not equal, suggesting differential water use of the three genotypes.

This phenotyping platform enables the monitoring of real-time growth measurement of leaf tissue and is applicable to other abiotic stresses, for example salinity or heat. It can also be applied to other grass species. Moreover, the Koubachis sensors can be used to determine the water availability in different soil substrates and/or media. The platform can collect data with high resolution and frequency (minutes) with little

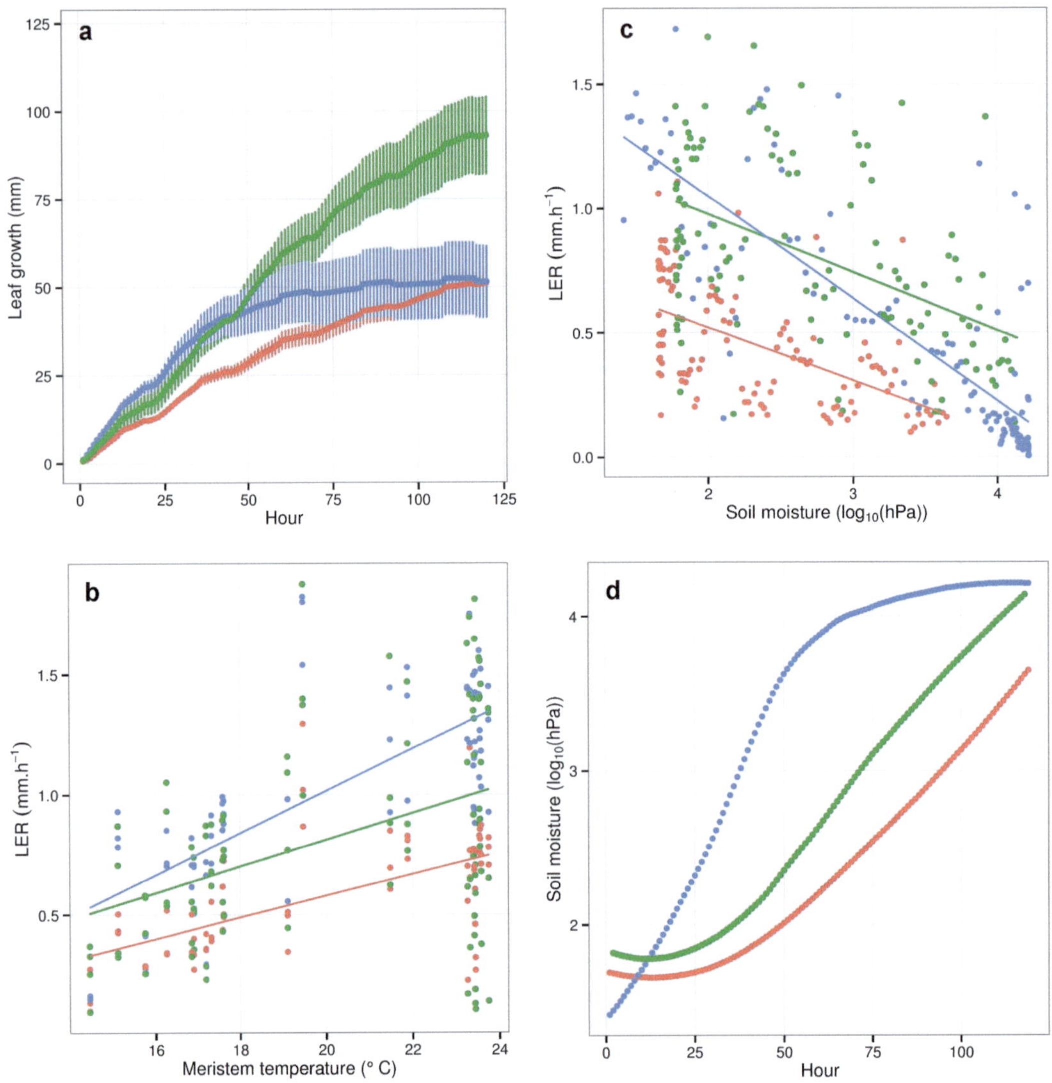

Fig. 15.2 Real-time growth analysis of three perennial ryegrass genotypes; during 5 days of increasing soil water deficit. The Figure shows three genotypes; 1299 in red, 3380 in blue and 3809 in green. (**a**) illustrates the leaf growth over time (in hours), with the mean values for each genotype including standard error bars. (**b**) shows LER with respect to meristem temperature for all genotypes and clones with linear regression fits in solid lines. (**c**) shows LER with respect to soil moisture, the mean is shown per hour per genotype to reduce over plotting. (**d**) shows soil moisture with respect to time as calculated by Loess fit

labor. The data obtained can be utilized in a variety of ways such as 'thermal time' (Auzanneau et al. 2011), and thus can be integrated into ecophysiological models to determine the genotypic response to environmental variables, as described by Reymond et al. (2003). Moreover, such parameters can be used for association mapping, to identify genetic components attributable to LER under adverse environmental conditions.

Conclusion

Here we demonstrate the application of this platform, which is able to measure LER with high temporal resolution and has advantages of being non-labor intensive and low cost.

This platform could also be used in other grass species and the data can be integrated into ecophysiological models for trait dissection. We are currently developing ecophysiological models to analyze this data, in a diverse population. We envision using the system to study other abiotic stresses such as salinity by dosing the plants with increasing salt concentrations.

References

Auzanneau J, Huyghe C, Escobar-Gutiérrez A J, Julier B, Gastal F, Barre P (2011) Association study between the gibberellic acid insensitive gene and leaf length in a *Lolium perenne* L. synthetic variety. BMC Plant Biology 11:183.

Laidlow AS (2009) The effect of soil moisture content on leaf extension rate and yield of perennial ryegrass. Irish Journal of Agricultural and Food Research 48: 1–20.

Liebisch F, Keller F, Huguenin-Elie O, E Frossard E, Oberson A (2014) Seasonal dynamics and turnover of microbial phosphorus in permanent grassland. Biology and fertility of soils 50: 465–475.

Mielewczik B, Friedli M, Kirchgessner N, Walter A (2013) Diel leaf growth of soybean: a novel method to analyze two-dimensional leaf expansion in high temporal resolution based on a marker tracking approach (Martrack Leaf). Plant Methods 9: 30.

Nagelmüller S, Kirchgessner N, Yates S, Hiltpold M, Walter A (2016) Leaf Length Tracker: a novel approach to analyse leaf elongation close to the thermal limit of growth in the field. Journal of Experimental Botany doi: 10.1093/jxb/erw003

Nelissen H, Moloney M, Inzé D (2014) Translational research: from pot to plot. Plant Biotechnology Journal 12: 277–285.

Norris I (1985) Relationships between growth and measured weather factors among contrasting varieties of *Lolium, Dactylis* and *Festuca* species. Grass and Forage Science 40:151–159.

Parent B, Boris, Suard B, Serraj R, Tardieu F (2010) Rice leaf growth and water potential are resilient to evaporative demand and soil water deficit once the effects of root system are neutralized. Plant, Cell & Environment 33: 1256–1267.

Reymond M, Muller B, Leonardi A, Charcosset A, Tardieu F (2003) Combining quantitative trait Loci analysis and an ecophysiological model to analyze the genetic variability of the responses of maize leaf growth to temperature and water deficit. Plant Physiology 131: 664–675.

Skirycz A, Inzé D (2011) More from less: plant growth under limited water. Current Opinion in Biotechnology 21: 197–203.

Tardieu F (2011) Any trait or trait-related allele can confer drought tolerance: just design the right drought scenario. Journal of Experimental Botany doi: 10.1093/jxb/err269.

16 UAV-Based Remote Sensing for Evaluation of Drought Tolerance in Forage Grasses

P. Lootens, W.H. Maes, T. De Swaef, J. Aper, K.C. Mertens, K. Steppe, J. Baert, and I. Roldán-Ruiz

Abstract

Global climate change is likely to create more and longer-lasting drought periods, which will have a negative effect on the performance of forage grasses. Existing varieties should be evaluated for drought tolerance and new, more drought tolerant varieties should be developed. Drought stress tolerance is currently assessed using visual scores, but new techniques and technologies such as imagery from airborne vehicles have the potential to make such assessments more automatic and objective. The aim of this study was to evaluate UAV-based remote sensing, using an unmanned aerial vehicles (UAV or drone) equipped with a visual and thermal camera as an alternative for scoring forage grasses for drought tolerance. A very good correlation (0.866) was found between the imagery-based indices and the visual scores assigned to individual plants by the breeder. This opens perspectives for using UAV-based remote sensing for breeding for drought tolerance in forage grasses.

Keywords

UAV-based remote sensing • UAV • Drone • Drought • *Festuca* • *Festulolium* • *Lolium* • Rainout shelter • Visual imagery • Thermal imagery

Introduction

Global climate change is likely to lead to more frequent and longer drought events, especially during spring and early summer (Olesen 2008). Consequently, periods during which the crop demand for water exceeds the supply (i.e., plant water deficit) will become more frequent (Blum 2011), causing production losses. The magnitude of the effects will depend on the crop and the region (Olesen et al. 2011), but the negative consequences

P. Lootens (✉) • T. De Swaef • J. Aper
J. Baert • I. Roldán-Ruiz
Plant Sciences Unit, Institute for Agricultural and Fisheries Research (ILVO), Melle, Belgium
e-mail: peter.lootens@ilvo.vlaanderen.be

K.C. Mertens
Technology and Food Sciences Unit, Institute for Agricultural and Fisheries Research (ILVO), Melle, Belgium

W.H. Maes • K. Steppe
Laboratory of Plant Ecology, Ghent University, Ghent, Belgium

I. Roldán-Ruiz et al. (eds.), *Breeding in a World of Scarcity: Proceedings of the 2015 Meeting of the Section "Forage Crops and Amenity Grasses" of Eucarpia*,
DOI 10.1007/978-3-319-28932-8_16

might be exacerbated in perennial cropping systems such as permanent grasslands if environmental hazards cause irreversible sward degradation. One strategy to mitigate possible yield losses is to breed drought tolerant varieties. This requires a thorough understanding of the eco-physiological responses of individual genotypes to drought and the availability of efficient and reliable screening tools.

A suitable approach for drought tolerance testing is the use of rainout shelters because these 'open rain shields' provide cover from the rain while limiting the impact on temperature or light level (Parra et al. 2012). In this way they provide a good approximation of the open-field situation during a dry period. Individual plant evaluation frequently consists of assigning visual scores of wilting, greenness or general appearance after a period of imposed drought and during subsequent recovery. Although this evaluation method is fast, it may become impractical when scoring large numbers of plants. Furthermore, bias can be introduced when several experts are needed to score a large trial or if several trials need to be scored over time. UAV-based remote sensing using an unmanned aerial vehicle (UAV or drone) equipped with a visual and thermal camera offers an alternative to visual scores, and can potentially deliver information that cannot be recorded by the human eye. Indeed, greenness indices derived from visual images have been developed and tested (e.g., Sonnentag et al. 2012), drought-related variables derived from thermal imagery have been described (e.g., Maes and Steppe 2012; Zia et al. 2013), and possibilities to monitor crops are increasing rapidly (Araus and Cairns 2014).

The aim of this study was to evaluate UAV-based remote sensing, using an UAV equipped with a visual and a thermal camera, as an alternative to visual scoring of drought tolerance trials of forage grasses in a breeding context.

Material and Methods

Experimental Setup

In the first half of April 2013, 2250 plants were planted (at a distance of 0.5 m between and within the rows) in a field equipped with three mobile rainout shelters in Melle, Belgium (50° 59.5′ N, 3°47.1′ E), on sandy loam soil. These 2250 plants comprised 10 families of *Festuca arundinacea*, 10 families of diploid and 10 families of tetraploid *Lolium perenne* (30 plants per family), and three clonal replicates of 450 genotypes selected from the breeding gene pool of Ghent University (*Festuca*) and the Institute for Agricultural and Fisheries Research (ILVO) (*Lolium–Festulolium*). The soil water content was monitored for the 10–80 cm depth profile during the drought period using TDR sensors (type CS616, Campbell Scientific Inc., UK) (Fig. 16.1).

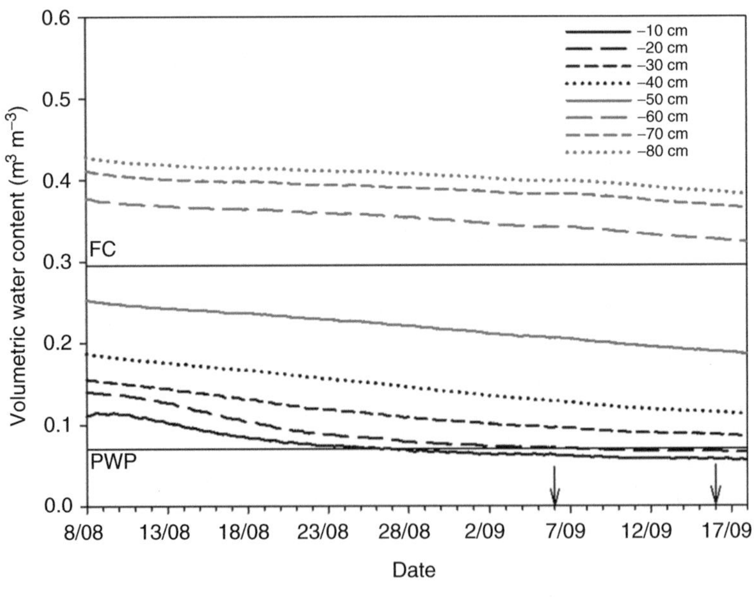

Fig. 16.1 Volumetric water content during the drought treatment, measured at eight depths. *Horizontal lines* indicate field capacity (*FC*) and permanent wilting point (*PWP*). *Arrows* indicate the timing of breeder visual scoring (6/09 and 16/09) and timing of the drone flight (6/09)

During the summer of 2013, two consecutive progressive drought periods were applied: from 5 July until 28 July, and from 8 August until 17 September. In this report we focus on the second drought period, which simulates a summer drought that lasted 41 days. The plants were mown and fertilized on 30 July and were able to regrow during 10 days before the drought treatment was applied. The plants were visually scored three times during the treatment (21 August, 6 September and 16 September). They were mown on 24 September, after removal of the rainout shelters, and scored again during the regrowth period, on 10 October. Visual plant scores varied from 1 to 9 and were an estimation of the amount of green biomass of the plant. Plants with score '9' had the highest amount of biomass (i.e., yield) that stayed green and healthy regardless of the drought applied. Plants with score '1' had only 0–20 % biomass of the plants with a score '9'. Each scoring round lasted approximately 6 h (n = 2250 plants).

UAV-Based Remote Sensing and Image Analysis

Aerial imagery was obtained with an unmanned aerial vehicle (UAV), the AT8 octocopter of AerialTronics (Scheveningen, The Netherlands), equipped with a thermal and a visual camera (Fig. 16.2). The thermal camera, a FLIR SC305 (FLIR Systems, Inc., Wilsonville, OR, USA) has a resolution of 320×240 pixels, a thermal accuracy of ±2 °C and a thermal sensitivity of 0.05 °C. It was equipped with a 10 mm lens, with a field of view (FOV) of 45° × 34°. The camera was connected with an Olimex on-board computer with data logged every 2–2.5 s. The visual camera (Canon S110, Canon, Japan) was programmed with CHDK, (Canon Hack Development Kit software which is available at http://chdk.wikia.com/wiki/CHDK) to log continuously every 0.5 s.

A UAV flight was carried out on 6 September 2013 at an altitude of 12–14 m, around 15 h. It was a sunny day with limited wind speed. The temperature ranged from 13.3 to 32.9 °C during that day with an average of 20.6 °C. The daily light integral (global radiation) was 1077.4 J cm^{-2}.

Georeferenced RGB orthophotos were created with AgiSoft PhotoScan Professional (AgiSoft LLC, St-Petersburg, Russia) based on the GPX-logfile of the UAV; orthophotos were registered manually in ArcGIS 10.1 (ESRI, Redlands, Ca, USA). The Green Chromatic Coordinate index (g_{cc} = 100 G/(R + G+ B)) and Excess Green index (ExG = 2G-(R + B)) were calculated. All thermal images were converted to canopy temperature (T_c) images as described by Maes et al. (2014).

A vegetation mask was created in ArcGIS 10.1 by combining g_{cc} and ExG of the RGB orthophoto of an earlier flight on 1 August 2013. Based on this mask, a polygon shapefile was generated automatically and checked manually if each polygon

Fig. 16.2 A drone equipped with a visual and thermal camera inspecting the drought tolerance trials at ILVO, Melle, Belgium

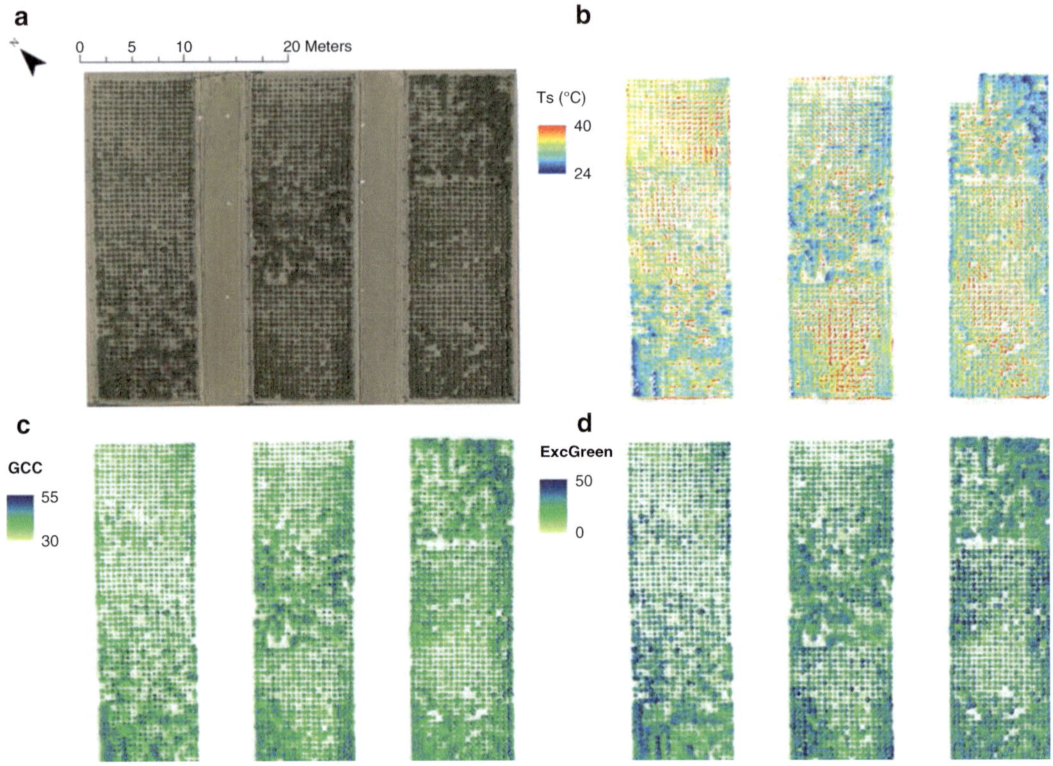

Fig. 16.3 Visual orthophoto (**a**), and visual representations of the temperature (***Tc***) (**b**), the Green Chromatic Coordinate index (***gcc***) (**c**) and the Excess Green index (***ExG***) (**d**) of the canopy of the forage grass plants in the drought experiment (All images were taken on 6 September 2012)

corresponded with a grass plant. This shapefile consisted of 2210 polygons, since 40 plants that had died off before the start of the summer treatment were not included. Row and column numbers were assigned to each polygon allowing to link each polygon to a single grass plant and to calculate g_{cc}, *ExG* and T_c per polygon. As an example a visual orthophoto and visual representations of the different parameters is shown in Fig. 16.3.

Statistics

Pearson correlations were calculated between the different variables. Furthermore, a general linear regression with the forward entry option was used to predict the scores of the breeder based on the drone-derived variables, using Statistica (Statsoft, Tulsa, OK, USA). A 0.05 level was used in all analyses to show significance.

Results and Discussion

The scores of the plants during the drought period and after regrowth were highly correlated (Table 16.1). The high correlation between observations taken after 29 (6 September) and 39 days (16 September) confirmed the consistency of the scoring.

The second column of Table 16.2 shows the correlations between the image analysis data and the visual score given the same day. All correlations were significant. Negative correlations were found for the T_c (canopy temperature) variables, indicating lower canopy temperature and lower temperature variability for more drought tolerant plants. The correlation between the visual score and the mean canopy temperature was higher than its correlation with the standard deviation of the canopy temperature. *ExG* and g_{cc} mean values correlated quite well with the visual scores, with

Table 16.1 Correlation between visual scores during the drought period (29 days = 6 September; 39 days = 16 September) and during regrowth (10 October)

	29 days of drought	16 days of regrowth after rain out shelter removal and cutting
29 days of drought		0.793
39 days of drought	0.919	0.800

Scores range from 1 to 9. All correlations were significant at the 0.05 level (n = 2210)

Table 16.2 Correlation between VIS (*ExG* excess green, g_{cc} green chromatic coordinate) and thermal (T_c canopy temperature, °C) image derived parameters estimated at 29 days of drought (6 September) and visual scores

	Scores	
Drone variables	29 days of drought	16 days of regrowth after rain out shelter removal and cutting
T_c (mean)	−0.619	−0.563
T_c (standard deviation)	−0.279	−0.289
ExG (mean)	0.809	0.679
ExG (standard deviation)	0.140	0.087
g_{cc} (mean)	0.802	0.738
g_{cc} (standard deviation)	0.603	0.611

All correlations were significant at the 0.05 level (n = 2210)

slightly higher values for *ExG* than for g_{cc}. In contrast, findings by Sonnentag et al. (2012) suggested better performance of g_{cc}. A good correlation was also found between the standard variation of g_{cc} and the visual scores, indicating that more tolerant plants showed more greenness color variation than less tolerant plants. When the mean *ExG* data was analyzed per species the correlations were 0.868, 0.844, and 0.692, respectively, for *Lolium perenne*, *Festolulium*, and *Festuca arundinacea*. For mean g_{cc} analyzed per species, the correlations were 0.843, 0.876, and 0.841, respectively, thus even higher correlations can be reached compared to a global correlation analysis. Very similar tendencies were found when the drone-derived data were compared to the visual scoring given after 16 days of regrowth, but the correlation coefficients were generally lower.

It is not surprising that parameters derived from images in the visual spectrum correlate better with visual scores than parameters derived from thermal images, as the scores assigned by the human eye are based on greenness and not on surface temperature. This suggests that thermal images provide complementary information about the eco-physiological responses of the plant.

A forward entry stepwise linear regression using all image analysis-derived parameters was used to find out whether it was possible to predict the regrowth score or the drought score of the breeder. The correlation (multiple r) between the drought score after 29 days of drought and the model was 0.866 (adjusted $R^2 = 0.750$) when all image analysis-derived variables were incorporated except the *ExG* standard deviation. Thus, the resulting correlation (0.866) was improved compared to the highest r value (for the *ExG* mean: 0.809). The correlation between the plant score assigned during regrowth and the second model based on the image analysis-derived variables was 0.772 (adjusted $R^2 = 0.594$). In this case, the resulting r (0.772) was only slightly improved compared to the highest correlation value of a single drone variable (for g_{cc} mean: 0.738).

Taken together, these results demonstrate that the use of UAVs and UAV-based remote sensing offer opportunities for the monitoring of large field trials of forage crops using visual and/or thermal imagery. More frequent observations and the use of plant physiology related parameters could enable improved selection progress. Here we demonstrate the use of these technologies for drought tolerance screening, but the possibilities are much broader, as a single flight in which visual, thermal and/or hyperspectral images are taken. Such combined data could allow the screening of thousands of plants or plots for multiple traits related to yield, quality and for tolerance to biotic and abiotic stresses.

Conclusion

Using only a few variables derived from UAV-based remote sensing, a highly significant correlation (0.866) was obtained between data obtained with a UAV and the visual scores assigned to individual plants by the breeder.

This opens perspectives for the use of UAV-based remote sensing for breeding for drought tolerance. Especially if a good workflow can be created to analyze the images and ultimately calculate a plant value in real time, UAV-based remote sensing may result in a more efficient plant evaluation for breeding purposes.

References

Araus JL, Cairns JE (2014) Field high-throughput phenotyping: the new crop breeding frontier. Trends in plant Science, 19(1), 52–61.

Blum A (2011) Plant breeding for water-limited environments. Springer, New York, 255p.

Maes WH, Steppe K (2012) Estimating evapotranspiration and drought stress with ground-based thermal remote sensing in agriculture: a review. Journal of Experimental Botany, 63(13), 4671–4712.

Maes WH, Minchin PE, Snelgar W, Steppe K (2014) Early detection of Psa infection in kiwifruit by means of infrared thermography at leaf and orchard scale. Functional Plant Biology, 41(12),1207–1220.

Olesen JE (2008) Climate change as a driver of European agriculture. Europe's rural areas in action - Facing the challenges of tomorrow. Limassol, Cyprus, October 16–17.

Olesen JE, Trnka M, Kersebaum KC, Skjelvag AO, Seguin B, Peltonen-Sainio P, Rossi F, Kozyra J, Micale F (2011) Impacts and adaptation of European crop production systems to climate change. European Journal of Agronomy, 34, 96–112.

Zia S, Romano G, Spreer W, Sanchez C, Cairns J, Araus JL, Müller J (2013) infrared thermal imaging as a rapid tool for identifying water-stress tolerant maize genotypes of different phenology. Journal of Agronomy and Crop Science, 199, 75–84.

17 Association Analysis of Candidate Genes with Tillering in Perennial Ryegrass

G. Statkevičiūtė, A. Aleliūnas, V. Kemešytė, and G. Brazauskas

Abstract

Because perennial ryegrass is propagated by seed, seed yield is particularly important for commercial breeding of ryegrass. Generative tillering is one of the phenotypic traits that determines seed yield potential. Associations between generative and vegetative tiller numbers and sequence polymorphisms in five perennial ryegrass genes with putative function in plant architecture were estimated in this study. One SNP in the *LpSHOOT1* gene was associated with generative tiller number while vegetative tiller weight was associated with two alternative SNPs in the same *LpSHOOT1* gene.

Keywords

Lolium perenne • Generative tillers • *LpSHOOT1*

Introduction

Perennial ryegrass (*Lolium perenne* L.) is the dominant grassland species in the temperate regions of the world. Ryegrass breeding objectives have mostly focused on improving herbage yield, feed quality and resistance to biotic and abiotic factors (Humphreys et al. 2006). However, seed production is crucial for commercial success of the new varieties. Seed yield is a complex trait that is affected by many factors and is generally difficult to predict (Elgersma 1990). The number of fertile generative tillers is an important component of seed yield potential. The number of fertile tillers has been shown to be closely related to seed yield in various grass species (Chastain et al. 1997; Entz et al. 1994).

The objectives of this study were to evaluate variation for generative and vegetative tiller number and weight in an association mapping population of perennial ryegrass and identify associations between these traits and sequence variants within five perennial ryegrass genes with putative function in plant architecture. Orthologs of selected genes

G. Statkevičiūtė (✉) • A. Aleliūnas • G. Brazauskas
Laboratory of Genetics and Physiology, Institute of Agriculture, Lithuanian Research Centre for Agriculture and Forestry, Akademija, Kėdainiai reg, Lithuania
e-mail: grazinastat@lzi.lt

V. Kemešytė
Department of Grass Breeding, Institute of Agriculture, Lithuanian Research Centre for Agriculture and Forestry, Akademija, Kėdainiai reg, Lithuania

I. Roldán-Ruiz et al. (eds.), *Breeding in a World of Scarcity: Proceedings of the 2015 Meeting of the Section "Forage Crops and Amenity Grasses" of Eucarpia*,
DOI 10.1007/978-3-319-28932-8_17

were shown to be involved in phytohormone signaling regulation and axillary meristem outgrouth in various plant species (Brazauskas et al. 2010).

Material and Methods

Plant Material and Phenotyping

From a collection of 96 perennial ryegrass (*Lolium perenne* L.) accessions consisting of 44 cultivars, 51 natural ecotype and one colchicine induced mutant (Pašakinskienė 2005), a single plant per accession was randomly taken and used in this study. Plants were established from seeds and clonally propagated in the greenhouse. A set of 96 genotypes went through a vernalization process in the greenhouse at the constant day/night temperature of 5 °C from November 2012 to February 2013 (4 months in total). Ramet-based field collection was established in April 2013. A set of non-vernalized genotypes was planted in the field in July 2013 and went through vernalization under natural conditions during winter 2013–2014. Four ramets of each genotype were planted at 50 × 50 cm distances using a randomized complete block design. Each plant was cut at full ripening stage and dried to a constant weight. Generative tiller number (GTN), generative tiller weight (GTW), vegetative tiller number (VTN) and vegetative tiller weight (VTW) were assessed in 2013 and 2014. Lattice analysis and broad sense heritability were estimated with PLABSTAT software package (Utz 2011). Repeatability was calculated by dividing the genotypic variance component σ^2_g by the sum of σ^2_g and the effective mean square of the error. Broad sense heritability was estimated by the formula $H^2 = \sigma^2_g / (\sigma^2 / re + \sigma^2_{ge} / e + \sigma^2_g)$, where r is the number of replicates and e is the number of environments.

Candidate Gene Amplification and Sequencing

Gene fragments of *LpIAA1*, *LpBRI1*, *LpRUB1*, and *LpSHOOT1* were amplified with primers described in Brazauskas et al. (2010), except for *LpTB1*, where sequences of forward and reverse primers were 5′-TGATCTGCTCCTGCTAGTCCT-3′ and 5′-TGCAGATTAGAATCCACGCAAGA-3′, respectively. PCR was carried out in 10 μl reaction mixtures, containing 1 × Phusion Flash High-Fidelity PCR Master Mix (Thermo Fisher Scientific, Vilnius, Lithuania), 0.5 μM of forward and reverse primers and 80 ng of DNA. PCR thermal profile was 98 °C for 10 s, followed by 30 cycles of 98 °C for 1 s, annealing temperature for 5 s, 72 °C for 40 s and final extension step of 72 °C for 60 s. The annealing temperature was set to 71 °C for *LpIAA1* and *LpBRI1*, 64 °C for *LpRUB1* and *LpSHOOT1* and 61 °C for *LpTB1*. The cloning of amplified fragments, transformation, plasmid DNA extraction and sequencing were performed as described in Aleliūnas et al. (2014). Each fragment was sequenced from both directions with forward and reverse pJET 1.2 sequencing primers. Sequence chromatograms were assembled into contigs using ChromasPro v.1.7.5 (Technelysium Pty Ltd, Australia). Sequence alignment was done with MEGA6 software (Tamura et al. 2013).

Marker–Trait Association Analysis

Population structure and relative kinship estimations were performed as described in Aleliūnas et al. (2014). Association analysis between SNPs and phenotypic traits was performed using a mixed linear model (MLM) implemented in the Tassel v.4.3 software (Bradbury et al. 2007). SNPs with minor allele frequency of less than 5 % were excluded from further analysis. Multiple testing correction of p values was performed by positive false discovery rate (q value) estimated using QVALUE package (Storey et al. 2004). Phenotypic means of marker genotype classes were compared using Tukey's honestly significant difference (HSD) method modified for unequal sample size (Unequal N test) in STATISTICA 7 (StatSoft Inc., Tulsa, Oklahoma, USA).

Results and Discussion

Phenotypic Variation

Genotypes of the first and the second planting went through different vernalization conditions. Differences in vernalization received may explain why the plants generated more vegetative than generative tillers in the first year whereas in the second experimental year, generative tillers were more abundant (Table 17.1). The total number of tillers was similar. Lattice analyses revealed significant ($p<0.01$) genotypic variance components for both traits.

Boelt and Studer (2010) state that the reproductive tiller number is strongly affected by the management practices as well as the environment and is thus a difficult breeding target. On the contrary, Şeker et al. (2014) estimated high heritabilities for both fertile tiller and total tiller numbers. The heritability of the traits in our study was generally low due to differential vernalization received over 2 years, yet the repeatability within years was higher (0.62–0.73 for generative and 0.49–0.82 for vegetative tiller traits). Marker-trait associations were estimated for 2013 and 2014 phenotyping separately.

Marker-Trait Associations

Associations were estimated between the phenotypic traits and 270 polymorphic sites in five perennial ryegrass genes with putative function in plant architecture. The analysis yielded three significant associations in *LpSHOOT1* (Table 17.2). However, these associations did not pass the threshold of multiple testing correction, although the post-hoc unequal N test confirmed the differences between the haplotype groups. Testing in a larger genotype set would

Table 17.1 Variation and heritability for phenotypic traits in 96 perennial ryegrass genotypes

	2013		2014		
Trait	Mean ± SD	Gen var	Mean ± SD	Gen var	Heritability
GTN	109.46 ± 67.52	4368.9**	330.63 ± 133.73	16316.0**	0.29
GTW, g	45.49 ± 34.32	747.7**	149.96 ± 75.99	5358.3**	0.43
VTN	303.36 ± 272.43	68815.2**	46.33 ± 46.92	1538.4**	0.26
VTW, g	39.30 ± 34.77	1158.0**	13.58 ± 8.03	58.7**	0.36

GTN generative tiller number, *GTW* generative tiller weight, *VTN* vegetative tiller number, *VTW* vegetative tiller weight, *Gen var* genotypic variance components

** $p<0.01$

Table 17.2 Significant associations between phenotypic traits and sequence polymorphism in *LpSHOOT1* gene

Locus	Coding/ noncoding	Trait	*p*	Genotype	Effect	n
SNP-602	C(s)	Generative tiller number 2014	0.00070	C:C	−0.67	78
				T:T	−0.44	3
				C:T	0	10
SNP-55	N	Vegetative tiller weight 2014	0.00090	C:C	−0.32	72
				C:A	12.66	6
				A:A	0	13
SNP-85	N	Vegetative tiller weight 2014	0.00090	C:C	−0.32	72
				C:T	12.66	6
				T:T	0	13

N non-coding region, *C(s)* synonymous substitution, *n* number of genotypes in the haplotype group help to elucidate the validity of these associations.

LpSHOOT1 gene is a putative regulator of outgrowth of axillary buds (Brazauskas et al. 2010) and a QTL for inflorescence length was detected at the *LpSHOOT1* gene-derived marker in perennial ryegrass VrnA mapping population (Studer et al. 2008). The association of *LpSHOOT1* with inflorescence length was not analyzed in this study, yet possible association with generative tiller number could be of great practical importance, as this trait is a component of seed yield.

Conclusion

Three marker-trait associations were identified between *LpSHOOT1* sequence polymorphisms and tillering traits. The associations with generative tiller number can be of particular interest as an indirect marker for seed yield. However, this marker requires a rigorous testing in a larger genotype set prior application in marker assisted selection.

Acknowledgments This study was funded by the Research Council of Lithuania, grant No. MIP-032/2012 (FUMAG).

References

Aleliūnas A, Jonavičienė K, Statkevičiūtė G, Vaitiekūnaitė D, Kemešytė V, Lübberstedt T, Brazauskas G (2015) Association of single nucleotide polymorphism in LpIRI1 gene with freezing tolerance traits in perennial ryegrass. Euphytica 204:523–534.

Boelt B, Studer B (2010) Breeding for grass seed yield. B. Boller et al. (eds.), Handbook of Plant Breeding 5. Springer Science + Business Media, Dordrecht. pp 161–174.

Bradbury PJ, Zhang Z, Kroon DE, Casstevens TM, Ramdoss Y, Buckler ES (2007) TASSEL: Sotware for association mapping of complex traits in diverse samples. Bioinformatics 23: 2633–2635.

Brazauskas G, Pašakinskienė I, Asp T, Lübberstedt T (2010) Nucleotide diversity and linkage disequilibrium in five Lolium perenne genes with putative role in shoot morphology. Plant Science 179: 194–201.

Chastain TG, Kiemenec GL, Cook GH, Garbacik CJ, Quebbeman BM, Crowe FJ (1997) Residue management strategies for Kentucky bluegrass seed production. Crop Science 37: 1836–1840.

Elgersma A (1990) Heritability estimates of spaced-plant traits in three perennial ryegrass (Lo-lium perenne L.) cultivars. Euphytica 51: 163–171.

Entz MH, Smith SR, Cattani DJ, Storgaard AK (1994) Influence of post-harvest residue management and cultivar on tiller dynamics and seed yield in timothy. Canadian Journal of Plant Science 74: 507–513.

Humphreys MW, Yadav RS, Caims AJ, Turner LB, Humphreys J, Skøt L (2006) A Changing climate for grassland research. New Phytologist 169: 9–26.

Pašakinskienė I (2005) Creeping stem and barren inflorescence: a new mutation in Lolium perenne 'VIROIZ'. Zwierzykowski Z, Kosmala S (eds.), Recent advances in genetics and breeding of the grasses. Institute of Plant Genetics, Poznan. pp 209–213.

Şeker H, Yazici A, Uysal P (2014) Analysis of variability, heritability, and genetic advance in seed yield and related traits of orchardgrass (Dactylis glomerata L.) populations. Turkish Journal of Agriculture and Forestry 38: 633–643.

Storey JD, Taylor JE, Siegmund D (2004) Strong control, conservative point estimation, and simultaneous conservative consistency of false discovery rates: A unified approach. Journal of the Royal Statistical Society: Series B (Statistical Methodology) 66: 187–205.

Studer B, Jensen LB, Hentrup S, Brazauskas G, Kölliker R, Lübberstedt T (2008) Genetic characterisation of seed yield and fertility traits in perennial ryegrass (Lolium perenne L.). Theoretical and Applied Genetics 117: 781–791.

Tamura K, Stecher G, Peterson D, Filipski A, Kumar S (2013) MEGA6: Molecular Evolutionary Genetics Analysis Version 6.0. Molecular Biology and Evolution 30: 2725–2729.

Utz HF (2011) PLABSTAT-A computer program for statistical analysis of plant breeding ex-periments. Version 3A. Germany, Universität Hohenheim.

Genomic Variation in the FT Gene Family of Perennial Ryegrass (*Lolium perenne*)

18

E. Veeckman, K. Vandepoele, T. Asp,
I. Roldán-Ruiz, and T. Ruttink

Abstract

The timing of flowering is of prime importance for several agronomic traits, and its genetic control is therefore of great interest to breeders. Several signaling pathways converge on FLOWERING LOCUS T (FT) gene family members, which act as central regulators of flowering, branching and seed dormancy. We identified the complete FT gene family in the *Lolium perenne* genome and performed phylogenetic analysis to delineate functional clades and to identify putative functionally redundant paralogs. Five FT genes of *L. perenne* were selected for targeted resequencing in a genepool of 746 accessions to describe genetic diversity in wild accessions, commercial cultivars and breeding material.

Keywords

Gene family • Flowering time • FLOWERING LOCUS T • TERMINAL FLOWER 1 • Perennial ryegrass (*Lolium perenne*)

Introduction

The transition from vegetative growth to flowering is one of the most important processes in plant development, and the timing is strictly controlled to coincide with conditions that enhance production of seeds and fruits. In perennial ryegrass (*Lolium perenne*), flowering is induced by a period of vernalization, followed by long days at higher temperatures. Because the timing and intensity of flowering affects agronomic traits such as biomass yield and quality and seed yield, the genetic control of flowering traits is of great interest to breeders. The regulatory network that controls flowering

E. Veeckman • I. Roldán-Ruiz • T. Ruttink (✉)
Plant Sciences Unit, Institute for Agricultural and Fisheries Research (ILVO), Growth and Development, Melle, Belgium
e-mail: tom.ruttink@ilvo.vlaanderen.be

K. Vandepoele
Department of Plant Systems Biology, VIB, Ghent, Belgium

Department of Plant Biotechnology and Bioinformatics, Ghent University, Ghent, Belgium

T. Asp
Department of Genetics and Biotechnology, Faculty of Agricultural Sciences, Research Centre Flakkebjerg, Aarhus University, Slagelse 4200, Denmark

I. Roldán-Ruiz et al. (eds.), *Breeding in a World of Scarcity: Proceedings of the 2015 Meeting of the Section "Forage Crops and Amenity Grasses" of Eucarpia*,
DOI 10.1007/978-3-319-28932-8_18

involves the perception and processing of a diverse range of environmental and internal signals and integrates them into a single decision: to flower or not to flower. This network slightly differs across species but always converges on genes of the PEBP (phosphatidylethanolamine-binding proteins) gene family, also known as the FLOWERING LOCUS T (FT) gene family. Phylogenetic analyses have revealed the complex evolutionary history of this gene family, with strong evidence of lineage specific expansion, creating the possibility for functional diversification between clades, as well as functional redundancy of paralogs within clades.

Two members of this gene family have been studied intensively: FLOWERING LOCUS T (FT), a positive regulator of flowering, and TERMINAL FLOWER 1, a negative regulator of flowering. MOTHER OF FT AND TFL1 (MFT) is another FT gene family member, that was found to act redundantly with other members of the FT gene family (Yoo et al. 2004). Only three members of this family, *LpFT3*, *LpTFL1* and *LpSFT*, have previously been characterized in *L. perenne* (Jensen et al. 2001; King et al. 2006; Fiil et al. 2011; Skøt et al. 2011). We have identified all FT genes in the *L. perenne* genome and classified them into the *MFT-like*, *TFL-like* or *FT-like* clade. Sequence variation was determined for five selected *LpFT* genes in 746 genotypes including wild accessions, commercial cultivars and breeding material. We investigated whether sequence variation affected residues that are critical for PEBP protein function, or may be associated with phenotypic variation in flowering time.

Material and Methods

Gene Family Annotation and Phylogenetic Analysis

Protein sequences of 18 *Brachypodium distachyon* genes were extracted from the PLAZA 3.0 Monocots database (Proost et al. 2014) and used for a tBLASTn search against the *L. perenne* reference genome sequence (Byrne et al. 2015), revealing 18 candidate orthologous loci. Overlapping gene models, predicted using MAKER v2.3 (Cantarel et al. 2008), were checked and corrected based on a multiple sequence alignment of the whole gene family (HOM03M000266, PLAZA 3.0 monocots) and RNA-seq data. Annotated gene sequences are deposited in GenBank under accession numbers KR706144–KR706161. For phylogenetic analysis, the set of 18 LpFT genes was complemented with homologs of *B. distachyon* (18), *Hordeum vulgare* (12), *Oryza sativa* spp. *japonica* (19), *Zea mays* (25) and *Arabidopsis thaliana* (6). A multiple sequence alignment was created using MUSCLE (Edgar 2004) and a phylogenetic tree was constructed using PhyML (Guindon et al. 2010) applying the JTT substitution model, and bootstrap values were calculated using 1000 replicates. Human phosphatidylethanolamine-binding protein 1 preproprotein (gi|4505621, NP_002558.1) was used to root the tree.

SNP and Indel Discovery

The collection comprised 746 genotypes, including wild accessions, breeding material and commercial varieties. From that collection, targeted resequencing of 548 genes of interest was performed using SureSelect probe capture enrichment of indexed genome shotgun libraries followed by Illumina HiSeq sequencing (PE 2×100). The raw reads were trimmed, mapped to the *L. perenne* genome sequence using BWA-mem (Li and Durbin 2009) and polymorphisms (SNPs and short indels) were identified using GATK v3.2–2 (McKenna et al. 2010), as described in Ruttink et al. (2015). Multiallelic SNPs and indels were filtered out and genotype calls were filtered on read depth (DP $\geq$6) and likelihood score (GQ $\geq$30). A total of 1645 SNPs and 505 indels are located in the five *LpFT* genes reported here.

Results and Discussion

Delineation of the FT Gene Family in *L. perenne*

BLAST searches with all 18 PEBP proteins of *B. distachyon* against the *L. perenne* genome sequence revealed 18 FT gene family members

(Fig. 18.1), consistent with a similar number of FT genes in other monocot species (see 18.2). *LpFT3, LpSFT,* and *LpTFL1* (here named *LpFT03*, *LpFT01* and *LpFT07*, respectively) have previously been described (Jensen et al. 2004; King et al. 2006; Fiil et al. 2011; Skøt et al. 2011).

Previously published phylogenetic analyses of the PEBP domains of the FT genes in *H. vulgare*, *O. sativa* spp. *japonica* and *A. thaliana* revealed that the FT gene family contains three major clades (Chardon and Damerval 2005; Faure et al. 2007). *LpFT05* and *LpFT18* belong to the MFT-like clade, *LpFT04*, *LpFT06* and *LpFT07* belong to the TFL1-like clade and the remaining 13 *LpFT* genes belong to the FT-like clade (Fig. 18.1). This last clade can be subdivided into two subclades: the FT-like clade I, containing *AtFT* and *AtTSF*, and the FT-like clade II, containing only grass and cereal-specific genes.

The number of PEBP genes in grasses and cereals is three to four times larger than that in *A. thaliana*, due to several whole-genome duplications and tandem duplications that are specific for the grass lineage. These ancient duplications are revealed by the consistent grouping of orthologs in clades containing members of various grass species. The most parsimonious hypothesis suggests that two *MFT-like*, two *TFL1-like* and at least eight *FT-like* genes were present in the ancestral grass genome (Chardon and Damerval 2005). The orthologous relationships within subfamilies are often difficult to deduce because genes likely evolved at least partially independently in each taxon by duplication and possible gene loss.

We selected five FT genes representing the four (sub) clades of the phylogenetic tree (indicated with an asterisk in Fig. 18.1) for detailed analysis (see below). One of them, *LpFT03*, has previously been shown to associate with heading date (Skøt et al. 2011). All *LpFT* genes showed the typical residues at positions that were conserved in the whole gene family (residues colored purple in Fig. 18.2). Moreover, positions that were only conserved within clades were also conserved in the *LpFT* gene family members (results not shown), suggesting that these paralogs share biochemical function.

Identification of Sequence Variation in Five Family Members

An average of 350 variant positions (SNPs and indels) were identified for each of the five target genes, but strong differences were found among genes (Table 18.1). As expected, coding sequences contain fewer variants than promoter, untranslated regions or intron sequences. Despite having sufficient read data for *LpFT04*, this gene shows remarkable little sequence variation.

Next, we investigated sequence variation in the external loop and key residues of the anion ligand binding site (Ahn et al. 2006; Danilevskaya et al. 2008) (Fig. 18.2). Projection of non-synonymous SNPs and indels onto the protein sequence alignment of the five selected genes (Fig. 18.2) reveals that most of them code for non-essential residues. In *LpFT03*, however, some variation in critical residues is observed, including the P136 in the external loop and the Y151 of the LYN triad, which are normally completely conserved in *FT-like* genes (Ahn et al. 2006). *TFL1-like* and *MFT-like* genes in other species also showed variation in the external loop, which may affect the surface charge around the ligand-binding pocket and thus may affect protein function or activity. On the other hand, no variation was found in the external loop of *LpFT04*. For *LpFT02*, a SNP introduces a premature stop-codon at W89 at the end of the second exon. This mutation occurs only in heterozygous state in 11 of the 746 resequenced genotypes, and can be considered a rare defective allele.

Ho and Weigel (2014) identified residues that were critical for the molecular function of FT and residues that are essential for interaction with FD or TCP proteins (Fig. 18.2, residue numbers refer to amino acid positions in *AtFT*). Only for two of these residues a non-synonymous SNP was detected: an alternative residue at D60 could lead to differential interactions with FD and sequence variation at L128 may affect the molecular function of *LpFT3*.

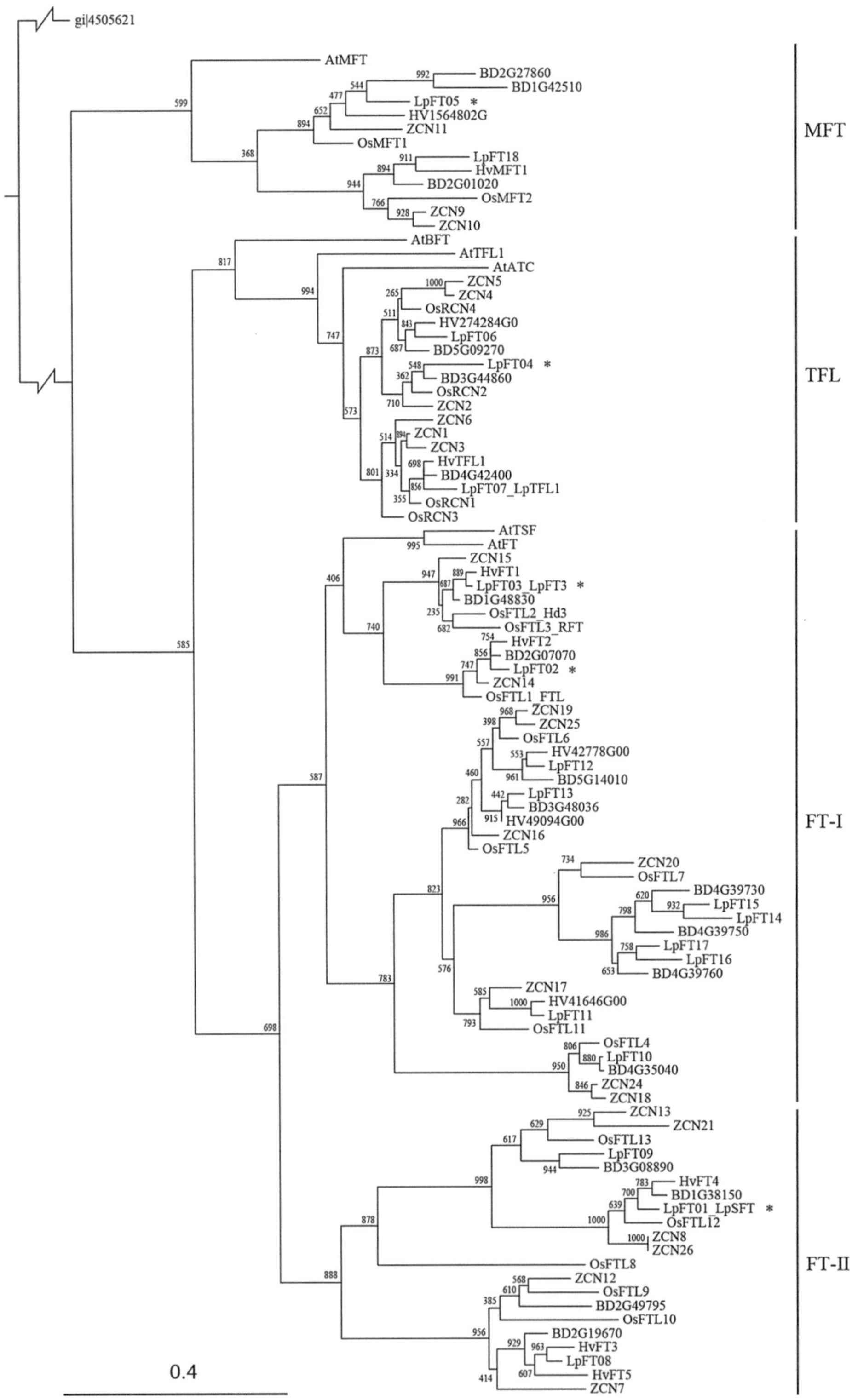

Fig. 18.1 Phylogenetic tree of the FT gene family of *L. perenne* (*Lp*), *B. distachyon* (*BD*), *H. vulgare* (*Hv*), *O. sativa* spp. *japonica* (*Os*), *Z. mays* (*ZCN*) and *A. thaliana* (*At*). Human PEBP1 (gi|4505621) was used to root the tree. Support values for branches are represented by bootstrap values (1000 replicates). Three major clades (MFT-like, TFL1-like, and FT-like) including two subclades (FT-like I and FT-like II) are shown. *LpFT* genes selected for targeted resequencing are indicated by an asterisk (*)

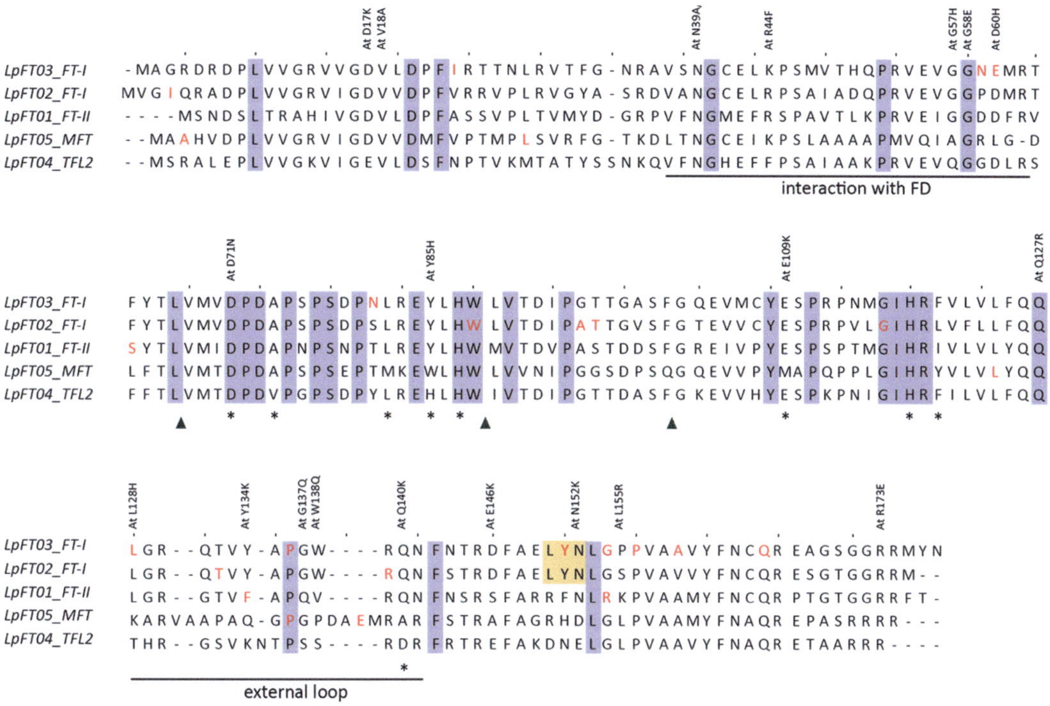

Fig. 18.2 Protein sequence alignment of five *LpFT* genes and delineation of conserved residues across the gene family (*purple*). Exon borders are indicated by *black triangles*. Key residues of the anion ligand binding pocket are indicated with an *asterisk* (Ahn et al. 2006; Danilevskaya et al. 2008). The LYN triad sequence of the FT-like clade I is colored *orange* (Ahn et al. 2006). Residues that are essential for molecular function of FT or for interaction with FD or TCP proteins are given on top of the alignment (Ho and Weigel 2014). Residues with non-synonymous substitutions found in the genepool collection are colored *red*

Table 18.1 Distribution of SNPs and indels for five *LpFT* genes in coding and non-coding regions at minor allele frequency (MAF) of >1 % across all 746 genotypes with positive observations. Non-synonymous SNPs in the CDS are given between brackets

		LpFT01	LpFT02	LpFT03	LpFT04	LpFT05
Coding	Length (bp)	519	528	531	513	534
	SNPs/indels	17(3)/–	24(7)/–	27(11)/–	–/–	23(5)/–
Non-coding	Length	2396	3032	1537	2365	1759
	SNPs/indels	223/62	253/65	114/71	27/3	158/41

Conclusion

We have completed the FT gene family of *L. perenne* by identifying 18 family members in the draft genome. All newly characterized genes contain amino acid residues that are conserved throughout the whole gene family, suggesting that all encode functional PEBP proteins. Furthermore, *L. perenne* members contain the clade-specific residues that differentiate between MFT, TFL and FT functions. In this way we have identified several novel paralogs that may be functionally redundant with previously described genes in the TFL1-like clade and the LpFT-like clade I.

Five FT family members were resequenced in 746 genotypes, revealing high degrees of genetic diversity. The sequence variation across the *L. perenne* genepool may be used in two complementary ways. First, we have identified several genotypes with sequence variation at critical residues. Non-synonymous substitutions that reside in the external loop may cause changes in the surface charge and have consequences for the biochemical function. Furthermore, the defective W89* allele of *LpFT02* could be useful to study the role of *LpFT02* and its putative functional redundancy with *LpFT03*. In addition, many deletions occur in the promoter and UTR, which may affect transcript expression or stability, but the effects are more difficult to predict by bioinformatics alone. Haplotype reconstruction could be exploited to combine sequence variation in coding and non-coding regions, and to perform association mapping with reduced multiple testing correction. The distribution of haplotypes across wild accessions, breeding populations, and cultivars may further reveal whether breeding has selected for particular alleles.

Acknowledgments KV acknowledges the Multidisciplinary Research Partnership "Bioinformatics: from nucleotides to networks" Project (no 01MR0310W) of Ghent University

References

Ahn JH, Miller D, Winter VJ, et al (2006) A Divergent External Loop Confers Antagonistic Activity on Floral Regulators FT and TFL1. EMBO 25(3):605–14. doi:10.1038/sj.emboj.7600950

Byrne SL, Nagy I, Pfeifer M et al (2015) A synteny-based draft genome sequence of the forage grass Lolium perenne. Plant J 84(4):816–26. doi: 10.1111/tpj.13037

Cantarel BL, Korf I, Robb SMC, et al (2008) MAKER: an easy-to-use annotation pipeline designed for emerging model organism genomes. Genome Res 18:188–96. doi: 10.1101/gr.6743907

Chardon F, Damerval C (2005) Phylogenomic analysis of the PEBP gene family in cereals. J Mol Evol 61:579–590. doi: 10.1007/s00239-004-0179-4

Danilevskaya ON, Meng X, Hou Z, et al (2008) A Genomic and Expression Compendium of the Expanded PEBP Gene Family from Maize. Plant Physiol 146(1): 250–64. doi:10.1104/pp.107.109538

Edgar RC (2004) MUSCLE: multiple sequence alignment with high accuracy and high throughput. Nucleic Acids Res 32:1792–7. doi: 10.1093/nar/gkh340

Faure S, Higgins J, Turner A, Laurie D a. (2007) The FLOWERING LOCUS T-like gene family in barley (Hordeum vulgare). Genetics 176:599–609. doi: 10.1534/genetics.106.069500

Fiil A, Lenk I, Petersen K, et al (2011) Nucleotide diversity and linkage disequilibrium of nine genes with putative effects on flowering time in perennial ryegrass (Lolium perenne L.). Plant Sci 180:228–237. doi: 10.1016/j.plantsci.2010.08.015

Guindon S, Dufayard J-F, Lefort V, et al (2010) New algorithms and methods to estimate maximum-likelihood phylogenies: assessing the performance of PhyML 3.0. Syst Biol 59:307–21. doi: 10.1093/sysbio/syq010

Ho WWH, Weigel D (2014) Structural Features Determining Flower-Promoting Activity of Arabidopsis FLOWERING LOCUS T. The Plant Cell 26 (2): 552–64. doi:10.1105/tpc.113.115220

Jensen CS, Salchert K, Nielsen KK (2001) A TERMINAL FLOWER1-like gene from perennial ryegrass involved in floral transition and axillary meristem identity. Plant Physiol 125:1517–1528. doi: 10.1104/pp.125.3.1517

Jensen CS, Salchert K, Gao C, et al (2004) Floral inhibition in red fescue (Festuca rubra L.) through expression of a heterologous flowering repressor from Lolium. Mol Breed 13:37–48. doi: 10.1023/B:MOLB.0000012327.47625.23

King RW, Moritz T, Evans LT, et al (2006) Regulation of flowering in the long-day grass Lolium temulentum by gibberellins and the FLOWERING LOCUS T gene. Plant Physiol 141:498–507. doi: 10.1104/pp.106.076760

Li H, Durbin R (2009) Fast and accurate short read alignment with Burrows-Wheeler transform. Bioinformatics 25:1754–60. doi: 10.1093/bioinformatics/btp324

McKenna A, Hanna M, Banks E, et al (2010) The Genome Analysis Toolkit: a MapReduce framework for analyzing next-generation DNA sequencing data. Genome Res 20:1297–303. doi: 10.1101/gr.107524.110

Proost S, Van Bel M, Vaneechoutte D, et al (2014) PLAZA 3.0: an access point for plant comparative genomics. Nucleic Acids Res 1–8. doi: 10.1093/nar/gku986

Ruttink T, Haegeman A, van Parijs F, et al (2015) Genetic diversity in candidate genes for developmental traits and cell wall characteristics in perennial ryegrass (Lolium perenne). Budak Spangenb. Proc. 8th Int. Symp. Mol. Breed. Forage Turf.

Skøt L, Sanderson R, Thomas A, et al (2011) Allelic variation in the perennial ryegrass FLOWERING LOCUS T gene is associated with changes in flowering time across a range of populations. Plant Physiol 155:1013–22. doi: 10.1104/pp.110.169870

Yoo SY, Kardailsky I, Lee JS, et al (2004) Acceleration of Flowering by Overexpression of MFT (MOTHER OF FT AND TFL1). Mol Cells 17:95–101

De novo Genome Sequencing and Gene Prediction in *Lolium perenne*, Perennial Ryegrass

19

E.M.B. Mollison, S. Barth, D. Milbourne, L. Milne, C. Halpin, M. McCabe, C. Creevey, and D.F. Marshall

Abstract

A 1.11Gbp *de novo* assembly of the *Lolium perenne* genome was generated, containing 424,745 scaffolds and with N50 of 25,193. Gene prediction on genomic, mitochondrial and chloroplast scaffolds was carried out using both *ab initio* and RNA-Seq based methods. *Ab initio* gene prediction, carried out using wheat-based gene models, identified a total of 188,822 potential gene models from genomic scaffolds and 109 from mitochondrial. Mapping of reads from a broad-based RNA-Seq study identified 67,706 potential genes from genomic scaffolds, 90 from mitochondrial and 18 from chloroplast. Comparison of *ab initio* predicted genes with RNA-Seq genes identified 44,252 predicted gene models from genomic scaffolds and three from mitochondrial that overlapped with RNA-Seq derived transcripts by more than 20 % of their length.

Keywords

Perennial ryegrass • Genome sequencing • RNA-Seq • Transcriptome • Gene prediction

Introduction

As one of the most widely grown fodder/forage crops for livestock in Europe *Lolium perenne* – perennial ryegrass – is an important crop species with implications for both animal and human

E.M.B. Mollison • L. Milne • D.F. Marshall (✉)
ICS Group, James Hutton Institute,
Errol Road, Invergowrie, Dundee, UK
e-mail: david.marshall@hutton.ac.uk

S. Barth (✉) • D. Milbourne
CELUP Crop Science Department, Teagasc, CELUP
Oak Park Research Centre, Carlow, Ireland
e-mail: susanne.barth@teagasc.ie

C. Halpin
Division of Plant Sciences,
University of Dundee at the James Hutton Institute,
Errol Road, Invergowrie, Dundee, UK

M. McCabe • C. Creevey
Teagasc Animal and Bioscience Research
Department, Animal & Grassland Research and
Innovation Centre, Teagasc, Grange,
Dunsany Co. Meath, Ireland

I. Roldán-Ruiz et al. (eds.), *Breeding in a World of Scarcity: Proceedings of the 2015 Meeting of the Section "Forage Crops and Amenity Grasses" of Eucarpia*,
DOI 10.1007/978-3-319-28932-8_19

nutrition. As a member of the *Poaceae* family it is closely related to major crop species such as barley, wheat and rice, as well as the model grass species *Brachypodium distachyon*, all of which have been subjects of major genome sequencing projects.

Many grass genomes are extremely large and complex, and thus challenging to sequence, even with current next generation sequencing (NGS) technology. However the low copy number regions of a genome tend to assemble well and an assembly of an organism's "gene-space" can be generated much more readily. Such an assembly will contain the bulk of the gene-coding regions and this gene-space approach to assembly was taken with *Lolium perenne*.

Genome sequencing can be complemented by RNA-Seq, which involves sequencing mRNA expressed in an organism's tissues, to identify genes expressed at the time and conditions of sampling. By sampling from a mixture of tissues and under a range of conditions the number of genes identified can be maximized.

Here the aim is to contribute towards a genomic resource for *Lolium perenne*, based on genome sequencing, RNA-Seq, and *ab initio* gene prediction to identify gene-like regions based on intrinsic sequence features. This genomic resource can be used to aid the development of novel tools and approaches in ryegrass breeding and as a basis for comparative genomic studies with other crop species.

Material and Methods

Plant Material

Sequencing for the genome assembly was performed on an inbred line that was originally a maintainer line in a cytoplasmic male sterility (CMS) program at Teagasc. The line originated from an inter-specific cross between meadow fescue (*Festuca pratensis*) and perennial ryegrass (*L. perenne*) followed by backcrossing the ryegrass parent and selfing for ten generations. The background of the *Lolium* contribution in the pedigree of the inbred line was the ryegrass cultivar 'Premo' (Mommersteeg International BV).

The genotype used for genome assembly is the paternal grand parent of an F2 mapping population recently subject to genotyping by sequencing. On the basis of almost 2000 SNP markers mapped in this population, the parent was shown to be 99 % homozygous (J. Velmurugan, pers. comm.) For RNA-Seq, both the inbred line used above and a second partially inbred F2 individual from the same CMS program were used.

Genome Assembly

A gene-space assembly for the *Lolium* genotype described above was generated based on Illumina HiSeq sequencing of a range of paired-end (PE), mate-pair (MP) and long-jumping distance (LJD) libraries with insert sizes of <500 bp (PE), 3 kbp (MP), 8 kbp (LJD), 20 kbp (LJD) and 40 kbp (LJD).

Approximately 207-fold raw coverage was generated; read pair de-duplication with *FastUniq* (Xu et al. 2012) and additional quality trimming with ***Sickle*** (Joshi and Fass 2011) to Phred quality score 30 (99.9 % base call confidence) reduced this to 105-fold coverage.

Reads were assembled using *CLC Assembly Cell* (CLC Bio 2008–2015) using *k*-mer length 41 and scaffolded with *SSPACE* (Boetzer et al. 2011).

Ab Initio Gene Prediction

Ab initio gene prediction was carried out using ***Augustus*** (Stanke and Waack 2003), with wheat-based gene structure models.

RNA-Seq Analysis

Data from Illumina sequencing of 22 RNA-Seq experiments carried out on *Lolium* genotypes, based on a range of tissues and experimental conditions, was used: Tissue type: whole plant, green parts; Temperature stress: cold, freezing; Nutrient stress: nitrogen starvation, phosphorus starvation; Drought stress; Growth substrate: soil-based, hydroponics; Life cycle: flowering, non-flowering. This breadth of tissues and conditions allows for a comprehensive transcriptome to be assembled from the RNA-Seq reads.

Transcriptome Assembly

From 22 RNA-Seq experiments, 860,480,042 reads were generated as 100 bp read-pairs, with approximately 500 bp insert size and 190,914,299 reads as 100 bp shotgun (unpaired) reads. RNA-Seq reads from each experiment were mapped to the *Lolium* genome assembly using *Tophat2* (Kim et al. 2013). The resulting alignments were pooled and merged using *Cufflinks* (Trapnell et al. 2010, 2012) to produce a single transcriptome assembly for *L. perenne*.

Comparison of Gene Predictions with Transcripts

Augustus gene predictions and RNA-Seq transcripts were compared using *BEDTools intersect* (Quinlan and Hall 2010) to identify overlapping features, based on base-pair locations within a set of scaffolds; features overlapping by more than 20 % of their length were considered strongest candidates for legitimate genes.

Completeness of Coverage Estimation

The *CEGMA* pipeline (Parra et al. 2007) searches a genome assembly for 458 core proteins that are conserved across eukaryotes, with a more highly conserved subset of 248 used to indicate completeness of coverage. This approach was used to estimate completeness of coverage of the *Lolium* gene-space assembly.

Results and Discussion

Genome Assembly

A final assembly of 1.11 Gbp in size was obtained. This contained 424,750 scaffolds, with N50 of 25,212 and a GC content of 44.16 %. This GC content is consistent with that of barley (Rostoks et al. 2002) and the assembled genome size captures approximately 40 % of the 2.6 Gbp *Lolium* genome (Pfeifer et al. 2013).

Comparison of the scaffolded assembly with published mitochondrial and chloroplast sequences for *Lolium* (Islam et al. 2013; Diekmann et al. 2008) identified three scaffolds that were mitochondrial in origin and two that were chloroplast in origin. Removal of mitochondrial and chloroplast scaffolds resulted in a genomic assembly containing 424,745 scaffolds and with N50 of 25,193.

Gene Identification from ab Initio and RNA-Seq Predictions

Ab initio prediction identified 188,822 gene-like regions from 59,900 genomic scaffolds. This is likely to greatly overestimate the gene numbers and reflects a number of confounding factors including retroelements, pseudogenes, genes with multiple open reading frames (ORFs), gene fragments and sequencing errors. The three mitochondrial scaffolds contained 20 predictions; Augustus did not predict gene models for scaffolds known to be associated with the chloroplast.

RNA-Seq identified 67,706 potential genes from 33,212 *Lolium* genomic scaffolds, with 111,464 transcripts; this is likely an overestimation, reflecting factors such as fragmented genes appear as multiple genes. Mitochondrial scaffolds contained 109 genes, with 109 transcripts; chloroplast scaffolds contained 12 genes, with 18 transcripts. The elevated estimate of gene numbers is unlikely to be affected by artefacts arising from interspecific crossing as a separate genomic *in situ* hybridisation (GISH) study carried out on this line identified minimal contribution from the *F. pratensis* parent (Anhalt et al. 2008).

A summary of gene identification results by *ab initio* and RNA-Seq methods appears in Table 19.1.

Comparing Augustus predictions with RNA-Seq genes identified 53,424 genomic and 12

Table 19.1 Genes identified by Augustus and RNA-Seq analysis

	RNA-Seq			Augustus		
	Genomic	Mt.	Ch.	Genomic	Mt.	Ch.
Predicted genes	67,706	109	12	188,822	20	0
Predicted transcripts	111,464	109	18	n/a	n/a	n/a
Scaffolds containing predictions	33,212	3	2	59,900	3	0
Genes per kilobase[a]	0.051	0.209	0.095	0.230	0.038	0

Mt. Mitochondrial, *Ch.* Chloroplast
[a]Genes per kilobase of gene-containing scaffolds

mitochondrial predictions that overlapped with RNA-Seq to any degree; with 44,252 genomic and 3 mitochondrial having a reciprocal overlap of at least 20 %. A reciprocal overlap of 20 % was chosen to allow for differing feature sizes between *ab initio* and RNA-Seq gene prediction methods.

Estimate of Completeness of Coverage

Of the 248 core proteins used by CEGMA to estimate completeness of coverage, 239 (96.37 %) were found to have "complete" alignment and 246 (99.19 %) were found to have either "complete" or "partial" alignment (complete alignment is defined by CEGMA as aligning over at least 70 % of the protein's length and partial is defined as a shorter alignment that is still significant).

Conclusions

A comprehensive transcriptome for *Lolium perenne* was generated from a broad RNA-Seq study. Comparing transcripts with *ab initio* gene predictions identified 44,252 gene models from genomic scaffolds with supporting evidence from RNA-Seq and three gene models from mitochondrial scaffolds with supporting RNA-Seq evidence. *Ab initio* gene finding alone leads to an overestimate of gene models and RNA-Seq is limited to what is being expressed at the time of sampling, therefore a broad range of tissues and conditions is needed. Combining RNA-Seq with *ab initio* allows a set of higher-confidence gene predictions to be identified. However it should be noted that some *ab initio* predictions without supporting RNA-Seq may be legitimate if those particular genes were not expressed at the time/conditions of sampling – comparison with related species will further refine this core set.

Identifying 246/248 CEGMA proteins at partial or complete level suggests that a very good degree of coverage of the *Lolium* gene-space has been achieved.

References

Anhalt UC, Heslop-Harrison PJ, Byrne S, Guillard A, Barth S (2008): Segregation distortion in Lolium: evidence for genetic effects. Theoretical and Applied Genetics 117 (2): 297–306

CLC Bio (2008 – 2015): CLC Assembly Cell [Software]. Available at http://www.clcbio.com/products/clc-assembly-cell/

Boetzer M, Henkel CV, Jansen HJ, Butler D, Pirovano W (2011): Scaffolding pre-assembled contigs using SSPACE. Bioinformatics 27 (4): 578–579

Diekmann K, Hodkinson TR, Fricke E, Barth S (2008): An optimized chloroplast DNA extrac-tion protocol for grasses (Poaceae) proves suitable for whole plastid genome sequencing and SNP detection. PLoS One 3 (7): e2813

Islam MS, Studer B, Byrne SL, Farrell JD, Panitz F, Bendixen C, Møller IM, Asp T (2013): The genome and transcriptome of perennial ryegrass mitochondria. BMC Genomics 14: 202

Joshi NA, Fass JN (2011): Sickle: A sliding-window, adaptive, quality-based trimming tool for FastQ files (Version 1.33) [Software]. Available at https://github.com/najoshi/sickle.

Kim D, Pertea G, Trapnell C, Pimentel H, Kelley R, Salzberg SL (2013) TopHat2: accurate alignment of transcriptomes in the presence of insertions, deletions and gene fusions. Ge-nome Biology 14: R36.

Parra G, Bradnam K, Korf I (2007): CEGMA: a pipeline to accurately annotate core genes in eukaryotic genomes. Bioinformatics 23: 1061–1067

Pfeifer M, Martis M, Asp T, Mayer KFX, Lübberstedt T, Byrne S, Frei U, Studer B (2013): The Perennial Ryegrass GenomeZipper: targeted use of genome resources for comparative grass genomics. Plant Physiology 161 (2): 571–582

Quinlan AR, Hall IM (2010): BEDTools: a flexible suite of utilities for comparing genomic features. Bioinformatics 26 (6): 841–842

Rostoks N, Park Y, Ramakrishna W, Ma J, Druka A, Shiloff B, SanMiguel P, Jiang Z, Brueggeman R, Sandhu D, Gill K, Bennetzen J, Kleinhofs A (2002): Genomic sequencing reveals gene content, genomic organization, and recombination relationships in barley. Functional & Integrative Genomics 2: 51–59

Stanke M, Waack S (2003): Gene prediction with a hidden Markov model and a new intron submodel. Bioinformatics 19 (suppl 2): ii215–ii225

Trapnell C, Williams BA, Pertea G, Mortazavi A, Kwan G, van Baren MJ, Salzberg SL, Wold BJ, Pachter L (2010): Transcript assembly and quantification by RNA-Seq revels unanno-tated transcripts and isoform switching during cell differentiation. Nature Biotechnology 28: 511–515

Trapnell C, Roberts A, Goff L, Pertea G, Kim D, Kelley DR, Pimentel H, Salzberg SL, Rinn JL, Pachter L (2012): Differential gene and transcript expression analysis of RNA-Seq experi-ments with TopHat and Cufflinks. Nature Protocols 7: 562–578

Xu H, Luo X, Qian J, Pang X, Song J, Qian G, Chen J, Chen S (2012): FastUniq: a fast de novo duplicates removal tool for paired short reads. PLoS One 7 (12): e52249

Genotyping by Sequencing in Autotetraploid Cocksfoot (*Dactylis glomerata*) without a Reference Genome

20

B.S. Bushman, M.D. Robbins, S.E. Larson, and J.E. Staub

Abstract

Quantitative trait locus mapping is often useful for understanding the number and the direction of effects for traits in germplasm of interest. New genotype-by-sequencing (GBS) methods are available for genetic mapping, but their application toward autopolyploid plants is limited. In this paper we discuss adapting semi-conductor sequencing to GBS of autotetraploid cocksfoot (*Dactylis glomerata*). Based on empirical results from a previous study, we selected 48 samples to pool on an Ion Torrent P1 sequencing chip. Whereas the Illumina-based SNP detection pipeline was designed to identify SNP markers in 64 bp sequence reads, the longer length of semi-conductor sequences was mined for additional SNPs. Using a small subset of plants from an F_1 cocksfoot mapping population, a genetic linkage map was constructed with GBS-derived markers and previously mapped SSR markers, and between 25 and 29 homologous linkage groups were detected with high LOD probability. These data provide evidence of successful implementation of existing GBS pipelines when using semi-conductor sequencing on complex autotetraploid plants.

Keywords

Cocksfoot • Genotyping-by-sequencing • Heading date

Introduction

Quantitative trait locus (QTL) mapping in outcrossing grasses has limited but specific utility in molecular breeding. Foremost is the ability to use QTL mapping to understand an approximate number of loci affecting a trait, and the direction of genetic effects of those loci. Our laboratory previously generated a genetic map of autotetraploid

B.S. Bushman (✉) • M.D. Robbins • S.E. Larson
J.E. Staub
USDA-ARS Forage and Range Research Laboratory, Logan, UT 84322-6300, USA
e-mail: shaun.bushman@ars.usda.gov

I. Roldán-Ruiz et al. (eds.), *Breeding in a World of Scarcity: Proceedings of the 2015 Meeting of the Section "Forage Crops and Amenity Grasses" of Eucarpia*,
DOI 10.1007/978-3-319-28932-8_20

cocksfoot (Xie et al. 2012) with seven base chromosomes and 28 homeologous linkage groups. Several QTLs for heading date were detected on linkage groups that overlapped with known vernalization and heading date genes in other grasses, and thus provided candidate genes from which association studies can be conducted. However, the genetic map was constructed of AFLP and SSR markers, and had poor resolution in several key linkage groups.

Genotyping by sequencing (GBS) has recently been implemented to develop dense genetic maps in forage crops, including autotetraploid species (Hackett et al. 2013; Li et al. 2014). All GBS pipelines have been based on Illumina sequencing of short reads and have reference genomic sequence to which reads can be mapped (Mascher et al. 2013). On the other hand, *de novo* GBS has no reference sequence and must rely on homology of shared sequence segments, called 'tags', to infer single nucleotide polymorphisms (SNPs). Additionally, semi-conductor platforms provide fewer sequences than Illumina Hi-Seq flow cells, but twofold longer sequences (up to 200 bp), such that *de novo* GBS pipelines must be further modified to take advantage of the longer sequences (Mascher et al. 2013).

In this paper we document the use of Ion Torrent semi-conductor platform to conduct GBS on a subset of plants from the previous cocksfoot mapping population. We use a two-enzyme system with size selection of fragments prior to sequencing. We report the number of reads obtained from barcoded samples, their distribution, average sizes, and utility for genetic mapping. Results indicate that the Ion Torrent reads can be useful for mapping and that the extra sequence length can be used to extract more markers, but that fewer samples should be pooled per sequencing run to ensure sufficient sequencing depth.

Materials and Methods

A cocksfoot F_1 genetic mapping population from a cross between PI 372621 and PI 295271 was previously used to generate a genetic map using TetraploidMap (Xie et al. 2012). Forty-five plants from that population, each parent and a blank lane were submitted for GBS. For GBS, a two-enzyme system was implemented with Pst1 and MspI enzymes as per Mascher et al. (2013). Two samples of each plant were included with separate barcodes to increase the read depth per sample and avoid potential issues with problematic barcodes. All samples were pooled and the fragment library was size selected between 200 and 300 bp using a Blue Pippin (Sage Science, Beverly, MA, USA). Emulsion PCR and sequencing used the Ion Torrent P1 Template OT200 Kit v3 (Thermo Fisher, Waltham, MA, USA).

Trimmed sequences were inputted into the UNEAK pipeline (Lu et al. 2013). UNEAK is specifically designed for the Illumina platform, and only utilizes the first 64 bp of a sequence read. Sequences less than 64 bp will cause the UNEAK pipeline to terminate prematurely; however, it is also a feature of UNEAK to remove any sequence reads that contain an "N" in the first 64 bp. Thus, if an "N" is placed at the end of sequence reads less than 64 bp, UNEAK will remove those individual reads rather than terminate. As Ion Torrent sequences are of varying sizes, a custom perl script was employed to add an "N" to the 3′ end of each sequence read. The following criteria were used to determine if a SNP identified by UNEAK was useful: sequence 'tags' (unique 64 bp sequences representing multiple identical reads) comprising less than five reads were discarded, a 5 % error tolerance rate was selected when detecting SNPs across the 47 plants to account for sequencing errors, homozygous loci within a sample were called as missing values if less than 11 reads constituted that tag, SNPs with a minor allele frequency less than 10 % in the population were discarded, and SNPs with more than 30 % missing values were discarded.

UNEAK only uses the first 64 bp of a sequence read, yet most of the semi-conductor sequence reads were longer. Therefore a custom perl script was developed to remove the first 64 bp of each sequence, add another "N" onto the 3′ end of newly shortened sequences, and submit the sequences to another round SNP detection for bases 65–129. A third iteration of SNP detection was also conducted. Resulting SNP genotypes were screened for redundancy by identifying identical genotypic calls among the 45 segregating progeny using JoinMap v4.0 (Van Ooijen 2011). The non-redundant SNPs, along with select SSR markers from previous mapping efforts (Xie et al. 2012), were used to test for linkage on a rough scale in JoinMap. The double haploid (DH) model was used and only markers with a chi-square value fitting a 1:1 ratio were included.

Results and Discussion

Using an Ion Torrent Proton P1 sequencing chip, 63,233,799 raw sequences were generated across all 47 samples providing approximately 1.3 million sequence reads per sample. The average length of trimmed sequences was 117 bp, with a maximum of 234 bp. After sequencing, removal of the adaptor sequences, and quality trimming, a bimodal distribution of lengths was observed (Fig. 20.1). As semi-conductor sequencing can generate base-calling INDELs in homopolymer regions, the quality control process may have trimmed longer sequence segments that contained low quality regions, which resulted in the bimodal distribution.

The coverage of sequence reads of each sample meeting the length criteria of UNEAK ranged from 239,797 to 483,439; this provided approximately six million shared sequence tags. From those tags, 48,548 SNPs were detected and 3,844 met our criteria of minor allele frequencies and missing data (Table 20.1).

As Ion Torrent Proton sequence reads can reach in excess of 200 bp but the UNEAK pipeline only uses 64 bp sequence reads, we tested if the additional length could be used to identify more SNP markers. The first pass of the UNEAK pipeline detected six million useable and shared sequence tags that translated into the 3,844 SNPs that met the strict criteria (Table 20.1). After removing the first 64 bp from each sequence, an additional 4.8 million shared sequence tags and 1,398 useable SNPs were detected. A third pass of this process (bases 130–194 of each sequence read) led to an additional 1.9 million shared sequence tags and 332 useable SNPs (Table 20.1). The attenuation of the number of SNPs detected between the first and second passes was 64 %, and between the second and third passes was 77 %. It is possible that the observed attenuation was a result of the modal peak of shorter sequence reads

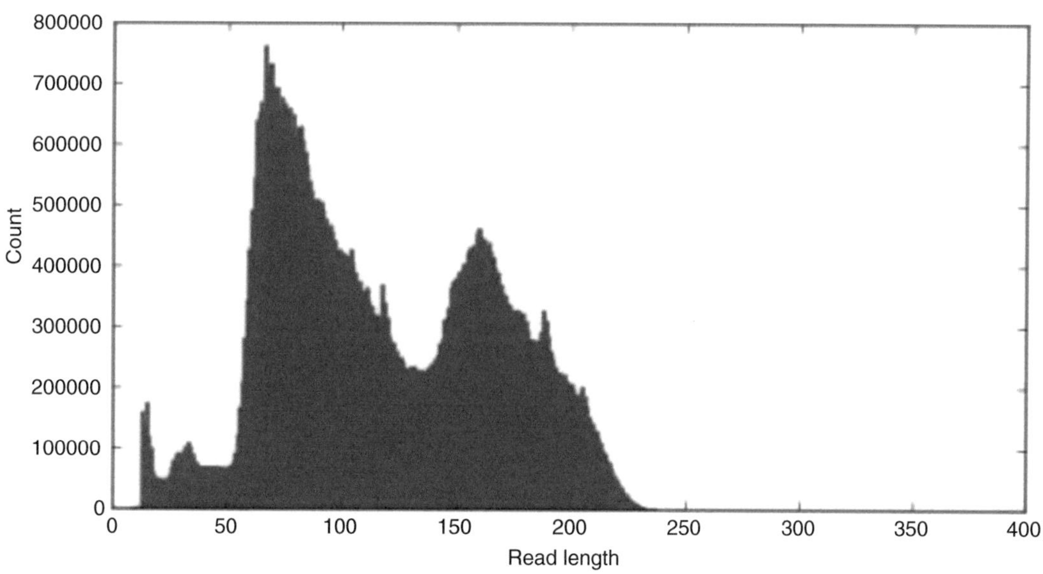

Fig. 20.1 The distribution of trimmed sequence lengths for 47 plants of a cocksfoot F_1 mapping population

Table 20.1 Number of sequence tags, SNPs, usable SNPs that met the criteria for genetic mapping and SNPs with 1:1 segregation ratios from three passes through the UNEAK pipeline after removing 64 bp segments. Data encompass parents and 45 plants in an F_1 cocksfoot mapping population

	Pass 1	Pass 2	Pass 3
Shared tags (million)	6	4.8	1.9
Total SNPs	48,548	33,971	10,708
Useable SNPs	3,844	1,398	332
1:1 segregating SNPs	1,433	514	109

(Fig. 20.1). Regardless, a total of 5,574 SNPs were detected across the samples in this study.

Of the useable SNPs, 2,056 had 1:1 segregation ratios suitable for genetic mapping, and 148 were redundant. The redundant SNPs could possibly result from multiple SNP identification within the same sequence fragment or from tightly linked loci in the same linkage phase that would not be resolved in the relatively small subset of 45 segregating progeny. A total of 1,267 were assigned linkage groups with LOD scores greater than five (Table 20.2). When combined with SSR markers from Xie et al. (2012), 25 of

Table 20.2 Linkage groups (*LG*), the number of total markers and SSR markers in each group, and the genetic distance (*cM*) in each group for two parental maps on 45 plants of an F_1 cocksfoot population

	PI 372621			PI 295271		
LG	Markers	SSRs	cM	Markers	SSRs	cM
1	40	6	41.91	41	4	46.06
2	28	0	27.34	45	2	32.77
3	10	1	26.60	34	1	38.81
4	33	5	35.83	31	0	34.35
5	23	3	43.42	49	3	73.02
6	22	3	31.80	16	2	45.37
7	14	4	31.65	36	2	32.11
8	20	4	19.12	20	1	30.49
9	20	0	36.30	17	1	44.48
10	27	1	35.78	45	4	50.19
11	27	4	55.54	30	4	37.85
12	18	4	31.74	39	3	39.55
13	19	1	35.74	17	2	43.20
14	16	3	52.46	40	1	59.58
15	9	4	41.06	41	1	36.59
16	34	1	45.35	46	1	47.53
17	32	5	50.93	45	3	45.03
18	39	4	60.47	32	5	48.28
19	17	1	35.22	32	1	46.19
20	16	3	36.47	32	0	37.14
21	23	5	50.38	33	2	61.31
22	19	3	18.79	24	0	22.95
23	13	0	19.08	23	1	98.77
24	9	1	37.50	20	2	37.49
25	6	1	9.24	15	0	30.56
26				11	2	30.26
27				6	0	43.49
28				20	3	42.33
29				12	1	29.79
Total	534	67	909.7	852	52	1265.5

the 28 linkage groups were detected in the PI 372621 map and 29 were detected in the PI 295271 parental map (Table 20.2). Several hundred markers were just under the LOD threshold for mapping suitability within the JoinMap program, and it is highly likely that genotyping the remaining plants in our genetic mapping population will provide genetic maps of 28 linkage groups from both parents with high LOD support. Additionally, pooling fewer samples on an Ion Torrent P1 chip would increase read-depth and lead to more useable SNPs.

Conclusion

The *de novo* GBS pipeline of UNEAK was successfully adapted by pooling 48 samples per Ion Torrent P1 sequencing chip and provided over 1,000 SNPs that could be mapped in a subset of a cocksfoot genetic mapping population. These results indicate that GBS protocols using all of the sequence length from semi-conductor sequencing methods are efficacious for higher density genetic mapping. Use of the remaining plants in the F_1 population will further indicate the map position of loci that affect heading date in cocksfoot, and allow us to hone in on candidate genes for association studies.

References

Hackett CA, McLean M, Bryan GJ (2013) Linkage analysis and QTL mapping using SNP do-sage data in a tetraploid potato mapping population. PLoS ONE 8: e63939

Li X, Wei Y, Acharya A, Jiang Q, Kang J, Brummer EC (2014) A saturated genetic linkage map of autotetraploid alfalfa (Medicago sativa L.) developed using genotype-by-sequencing is highly syntenous with the Medicago truncatula genome. Gene Genome Genet 4:1971–1979

Lu F, Lipka AE, Glaubitz J, Elshire R, Cherney JH, et al. (2013) Switchgrass genomic diversity, ploidy, and evolution: Novel insights from a network-based SNP discovery protocol. PLoS Genet 9: e1003215

Mascher M, Wu S, Amand PS, Stein N, Poland J (2013) Application of genotyping-by-sequencing on semiconductor sequencing platforms: A comparison of genetic and reference-based marker ordering in barley. PLoS ONE 8: e76925

Van Ooijen JW (2011) Multipoint maximum likelihood mapping in a full-sib family of an out-breeding species. Genetics Research 93:343–349

Xie W, Robins JG, Bushman BS (2012) A genetic linkage map of autotetraploid cocksfoot and quantitative trait loci for heading date. Genome 55:360–369

Development of Genomic Selection for Perennial Ryegrass

21

M.J. Faville, S. Ganesh, R. Moraga, H.S. Easton, M.Z.Z. Jahufer, R.E. Elshire, T. Asp, and B.A. Barrett

Abstract

Mixed-species pasture based on perennial ryegrass (*Lolium perenne*) and white clover (*Trifolium repens*) is the foundation for profitable production from temperate grasslands. In theory, genomic selection (GS) offers an opportunity to lift the rate of genetic gain in these species. Critical research questions include to what extent theoretical expectations for GS can be realized in practice, and understanding the genetic and economic implications of GS in breeding programs and on farm. We describe a limited experiment to derive and cross-validate genomic estimated breeding values (GEBVs) from 211 perennial ryegrass plants evaluated for plant herbage dry weight (DW) and days-to-heading (DTH), using field phenotypic data and up to 10,885 markers typed via genotyping-by-sequencing (GBS). Using Ridge Regression-BLUP and Random Forest regression, cross validation prediction accuracies ranged from $r = 0.16$–0.34 (DW) and $r = 0.52$–0.56 (DTH). Accuracy was not influenced by marker density, but there was an interaction between statistical model and trait. The data indicate that, in these elite breeding populations, low marker densities in a limited training population dataset may be viable for generation of GEBVs in perennial ryegrass. Variance attributable to population structure, rather

M.J. Faville (✉) • H.S. Easton • M.Z.Z. Jahufer
R.E. Elshire • B.A. Barrett
Forage Improvement, AgResearch Limited, Grasslands Research Centre, Palmerston North, New Zealand
e-mail: marty.faville@agresearch.co.nz

S. Ganesh • R. Moraga
Bioinformatics Maths and Statistics, AgResearch Limited, Grasslands Research Centre, Palmerston North, New Zealand

T. Asp
Department of Genetics and Biotechnology, Aarhus University, Slagelse, Denmark

I. Roldán-Ruiz et al. (eds.), *Breeding in a World of Scarcity: Proceedings of the 2015 Meeting of the Section "Forage Crops and Amenity Grasses" of Eucarpia*,
DOI 10.1007/978-3-319-28932-8_21

than linkage disequilibrium, is likely the primary basis of GEBV accuracy in this study. Generation of a training set of increased size, scope and greater relevance for key economic traits is outlined in the context of developing a GS capability with an Australasian focus.

Keywords
Genomic selection • GBS • Perennial ryegrass • Random Forest • Ridge Regression

Introduction

Perennial ryegrass (*Lolium perenne* L.) is a species commonly used for grassland fodder and amenity purposes. In agriculture, rate of genetic improvement over time is a key performance indicator for forage breeding. Historically, rates of gain in forage species have been relatively low (Woodfield 1999; Brummer and Casler 2014). Given the importance of forage for profitability and sustainability in pastoral systems, the need for breeding strategies that lift the rate of genetic gain is critical.

Genomic selection (GS) enables a plant breeder to use a comprehensive DNA fingerprint to assess the genetic potential of an otherwise untested individual for target traits, and thereby to make a genomic prediction for use in selection. The GS concept (Meuwissen et al. 2001) is now enabled by gains in DNA marker efficiency (Davey et al. 2011; Elshire et al. 2011). Emerging theoretical (Resende et al. 2014; Heffner et al. 2010; Hayes et al. 2013) and empirical (Pryce et al. 2014; Poland et al. 2012) evidence indicates that relative to phenotypic selection, GS can provide a substantial increase in rate of genetic gain when integrated with existing plant breeding systems. This creates options to shorten the breeding cycle, improve accuracy of selection, use within-family variation, and realize gain for low heritability traits, thereby improving the rate of genetic gain in forage species (Heffner et al. 2010; Resende et al. 2014).

While GS theoretical models indicate marked rates of improvement and scope to breed for previously intractable traits, they are largely untested in forages: critical parameters have yet to be elucidated. Within our general aim to develop and deploy a GS system for perennial ryegrass, we report here on a pilot experiment to explore relationships between marker density, statistical model and contrasting traits in elite perennial ryegrass breeding populations.

Material and methods

Plant material

Clones (n = 238) from two late-flowering perennial ryegrass multi-parent breeding populations were assessed for several traits, including individual plant herbage dry weight (DW, after cutting, drying and weighing) at four seasonal time points over 2 years and days to heading (DTH, number of days after 12 October to five fully-emerged flower heads), over 2 years at Palmerston North, New Zealand (40° 21′ S, 175° 37′ E). The trial was a spaced plant row-column design with three replicates. Clonal check plants were repeated within and across replicate blocks to adjust plant trait means for spatial variation within rows and columns. Superphosphate fertilizer (8.8 kg P ha^{-1}) was applied in late autumn each year and 15–20 kg ha^{-1} of nitrogen was applied two or three times each year. Plants were defoliated at intervals based on growth pattern (including after each DW evaluation) by sheep grazing and then mechanical mowing to defoliate under-grazed areas to a residual height of 5 cm.

Genotyping

Genotyping-by-sequencing (GBS) marker data for 211 clones were obtained from the Institute for Genomic Diversity, Cornell University,

Ithaca, NY, USA. GBS libraries were based on *PstI* digestion of DNA samples and SNP-calling was conducted using the Universal Network-Enabled Analysis Kit (UNEAK) computational pipeline (Lu et al. 2013). Total, unfiltered SNP data were subsequently re-analyzed using the TASSEL 3.0 pipeline in conjunction with ryegrass (Byrne et al. 2014) pseudo-reference genome data from Aarhus University, Aarhus, DK, with assembly sorted into 12 pseudo-molecules based on macrosynteny with the rice genome. Missing SNP data were imputed using a Random Forest (RF) algorithm.

Statistical analysis

Phenotypic data were analyzed as mixed linear models, using the variance component analysis procedure residual maximum likelihood (REML) option in GenStat 7.1 (GenStat 2003). Repeated clonal checks and years were considered as fixed effects. Final phenotypic means were based on best linear unbiased predictors (BLUPs) (White and Hodge 1989), enabling adjustment for random error across replicates, columns and rows within replicates, and trial spatial effects estimated using repeated checks data (Gleeson 1997; Gleeson and Kempton 1997). Accuracy of Ridge Regression-best linear unbiased prediction (RR-BLUP) and RF models for genomic prediction was assessed by tenfold cross-validation (CV) at two SNP densities for DW (BLUP estimate from four observations across 2 years) and DTH (BLUP estimate across 2 years). In both cases, data were combined from the two populations. Accuracy of GEBV was the mean Pearson correlation coefficient (*r*) between GEBV and the phenotypic BLUP, computed from 10 CV repetitions.

Results and discussion

Significant genotypic effects ($p < 0.05$) and normal distributions were observed for both traits. Population I DTH was mean 33 days ± 12.2 (SD) and population II mean 38 days ± 7.3. For populations I and II, respectively, DW was 8.3 g ± 5.23 and 9.5 g ± 4.87. The GBS dataset yielded 3141 SNPs from the UNEAK pipeline and 10,885 from TASSEL. Prediction accuracies for GEBV's from this limited dataset varied from 0.16 to 0.56 and were affected most by trait and model applied, with marker density less influential (Table 21.1). Accuracy was higher for DTH than for DW and RR-BLUP was only marginally more accurate than RF for DTH. In contrast, prediction accuracy with RF was superior for DW. These differences may reflect contrasting genetic architectures, with RF potentially better able to assess effects for genetically complex traits, possibly due to capture of non-additive effects (Heslot et al. 2012).

Increasing marker number by nearly threefold did not improve estimated accuracy, with the exception of DW/RF (Table 21.1). This suggests prediction accuracy in this dataset may be principally driven by population structure, rather than linkage disequilibrium (LD) between markers and quantitative trait loci (Daetwyler et al. 2012), although LD has yet to be estimated in these populations.

Overall, these data indicate that moderate GS accuracy might be achievable in small popula-

Table 21.1 Summary table of cross-validation (*CV*) prediction accuracies for days-to-heading (*DTH*) and individual plant herbage dry weight (*DW*) among 211 advanced breeding population plants, based on GEBVs estimated by Ridge Regression BLUP (*RR-BLUP*) or Random Forests (*RF*) approaches

			CV prediction accuracy, *r* (SE)	
Trait	Repeatability	Model	3141 SNPs	10,885 SNPs
DTH	0.93	RF	0.55 (0.0040)	0.52 (0.0051)
DTH		RR-BLUP	0.56 (0.0035)	0.53 (0.0043)
DW	0.82	RF	0.29 (0.0054)	0.34 (0.0067)
DW		RR-BLUP	0.18 (0.0042)	0.16 (0.0057)

SE standard error of the mean. Repeatability, the upper limit of the degree of genetic determination (Falconer 1989), is also shown for each trait tions and with relatively low numbers of markers. However, use of GEBVs with accuracy dependent on genetic relationships may limit transferability of GS models and the persistence of GEBV accuracy through more than one cycle of selection (Habier et al. 2007).

We have now established a multi-population training set of 1000 elite plants and their half-sibling families. These are under evaluation in sown rows to estimate breeding values based on multi-environment yield and quality trait observations. This training set is expected to produce data that will allow assessment of GS for traits that are more industry-relevant. Selections based on GS models from these data will also be used to assess critical parameters that may influence the efficacy of GS in perennial ryegrass, including model portability, rate of decay in accuracy over generations and rate of inbreeding.

Conclusions

The extent to which GS can improve the rate of gain in perennial ryegrass and the genetic and economic implications of doing so are critical research questions. Early data indicate that low marker densities generated by GBS may provide reasonable assessment of GEBVs in elite breeding populations, with relatedness rather than linkage disequilibrium the likely basis for prediction accuracy. Research is underway to expand the training population and establish a genotype × environment component across forage yield and quality related traits of relevance for Australasia.

Acknowledgments The authors acknowledge research support from the NZ Ministry of Business, Innovation, and Employment, Pastoral Genomics, Grasslands Innovation Limited, and AgResearch Core funds.

References

Brummer EC, Casler MD (2014) Cool-Season Forages. In: Smith S, Diers B, Specht J, Carver B (eds) Yield Gains in Major U.S. Field Crops. CSSA Special Publications. American Society of Agronomy, Inc., Crop Science Society of America, Inc., and Soil Science Society of America, Inc., pp 33–52. doi:10.2135/cssaspecpub33.c3

Byrne S, Armstead I, Swain SC, Panitz F, Bendixen C, Hedegaard J, Nagy I, Farrell JD, Studer B, Asp T, Cassamo M A draft sequence assembly of the perennial ryegrass (Lolium perenne) genome. In: Proceedings of Plant & Animal Genome XXII: The International Conference on the Status of Plant & Animal Genome Research, San Diego, USA, 2014. p 161

Daetwyler HD, Kemper KE, van der Werf JHJ, Hayes BJ (2012) Components of the accuracy of genomic prediction in a multi-breed sheep population1. J Anim Sci 90 (10):3375–3384. doi:10.2527/jas.2011-4557

Davey JW, Hohenlohe PA, Etter PD, Boone JQ, Catchen JM, Blaxter ML (2011) Genome-wide genetic marker discovery and genotyping using next-generation sequencing. Nature Review Genetics 12 (7):499–510

Elshire RJ, Glaubitz JC, Sun Q, Poland JA, Kawamoto K, Buckler ES, Mitchell SE (2011) A robust, simple genotyping-by-sequencing (GBS) approach for high diversity species. PLoS ONE 6 (5):e19379. doi:10.1371/journal.pone.0019379 PONE-D-10-04702 [pii]

Falconer DS (1989) Introduction to quantitative genetics. Longman Scientific and Technical, New York

GenStat (2003) GenStat for Windows: Release 7.1., vol 6th. VSN Int, Oxford, UK

Gleeson AC (1997) Spatial analysis. In: Kempton RA, Fox PN (eds) Statistical methods for plant variety evaluation. Chapman and Hall, London, pp 68–85

Gleeson AC, Kempton RA (1997) Unreplicated trials. In: Kempton RA, Fox PN (eds) Statistical methods for plant variety evaluation. Chapman and Hall, London, pp 86–100

Habier D, Fernando RL, Dekkers JCM (2007) The impact of genetic relationship information on genome-assisted breeding values. Genetics 177 (4):2389–2397. doi:10.1534/genetics.107.081190

Hayes BJ, Cogan NOI, Pembleton LW, Goddard ME, Wang J, Spangenberg GC, Forster JW (2013) Prospects for genomic selection in forage plant species. Plant Breeding 132:133–143. doi:10.1111/pbr.12037

Heffner EL, Lorenz AJ, Jannink J-L, Sorrells ME (2010) Plant Breeding with Genomic Selection: Gain per Unit Time and Cost. Crop Science 50 (5):1681–1690. doi:10.2135/cropsci2009.11.0662

Heslot N, Yang H-P, Sorrells ME, Jannink J-L (2012) Genomic Selection in Plant Breeding: A Comparison of Models. Crop Science 52 (1):146–160. doi:10.2135/cropsci2011.06.0297

Lu F, Lipka AE, Glaubitz J, Elshire R, Cherney JH, Casler MD, Buckler ES, Costich DE (2013) Switchgrass genomic diversity, ploidy, and evolution: novel insights from a network-based SNP discovery protocol. PLoS Genetics 9 (1):e1003215. doi:10.1371/journal.pgen.1003215

Meuwissen THE, Hayes BJ, Goddard ME (2001) Prediction of Total Genetic Value Using Genome-Wide Dense Marker Maps. Genetics 157 (4):1819–1829

Poland J, Endelman J, Dawson J, Rutkoski J, Wu S, Manes Y, Dreisigacker S, Crossa J, Sánchez-Villeda H, Sorrells M, Jannink J-L (2012) Genomic Selection in Wheat Breeding using Genotyping-by-Sequencing. The Plant Genome 5 (3):103–113. doi:10.3835/plantgenome2012.06.0006

Pryce JE, Gonzalez-Recio O, Thornhill JB, Marett LC, Wales WJ, Coffey MP, de Haas Y, Veerkamp RF, Hayes BJ (2014) Short communication: Validation of genomic breeding value predictions for feed intake and feed efficiency traits. Journal of Dairy Science 97 (1):537–542. doi:10.3168/jds.2013-7376

Resende RMS, Casler MD, de Resende MDV (2014) Genomic selection in forage breeding: Accuracy and methods. Crop Science 54 (1):143–156

White TL, Hodge GR (1989) Predicting breeding values with applications in forest tree improvement. Forestry Sciences, vol 33. Kluwer Academic, Boston, MA

Woodfield DR (1999) Genetic improvements in New Zealand forage cultivars. Proceedings of the New Zealand Grasslands Association 61:3–7

22

Alfalfa Genomic Selection: Challenges, Strategies, Transnational Cooperation

P. Annicchiarico, N. Nazzicari, and E.C. Brummer

Abstract

Alfalfa (*Medicago sativa*) is the most commonly grown perennial forage legume worldwide. Some factors that constrain its genetic progress such as low breeding investment, long selection cycles and genotype-environment interactions, could be tackled more easily by genomic selection (GS), which is now conceivable using genotyping-by-sequencing approaches. Implementing such GS procedures can be cumbersome and quite expensive, however, at least initially. Establishing GS models by genotyping plants and phenotyping them according to their half-sib progenies is a suitable strategy. We found observed heterozygosity for SNP markers approaching closely the expected values for a tetraploid population in Hardy-Weinberg equilibrium, which reinforces the potential interest of assessing and imputing allele dosage information into GS models (currently too expensive to implement). GS models for broadly-based reference populations are less accurate but imply wider application and lower overall phenotyping costs relative to those for narrowly-based populations, and are compatible with their genetic structure as described by SNP markers. We provide examples of ongoing or possible future cooperation among breeding programs in different countries, which could be a strategic solution for sharing costs and technical challenges associated with GS implementation.

Keywords

Medicago sativa • Genetic structure • Genotyping-by-Sequencing • Heterozygosity • Reference population

P. Annicchiarico (✉) • N. Nazzicari
Centre for Fodder Crops and Dairy Productions, Council for Agricultural Research and Economics, Lodi, Italy
e-mail: paolo.annicchiarico@crea.gov.it

E.C. Brummer
Department of Plant Sciences, University of California, Davis, CA, USA

Introduction

Alfalfa (*Medicago sativa*) is the most commonly grown perennial forage legume worldwide (Annicchiarico et al. 2015a). Its contribution to

I. Roldán-Ruiz et al. (eds.), *Breeding in a World of Scarcity: Proceedings of the 2015 Meeting of the Section "Forage Crops and Amenity Grasses" of Eucarpia*,
DOI 10.1007/978-3-319-28932-8_22

more sustainable crop-livestock systems can be remarkable in terms of soil fertility, rate of nitrogen fixation, reduction of energy and greenhouse gas emissions, and increased self-sufficiency of feed proteins at the farm and regional level. However, higher rates of genetic progress for forage yield are necessary to safeguard its economic sustainability. Modest progress has been attributed to low breeding investment, long selection cycles, severe inbreeding depression, inability to breed real hybrids, tetrasomic inheritance, high extent of non-additive genetic variance, and large genotype-environment (GE) interactions (Annicchiarico et al. 2015a). Genomic selection (GS) based on low-cost, genotyping-by-sequencing (GBS) approaches (Elshire et al. 2011) may offer new opportunities for tackling some of these limitations such as long selection cycles, or GE interactions (by developing and exploiting GS models for wide or specific adaptation). In alfalfa, GBS can produce several thousand SNP markers (although with significant amounts of missing data) (Li et al. 2014; Annicchiarico et al. 2015b), which are expected to suffice for proper genome coverage (Li et al. 2015). However, the implementation of GS for higher yield in breeding programs can be cumbersome and quite expensive, at least initially. The aim of this paper is to discuss strategies to make its implementation more feasible. We also consider transnational cooperation as a means for increasing GS cost efficiency.

Selection Schemes Including Genomic Selection

A number of selection schemes are applicable for alfalfa and other forage crops. In the most straightforward method, a population of plants can be genotyped and cloned for phenotypic evaluation across target environments. In a GS experiment based on clonally evaluated phenotypes and GBS markers, we developed prediction models that explained over 40 % of the yield variation both across locations and across generations (Li et al. 2015).

Probably the most common selection scheme for perennial forages is to evaluate a large number of plants and reduce them by one stage of phenotypic selection to a manageable set of genotypes for selection based on their breeding value (as estimated from responses of their half-sib progenies). GS can reduce the duration of selection plus first-recombination stages to 1 year, which otherwise takes at least 3–5 years. Establishing GS models by genotyping plants and phenotyping them according to their half-sib progenies (obtained by poly-crossing all plants) has the advantage of tying GS models directly to breeding values (as opposed to plant genotypic values as in clonal evaluation). In addition, the production of half-sib seed enables the preservation of mapping/reference populations, and facilitates multi-environment experiments and evaluation under dense-sward conditions. Results for two reference populations indicated the possibility to develop moderately accurate GS models also for prediction of breeding values (Annicchiarico et al. 2015b).

Tetraploidy and Genomic Selection Modeling

GS modeling in alfalfa is complicated by autotetraploidy. Accurate GS models for prediction of plant breeding value (which relates essentially to additive genetic variation) can differentiate all genotypic classes at a locus, which in diploids includes homozygotes for either alleles or heterozygotes. In autotetraploids, the heterozygote class include three genotypes that differ in allele dosage, but in current model development, these are all lumped together into one class. Prediction model accuracy could be improved by imputing allele dosage, but the high number of sequencing reads per individual genotype required to do this effectively would dramatically reduce the number of usable SNP markers, at least at the current costs of genotyping. The loss of accuracy (as correlation between GS-predicted and actual genotype breeding values) due to this simplification is associated with the number of individuals that are heterozygous at a given locus, which is expected to be high in an outbred species.

We analyzed the proportion of heterozygous individuals (which included simplex, duplex and triplex genotypes) for each of 2,691 SNP markers obtained by GBS for a reference population of

124 plants. These individuals were selected phenotypically from 4,480 plants belonging to 14 Italian landraces or varieties (Annicchiarico 2006a). Despite the possible occurrence of relatively high bee-driven self-fertilization (Riday et al. 2013), nearly all heterozygosity values closely approached those expected for a tetraploid population in Hardy-Weinberg equilibrium (Fig. 22.1). On average, the mean difference between observed and expected proportion amounted to only −0.0032. This result reinforces the potential interest of imputing allele dosage information. Decreasing genotyping cost may make this option cost-efficient in the near future.

Genotype-Environment Interactions and Phenotyping

Alfalfa is characterized by large GE interaction, particularly between favorable and unfavorable (drought-prone, severely grazed, or saline) environments (Annicchiarico et al. 2015a). GS can help to cope with GE interactions by defining different GS models for yield response under specific phenotyping conditions, and applying them as a function of the target population of environments (possibly along with marker-assisted selection for oligogenic traits of specific interest) (Annicchiarico et al. 2015a). GS models could also be developed to breed for wide adaptation. Thus, GS models could be used to limit the costs of multi-environment testing when breeding for wide adaptation, or to facilitate the early allocation of genotypes to specifically-adapted germplasm pools. Another relevant type of GE interaction for biomass yield is that between spaced and dense planting (Annicchiarico 2006b), which can be minimized by phenotyping material under dense-sward conditions.

Genetic Structure and Reference Genetic Bases

GS models are established on a sample of the reference population (the training set of genotypes). Defining the reference population is of key importance. A broad-based population encompassing the main germplasm sources of possible interest would enable GS models to be applied for different traits throughout the entire genetic base of the breeding program. However, large numbers of markers would be needed to possibly allow for moderately high accuracy of GS models, because of the shorter linkage disequilibrium and greater

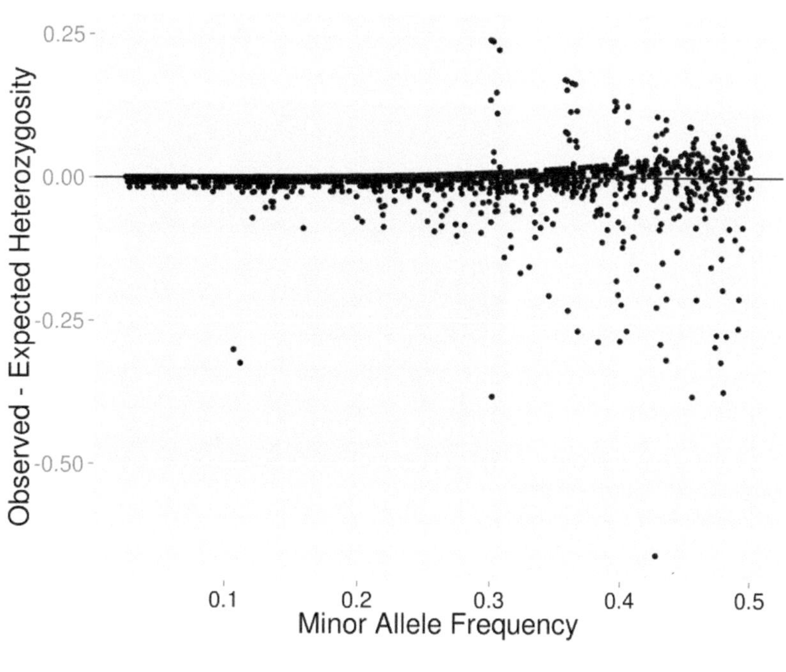

Fig. 22.1 Difference between observed and expected proportion of heterozygous individuals for 2,691 SNP markers ordered as a function of minor allele frequency, q ($q=(1-p)$), filtered at ≥2.5 %; p estimated as the fourth root of the observed population value; max. 10 % missing genotypes per SNP)

allele diversity compared with a narrowly-based reference population. The latter (e.g., established from few founding genotypes) could lead to higher accuracy, but also negligible value of GS models beyond the specific germplasm set. Non-transferable GS models would require phenotyping of more populations, thus increasing the cost. Some 124 elite genotypes derived from 15 Italian landraces or varieties that had showed outstanding diversity of adaptive responses in Annicchiarico and Piano (2005) displayed a limited genetic structure (Fig. 22.2). Therefore, these phenotypically different populations may share a large part of the overall genetic variation. Similarly, the population in Li et al. (2015) was derived from intercrossing 100 individuals of three cultivars, yet no population structure was identified. In general, these results agree with the much greater intra-population diversity relative to inter-population diversity that had emerged from molecular or morphophysiological studies (Annicchiarico et al. 2015a). Our findings suggest the possibility for developing one or very few broad-based reference populations instead of several narrowly-based populations, to limit the high investment required by phenotyping experiments. This strategy requires making several thousand SNP markers available using genotyping-by-sequencing techniques (Annicchiarico et al. 2015b).

Transnational Cooperation

Cooperation among alfalfa breeding programs in different countries can be strategic for sharing costs (especially the costs of phenotyping) and technical challenges associated with implementing GS. To be useful for individual programs, this prospect necessarily implies the adoption of a broad-based reference population composed by elite populations of different origin (preferably subjected to at least one generation of prior intercrossing, as in (Annicchiarico et al. 2015b)). Ideally, these component populations could be selected from best-performing populations in previous testing across target environments of mutual interest, as performed for non-dormant alfalfa in the western Mediterranean basin within the REFORMA project. A reference population was

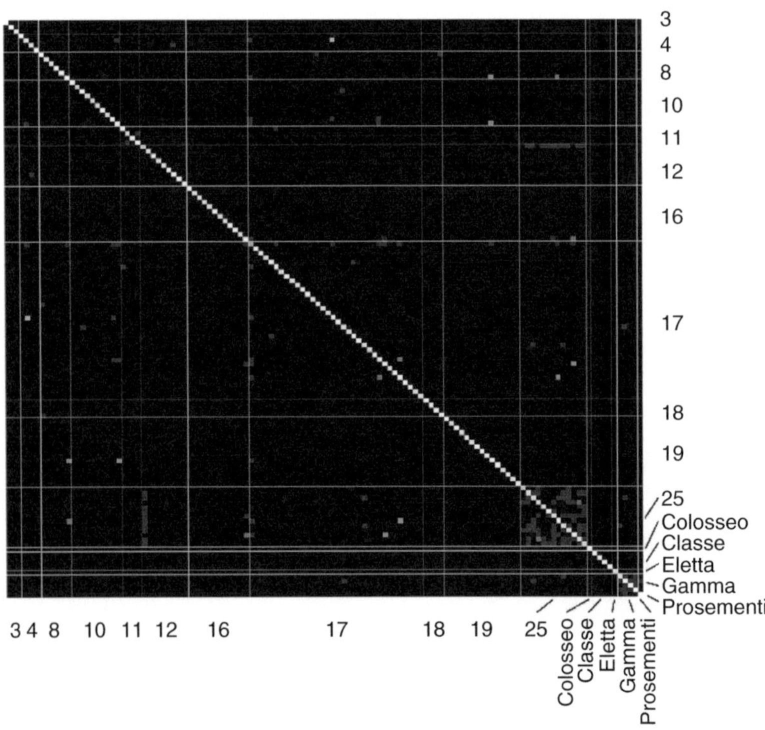

Fig. 22.2 Symmetric heat map genetic kinship matrix [according to Astle and Balding (2009)] between 124 alfalfa genotypes ordered within 15 Italian landraces or varieties, based on 3,491 SNP loci. Brighter color corresponds to greater genetic similarity between individuals (very high for the same individual). See Annicchiarico (2006a) for landrace numeric code and origin

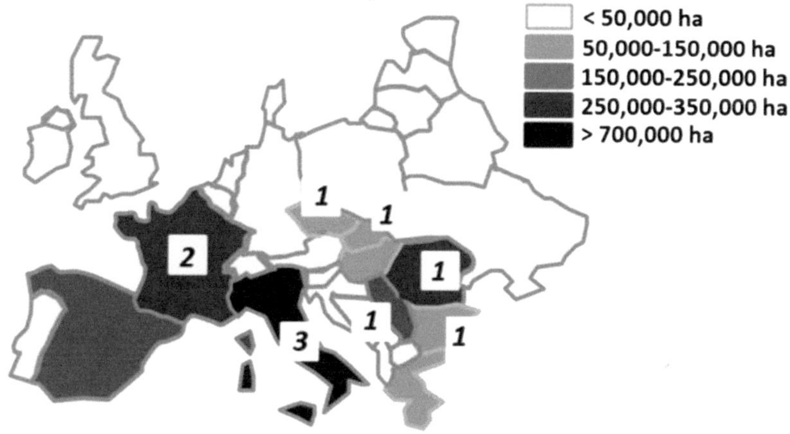

Fig. 22.3 Alfalfa cropping area of European countries, and number elite cultivars per major country contributing to the constitution of a European reference population of alfalfa

built by intercrossing one drought-tolerant Sardinian landrace, one widely-adapted Australian variety, and one Moroccan landrace that was moderately tolerant to drought and saline conditions (Annicchiarico et al. 2011).

A European semi-dormant reference population is under construction in Lodi, Italy, by intercrossing ten elite European cultivars that were selected according to the extent of alfalfa country cultivation and breeders' recommendations (Fig. 22.3). Each cultivar represents or includes elite genetic resources originated in a definite region. The expected inclusion of most key adaptive genes for European environments in the reference population may also allow to devise a GS-enabled introduction of novel genetic resources into European breeding programs.

Acknowledgments The experimental data were generated by the project 'Genomic selection in alfalfa (GENALFA)' funded by the Italian Ministry of Foreign Affairs in the framework of the Italy-USA scientific cooperation program

References

Astle W, Balding DJ (2009) Population structure and cryptic relatedness in genetic association studies. Statistical Science 24:451–471

Annicchiarico P (2006a) Diversity, genetic structure, distinctness and agronomic value of Italian lucerne (*Medicago sativa* L.) landraces. Euphytica 148: 269–282

Annicchiarico P (2006b) Prediction of indirect selection for lucerne seed and forage yield from space-planted evaluation. Plant Breeding 125: 641–643

Annicchiarico P, Piano E (2005) Use of artificial environments to reproduce and exploit genotype × location interaction for lucerne in northern Italy. Theoretical and Applied Genetics 110: 219–227

Annicchiarico P, Barrett B, Brummer EC, Julier B, Marshall AH (2015a) Achievements and challenges in improving temperate perennial forage legumes. Critical Reviews of Plant Science 34: 327–380

Annicchiarico P, Nazzicari N, Li X, Wei Y, Pecetti L, Brummer EC (2015b) Accuracy of genomic selection for alfalfa biomass yield in different reference populations. BMC Genomics 16:1020

Annicchiarico P, Pecetti L, Abdelguerfi A, Bouizgaren A, Carroni AM, Hayek T, Bouzina MM, Mezni M (2011) Adaptation of landrace and variety germplasm and selection strategies for lucerne in the Mediterranean basin. Field Crops Research 120: 283–291

Elshire RJ, Glaubitz JC, Sun Q, Poland JA, Kawamoto K, Buckler ES, Mitchell SE (2011) A robust, simple genotyping-by-sequencing (GBS) approach for high diversity species. PLoS ONE 6: e19379

Li X, Wei Y, Acharya A, Hansen JL, Crawford JL, Viands DR, Michaud R, Claessens A, Brummer EC (2015) Genomic prediction of biomass yield in two selection cycles of a tetraploid alfalfa breeding population. Plant Genome (in press: doi:10.3835/plantgenome2014.12.0090)

Li X, Wei Y, Acharya A, Jiang Q, Kang J, Brummer EC (2014) A saturated genetic linkage map of autotetraploid alfalfa (*Medicago sativa* L.) developed using genotyping-by-sequencing is highly syntenous with the M. truncatula genome. Genes Genomes Genetics 4:1971–1979

Riday H, Johnson D, Heyduk K, Raasch J, Darling M, Sandman J (2013) Paternity testing in an autotetraploid alfalfa breeding polycross. Euphytica 194: 335–349

Efficient Doubled Haploid Production in Perennial Ryegrass (*Lolium perenne* L.)

23

R.F. Begheyn, K. Vangsgaard, N. Roulund, and B. Studer

Abstract

Hybrid breeding has contributed significantly to the enormous yield increases that many major crops have undergone during the previous century. Its success relies on the exploitation of heterosis, the superior performance of an F1 hybrid compared to its inbred parents. Attempts to implement hybrid breeding in forage grasses, such as perennial ryegrass (*Lolium perenne* L.), are hampered by its highly effective self-incompatibility system as well as its sensitivity to inbreeding depression. Homozygous inbred lines are therefore difficult to create using the classical method of repeated selfing. Here, we report an efficient method to obtain homozygous genotypes of perennial ryegrass using doubled haploid (DH) induction. By means of anther culture, completely homozygous lines were obtained within one generation cycle. A highly genotype dependent response was observed for traits such as the number of embryos/calli per 100 cultured anthers and the percentages of green and albino plants regenerated. Transgressive segregation, indicative of heritable and polygenic control of the traits, was also found. Our general aim is to develop a molecular marker system to select for high responsiveness and to facilitate the introgression of this trait into advanced breeding germplasm. Segregating mapping populations will be phenotyped during anther culture and genotyped via a genotyping-by-sequencing (GBS) approach. Family-based association mapping will be used to identify marker-trait associations. In this way, an efficient breeding tool to screen germplasm for DH induction capacity will be developed. Our work will significantly accelerate forage grass breeding and constitute the first step towards efficient production of grass hybrids.

R.F. Begheyn (✉) • B. Studer
Forage Crop Genetics Institute of Agricultural Sciences, Forage Crop Genetics, ETH Zurich, Zurich, Switzerland
e-mail: rachel.begheyn@usys.ethz.ch

K. Vangsgaard • N. Roulund
DLF Trifolium A/S, Højerupvej 31, Store Heddinge DK-4660, Denmark

I. Roldán-Ruiz et al. (eds.), *Breeding in a World of Scarcity: Proceedings of the 2015 Meeting of the Section "Forage Crops and Amenity Grasses" of Eucarpia*,
DOI 10.1007/978-3-319-28932-8_23

Keywords

Hybrid breeding • Doubled haploids • Androgenesis • Molecular markers • Breeding tool • Perennial ryegrass

Introduction

To maintain and improve the productivity and sustainability of grasslands, novel tools and innovative breeding strategies are needed (Smith and Spangenberg 2014). The effective method of hybrid breeding can currently not be applied to forage grasses such as perennial ryegrass (*Lolium perenne* L.), because the self-incompatibility (SI) system of this species hampers the development of homozygous lines. Usually created by repeated selfing, inbred lines are used by breeders as parents to generate F1 hybrids. The best performing offspring of all tested parental combinations may then be released as a new cultivar. Hybrids often display hybrid vigor (heterosis), enabling them to outperform their parents for important agronomical traits (Shull 1948). This phenomenon has also been documented in perennial ryegrass (Posselt 2010). Implementing hybrid breeding therefore has great potential to contribute to the acceleration of forage grass breeding.

Doubled haploid (DH) induction is an efficient way to achieve 100 % homozygosity in one generation. This method not only bypasses the obstacle of SI, but also that of inbreeding depression, which increases in severity with every additional cycle of selfing. DH production in perennial ryegrass through anther culture has been reported (Olesen et al. 1988). However, the response to androgenesis, for example the percentage of albinos obtained, is highly variable and depends on the genotype used (Madsen et al. 1995). Nevertheless, androgenetic capacity has successfully been transferred to less responsive material by crossing it with so-called inducers, extremely rare genotypes that respond well to DH induction (Halberg et al. 1990; Madsen et al. 1995). Although the genetics underlying androgenetic capacity have not been studied in detail in perennial ryegrass, it is generally thought that a limited number of recessive genes is responsible (Madsen et al. 1995).

Here we report the results of a pilot experiment in which we implemented an efficient DH induction method in perennial ryegrass and phenotyped a segregating population for androgenetic capacity. The next step will be to apply the DH induction protocol to a set of segregating mapping populations, which will also be genotyped using a genotyping-by-sequencing (GBS) approach (Elshire et al. 2011). Family-based association mapping can then be used to identify marker-trait associations for traits of importance in adrogenesis (Guo et al. 2013).

Our general aim is to develop a molecular marker based breeding tool to select for high responsiveness to anther culture and to facilitate the introgression of this trait into advanced breeding germplasm. Research on forage grass cytoplasmic male sterility (CMS) has recently shown significant progress (Studer et al. 2012; Islam et al. 2014). Practical and economical systems for pollination control as well as homozygous line production are now within reach. Our work will therefore significantly accelerate forage grass breeding and will constitute the first step towards the realization of hybrid grass cultivars.

Material and Methods

Plant Material

Three so-called inducer genotypes, known to exhibit a good response to DH induction, were used. Offspring of a polycross between a larger set of inducer genotypes, including the three used in this experiment, was also made available to us and we used the eighteen genotypes that flowered at the convenient time. The plants were vernalized and grown in an unheated greenhouse in 2013/2014. Spikes with microspores in the late uninucleate stage, as determined by microscopy, were harvested in May-June 2014 for use in the doubled haploid induction experiment.

Doubled Haploid Induction

The method used was adapted from the androgenesis protocols based on 190–2 media for perennial ryegrass described by Olesen et al. (1988) and Opsahl-Ferstad et al. (1994). Kinetin and 2,4-dichlorophenoxyacetic acid (2,4-D) were included in the induction medium. Kinetin and 1-naphthaleneacetic acid (NAA) were included in the regeneration medium, but the shoot rooting medium did not contain any hormones. Maltose instead of sucrose was used as the carbon source in the induction medium. The gelling agent was Gelrite at 3.5 g L^{-1}. A cold pre-treatment of the spikes of 1–5 days was applied. Spikes were sterilized by a quick rinse in 70 % ethanol, followed by a 10-min soak in a 2.5 % hypochlorite solution.

Depending on spike availability, 84–168 anthers of the three inducer and eighteen polycross progeny genotypes were subjected to the *in vitro* DH induction procedure. Component traits of androgenetic capacity, such as the numbers of responsive anthers, embryo-like structures or calli and green or albino plants that were obtained were recorded. A representative subset of regenerated green putative DH plantlets was finally transplanted into soil and grown in the greenhouse.

Flow Cytometry

Leaf material of a subset of the green putative DH plants was harvested to estimate DNA content using flow cytometry. A genotype of the cultivar ‘Arara’ was used as a diploid control to determine the ploidy level of the putative DH plants.

Results and Discussion

DH Induction

Marked differences in androgenetic capacity were observed between the genotypes for the number of embryo-like structures (ELS) and plants produced per 100 anthers as well as for the percentages of albino and green regenerants (Table 23.1). Some genotypes did not produce any ELS, while for other genotypes the ELS did not regenerate into plants. Genotypes that yielded plants showed distinct differences in ratios between albino or green plants. Two genotypes could only be induced to regenerate albinos.

Several of the PP genotypes performed better than the best performing inducer for several or even all recorded traits, as was the case for PP 078. Such transgressive segregation is an indication of polygenic control of the component traits of androgenic capacity, confirming similar reports in literature (Madsen et al. 1995; Wang et al. 2005).

Flow Cytometry

In perennial ryegrass anther culture, the chromosome doubles spontaneously in the early phases of embryogenesis, with a reported efficiency of 50–70 % (Olesen et al. 1988; Bante et al. 1990; Halberg et al. 1990). Our results confirm that spontaneous chromosome doubling takes place. Interestingly, variation between genotypes was observed, suggesting genetic control of this characteristic (Table 23.2). It is also interesting that some genotypes that display similar performances for plantlet regeneration per 100 anthers and percentage of green regenerants differed in the percentages of diploids they yield. PP 007 and PP 012 produced 98 and 114 plants per 100 cultured anthers, respectively, approximately 70 % percent of which were green (Table 23.2). However, PP 007 had a chromosome doubling efficiency of 33 % compared to 50 % for PP 012.

Conclusions

An effective DH induction protocol based on anther culture was successfully established and could be used to screen perennial ryegrass genotypes for DH induction capacity. An intriguing variation for component traits of androgenetic capacity was observed between the genotypes. The ploidy level

Table 23.1 Overview of the number of embryo-like structures (*ELS*) and number of plants regenerated per 100 anthers, as well as the percentages of albino and green plants that were obtained by doubled haploid induction

Genotype	Number of ELS per 100 anthers	Number of plants per 100 anthers	Percentage albino plants	Percentage green plants
Inducer 002	277	173	81	19
Inducer 046	4	0		
Inducer 149	95	0		
PP 006	50	0		
PP 007	297	98	29	71
PP 012	323	114	34	66
PP 015	52	32	70	30
PP 017	24	0		
PP 023	67	32	100	0
PP 024	113	10	69	31
PP 031	23	36	67	33
PP 039	42	15	38	62
PP 046	1	0		
PP 051	0			
PP 056	102	70	41	59
PP 071	5	0		
PP 078	580	349	32	68
PP 079	40	42	29	71
PP 081	30	21	100	0
PP 086	2	0		
PP 089	0			

PP polycross progeny

Table 23.2 Ploidy levels, determined by flow cytometric DNA content estimation, of a subset of the green putative doubled haploid (*DH*) plantlets obtained in the DH induction experiment

Genotype	Number of plants checked	Percentage of haploids (n)	Percentage of diploids (2n)	Percentage of tetraploids (4n)
Inducer 002	11	36	64	
PP 007	9	56	33	11
PP 012	10	30	50	20
PP 015	10		100	
PP 024	1	100		
PP 031	10	10	90	
PP 056	10		100	
PP 078	10	10	70	20
PP 079	10		100	

PP polycross progeny

analysis still needs to be complemented with a marker assay to determine whether the diploid genotypes are actual DH plants. Those results will enable us to draw clear conclusions on the outcome of our experiment. These preliminary results are encouraging for our future work with mapping populations segregating for those same characteristics and inspire confidence that we will be able to find marker–trait associations useful for the development of a novel breeding tool.

Inbreeding depression is problematic when homozygosity is increased in outcrossing species such as perennial ryegrass, leading to DHs that are often weak and suffer from fertility issues (Opsahl-Ferstad 1993). However, reports of DH genotypes with a higher vigor and seed set than their parents do exist (Andersen et al. 1997). In effect, a selection against inbreeding depression takes place while raising DH plants. The fertile and vigorous genotypes are excellent candidates for parents of hybrid grass cultivars.

In conclusion, our future results and the molecular breeding tool we will develop, taken together with the foreseen developments in grass CMS research, will bring us closer to the reality of hybrid grass cultivars than ever before.

References

Andersen SB, Madsen S, Roulund N, et al (1997) Haploidy in ryegrass. In: Jain SM, Sopory SK, Veilleux RE (eds) In vritro haploid production in higher plants. Kluwer Academic Publishers, pp 133–147

Bante I, Sonke T, Tandler RF, et al (1990) Anther culture of Lolium perenne and Lolium multiflorum. In: Sangwan R., Sangwan-Norreel BS (eds) The impact of biotechnology in agriculture. Kluwer Academic Publishers, pp 105–127

Elshire RJ, Glaubitz JC, Sun Q, et al (2011) A robust, simple genotyping-by-sequencing (GBS) approach for high diversity species. PLoS One 6:e19379. doi: 10.1371/journal.pone.0019379

Guo B, Wang D, Guo Z, Beavis WD (2013) Family-based association mapping in crop species. Theor Appl Genet 126:1419–30. doi: 10.1007/s00122-013-2100-2

Halberg N, Olesen A, Tuvesson IKD, Andersen SB (1990) Genotypes of Perennial Ryegrass (Lolium perenne L.) with High Anther-Culture Response through Hybridization. Plant Breed 105:89–94.

Islam MS, Studer B, Møller IM, Asp T (2014) Genetics and biology of cytoplasmic male sterility and its applications in forage and turf grass breeding. Plant Breed 133:299–312. doi: 10.1111/pbr.12155

Madsen S, Olesen a., Dennis B, Andersen SB (1995) Inheritance of anther-culture response in perennial ryegrass (Lolium perenne L.). Plant Breed 114:165–168. doi: 10.1111/j.1439-0523.1995.tb00783.x

Olesen A, Andersen SB, Due IK (1988) Anther Culture Response in Perennial Ryegrass (Lolium perenne L.). Plant Breed 101:60–65.

Opsahl-Ferstad HG (1993) Androgenetic response in grasses. I. Anther culture in perennial ryegrass (Lolium perenne L.). II. Molecular studies of embryogenesis in barley (Hordeum vulgare L.). Ph. D. Thesis, Agricultural University of Norway

Opsahl-Ferstad H-G, Bjornstad A, Rognli OA (1994) Influence of medium and cold pretreatment on androgenetic response in Lolium perenne L. Plant Cell Rep 13:594–600.

Posselt UK (2010) Identification of heterotic patterns in perennial ryegrass. In: Huyge C (ed) Sustainable use of genetic diversity in forage and turf breeding. Springer Netherlands, pp 569–572

Shull GH (1948) What is "Heterosis"? Genetics 33:439–446.

Smith KF, Spangenberg G (2014) Forage breeding for changing environments and production systems: an overview. Crop Pasture Sci 65:i–ii. doi: 10.1071/CP12009

Studer B, Byrne S, Nielsen RO, et al (2012) A transcriptome map of perennial ryegrass (Lolium perenne L.). BMC Genomics 13:140. doi: 10.1186/1471-2164-13-140

Wang Z-Y, Ge Y, Mian R, Baker J (2005) Development of highly tissue culture responsive lines of Lolium temulentum by anther culture. Plant Sci 168:203–211. doi: 10.1016/j.plantsci.2004.07.036

Testing a Pollen-Parent Fecundity Distribution Model on Seed-Parent Fecundity Distributions in Bee-Pollinated Forage Legume Polycrosses

24

H. Riday and M.D. Peel

Abstract

Random mating (panmixis) is a fundamental assumption in quantitative genetics. In outcrossing bee-pollinated perennial forage legume polycrosses, mating is assumed by default to follow theoretical random mating. This assumption informs breeders of expected inbreeding estimates based on polycross size. In examining forage legume pollen-parent fecundity distributions (i.e., distribution of pollen-parent gamete contribution frequencies to the progeny generation in a polycross) it is obvious that the pollen-parent fecundity distribution is not uniform (i.e., panmixis expectation). Using an empirical approach, a Weibull distribution-based model has been developed to accurately predict a polycross' pollen-parent fecundity distribution depending only on polycross size. In this study we wanted to determine if this pollen-parent fecundity distribution model was able to predict seed-parent fecundity distributions in bee-pollinated forage legume polycrosses of varying sizes. We found that the pollen-parent fecundity distribution model worked very well in predicting observed seed-parent fecundity distributions with R^2 for eight polycrosses tested ranging from 0.83 to 0.99. A poor fit was observed for a ninth polycross tested ($R^2=0.30$). Preliminarily we attributed this poor fit to the fact that parent plants in this polycross were clonally replicated. Using the Weibull-based model, both expected pollen-parent and seed-parent fecundity distributions can be predicted for forage legume polycrosses knowing only polycross size. This model should allow breeders to estimate more realistic inbreeding expectations based on polycross size.

H. Riday (✉)
U.S. Dairy Forage Research Center (USDA-ARS),
Madison, WI, USA
e-mail: heathcliffe.riday@ars.usda.gov

M.D. Peel
Forage and Range Research Lab (USDA-ARS),
Logan, UT, USA

I. Roldán-Ruiz et al. (eds.), *Breeding in a World of Scarcity: Proceedings of the 2015 Meeting of the Section "Forage Crops and Amenity Grasses" of Eucarpia*,
DOI 10.1007/978-3-319-28932-8_24

Keywords
Fecundity distributions • Random mating • Polycross • Forage legumes • Inbreeding

Introduction

Random mating (panmixis) is a fundamental quantitative genetic assumption (Weismann 1883). Detailed mating examinations have revealed, however, that theoretical random mating is rarely observed (Clegg 1980). Allard (1999) cautions breeders on the random mating assumption: "It can be questioned whether the theoretical form of random mating is ever fulfilled exactly, because some form of selection, natural or human, is likely to intervene . . . environmentally induced differences . . . make it improbable that fertilizations are ever entirely random events, either in nature or in cultivation" (pg. 53, Allard 1999). Few if any studies have examined mating in bee-pollinated outcrossing perennial forage legume polycrosses, by default letting the theoretical random mating assumption stand. This default assumption has thus informed the expected inbreeding estimates based on polycross size taught to forage breeders (e.g., (Busbice 1969)). However, recently studies have reported pollen movement in bee-pollinated outbred forage legume polycrosses (Riday et al. 2013, 2015; Vleugels et al. 2014). Riday et al. (2015), in examining forage legume pollen-parent fecundity distributions (i.e., distribution of pollen-parent gamete contribution frequencies to the progeny generation in a polycross), determined that these distributions were not uniform distributions (in other words, they were contradictory to the panmixis expectation). Using an empirical approach, Riday et al. (2014) developed a model based on the Weibull distribution that was able to very accurately predict a polycross' pollen-parent fecundity distribution by knowing only the polycross size.

It is already known that seed-parent fecundity (i.e., distribution of seed-parent gamete contribution frequencies to the progeny generation in a polycross) distributions are not uniform distributions (i.e., not panmixis). Breeders know that seed yield from plant to plant in a polycross is not equal. This is why breeders often respond to this awareness through the common practice of bulking equal quantities of seed-parent seed (i.e., halfsib-seed) to form a synthetic. This practice artificially creates panmixis at least for seed-parents. However, attempting to model the plant to plant seed yield variation is likely thought to be complicated by stochastic processes and the polycross environment, making simple modeling of this distribution difficult. Because Riday et al. (2015) was able to demonstrate a simple model for pollen-parent fecundity distributions that require no knowledge of the polycross environment, in this study we wanted to try to determine if and how well this pollen-parent fecundity distribution model was able to predict seed-parent fecundity distributions in bee-pollinated perennial forage legume polycrosses.

Material and Methods

Polycrosses

This study utilized seven diploid red clover (*Trifolium pratense* L.), one autotetraploid alfalfa (*Medicago sativa* L.), and one autohexaploid kura clover (*Trifolium ambiguum* M. Bieb.) polycross. The size of the polycross (N) was assigned as the number of parent plants that had produced seed and were detected as pollen-parents in the progeny. The seven red clover polycrosses were: LP08 $N=22$; VIS09 $N=25$; C328WS $N=32$; YLD09 $N=72$; WI21 $N=91$ (Riday 2011); C584Y07 $N=92$ (Riday 2011); and C276 $N=93$ (Riday 2011). The alfalfa polycross was Alforex ($N=15$) (Riday et al. 2013). The kura clover polycross was KU09B ($N=26$). All red clover polycrosses were conducted at Prairie du Sac, WI, USA in a 1.8 m by 3.6 m by 1.8 m tall pollination enclosure as described in (Riday and Krohn 2010) using bumblebees (*Bombus impatiens* Cr.). Plants were spaced equidistant from each other within the crossing blocks. The alfalfa polycross was conducted in Woodland, CA, USA utilizing leafcutter bees (*Megachile rotundata* F.) (fully described in (Riday et al. 2013)). The kura clover polycross was conducted in Logan, UT, USA utilizing leafcutter bees in an open isolation with plants spaced 1 m apart in a 9 × 18 plant grid with parent plants replicated six times (randomized complete block design). One KU09B parent out of the original 27 parents was culled prior to pollination, thus $N=26$.

Following pollination, seed was harvested from individual parent plants and weighed. In polycrosses with replicated parental clones (i.e., Alforex and KU09B), an average seed yield per seed-parent was determined. Based on total seed yield of each polycross, the individual proportion or frequency that each seed-parent contributed to the progeny generation of the polycross was estimated (Pf_i). This methodology assumes that weight per individual seed was constant within and among half-sib seed lots within a polycross.

Fecundity Distribution Model

Based on empirically measured pollen-parent fecundity distributions of 10 polycrosses determined previously ((Riday et al. 2015) and Riday unpublished; pollen-parent identity was determined using DNA simple sequence repeat [SSR] molecular markers), we developed a model based on the Weibull distribution (Forbes et al. 2011; Weibull 1951) that very accurately predicts pollen-parent fecundity distributions in a polycross based on only knowing the polycross size (N). In this study we used model Eq. 24.1; however, we made Pf_i equal the ith seed-parent as opposed to ith pollen-parent. We also made Pp_i equal the ith seed-parent rank position in the array of all possible seed-parents (N) ranked from least frequent seed-parent to most frequent seed-parent. We compared the observed seed-parent fecundity distribution to the model predicted distribution using R^2 (i.e., 1 − [residual SS/Total SS]). Confidence limits of 95 % around the R^2 values were estimated using Fisher transformations (Kirk 1999).

$$Pf_i^* = \frac{\frac{1}{N}\left[-1+1.126\left(-\ln\left(1-\frac{Pp_i}{N+1}\right)\right)^{\frac{1}{1.524}}\right]+\frac{1}{N}}{\sum_{i=1}^{N} Pf_i} \tag{24.1}$$

Where,

N = Total number of possible seed-parents in polycross.

Pf_i = Frequency of progeny having the ith seed-parent.

Pp_i = ith seed-parent rank position in the array of all possible seed-parents (N) ranked from least frequent seed-parent of progeny in a polycross (Pf_i, Minimum) to most frequent seed-parent of progeny in a polycross (Pf_i, Maximum).

Pf_i^* = adjusted Pf_i such that

$$\sum_{i=1}^{N} Pf_i^* = 1$$

Results and Discussion

Although panmixis or random mating serves as a theoretical null hypothesis for mating in nature, theoretical random mating in nature is in fact rarely observed (Clegg 1980). Previously, Riday et al. (2015) demonstrated that pollen-parent fecundity distributions in bee-pollinated, isolated, physically small forage legume polycrosses could be very accurately modeled knowing only the polycross size using a Weibull distribution (see equation Eq. 24.1). Here we demonstrate that the same fecundity distribution model can be used to quite accurately predict seed-parent fecundity distributions (Fig. 24.1).

The goodness of fit between the predicted Pf_i values and the observed Pf_i values ranged from R^2 equals 0.301 to 0.991 (Fig. 24.1). Among the nine polycrosses the kura clover polycross

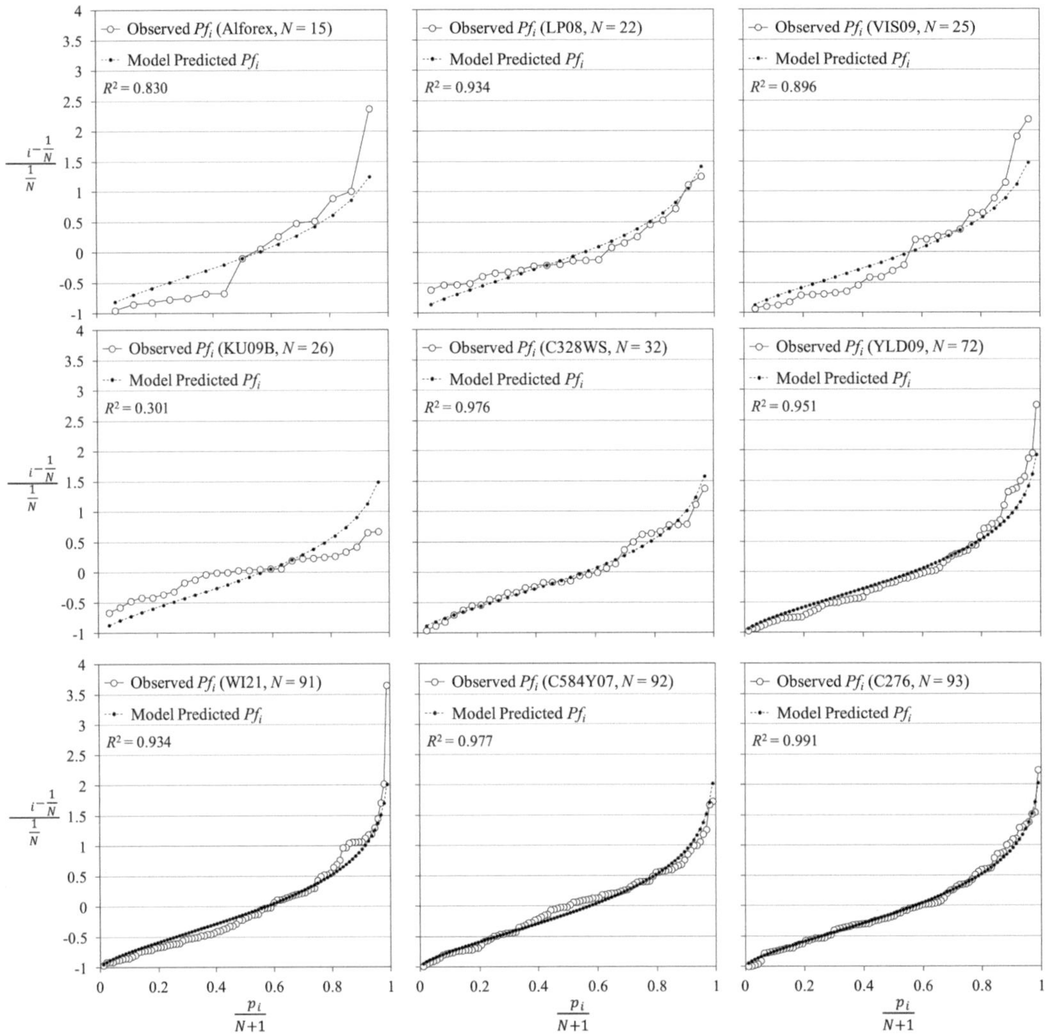

Fig. 24.1 Observed and model-predicted Pf_i's from least frequent to most frequent (Pp_i) for nine polycrosses (red clover [*Trifolium pratense* L.]: VIS09, LP08, C328WS, YLD09, WI21, C584Y07, and C276; alfalfa [*Medicago sativa* L.]: Alforex; and kura clover [*Trifolium ambiguum* M. Bieb.]: KU09B) ranging in polycross size (N) from 15 to 93; including R^2 goodness of fit between observed and model-predicted values

KU09B had an R^2 of 0.301 (95 % C.I. 0.042, 0.596) and clearly had a worse fit than any of the other polycrosses. The kura clover polycross was different from the other polycrosses in that parent plants were clonally replicated in the crossing block. This factor may be the cause of the poor model prediction for this polycross; although the Alforex polycross also had parent plants clonally replicated (R^2=0.830; 95 % C.I. 0.559, 0.942). Despite the poor model prediction for the KU09B seed-parent fecundity distribution, the model predicted that pollen-parent fecundity distribution had a good fit (Riday et al. 2015). Other than for the KU09B polycross, the other eight polycrosses had model-predicted seed-parent Pf_i values remarkably close to the observed values. In this group of eight polycrosses, the previously mentioned Alforex polycross had a somewhat lower R^2 value. The VIS09 polycross was the only other polycross whose goodness of fit was less than 0.90 (R^2=0.896; 95 % C.I. 0.776, 0.954). Six of the nine polycrosses had goodness of fits between model-predicted and observed Pf_i values higher than 0.90. Based on these observations, we conclude that seed-parent fecundity distributions can quite accurately be predicted using the pollen-parent fecundity distribution presented by Riday et al. (2015). We also conclude that pollen-parent and seed-parent fecundity distributions in small, isolated, bee-pollinated polycrosses similarly follow a Weibull distribution.

Conclusion

By default, the random mating assumption has stood in forage legume breeding (i.e., Pf_i values having a uniform distribution). This study demonstrates that the same, more accurate model can be used for expected pollen-parent fecundity distributions and for seed-parent fecundity distributions. The fecundity distribution deviation from uniformity indicates that actual inbreeding expected during polycross mating is more severe than assumed under random mating. If the breeder is concerned about minimizing inbreeding in their polycrosses, the common practice of bulking equal quantities of seed from seed-parent should be continued. Eliminating inbreeding associated with Weibull-distributed pollen-parent fecundity distributions, however, is more difficult to overcome. This study provides some preliminary evidence that clonally replicating seed-parents in polycrosses will lead to more uniform seed-parent fecundity distributions. However, Riday et al. (2015) did not observe a similar increased propensity towards a uniform distribution for pollen-parent fecundity distributions in polycrosses with clonally replicated parents.

Due to the simple nature of the fecundity distribution prediction model used in this study (i.e., Eq. 24.1), simulations and other quantitative theory studies should be examined in light of this distribution compared to the standardly assumed uniform distribution associated with panmixis or so-called random mating.

References

Allard RW (1999) Principles of plant breeding. John Wiley and Sons, Inc.: New York.

Busbice TH (1969) Inbreeding in synthetic varieties. Crop Sci. 9:601–604.

Clegg MT (1980) Measuring plant mating systems. BioScience 30:814–818.

Forbes C, Evans M, Hastings N, Peacock B (2011) Statistical distributions (4th ed.). John Wiley and Sons, Inc.: Hoboken, NJ, USA.

Kirk RE (1999) Statistics: an introduction (4th ed.). Harcourt Brace: Fort Worth, TX, USA.

Riday H, Krohn AL (2010) Increasing population hybridity by restricting self-incompatibility alleles in red Clover populations. Crop Sci. 50:853–860.

Riday H (2011) Paternity testing a non-linkage based marker assisted selection scheme for outbred forage species. Crop Sci. 51: 631–641.

Riday H, Johnson DW, Heyduk K, Raasch JA, Darling MM, Sandman JE (2013) Paternity testing in an autotetraploid alfalfa breeding polycross. Euphytica 194:335–349.

Riday H, Smith MA, Peel MD (2015) A simple model for pollen-parent fecundity distributions in bee-pollinated forage legume polycrosses. Theor Appl Genet 128:1865-1879.

Vleugels T, Cnops G, Roldán-Ruiz I (2014) Improving seed yield in red clover through marker assisted parentage analysis. Euphytica 200:305–320.

Weibull W (1951) A statistical distribution function of wide applicability. J. Appl. Mech.-Trans. ASME 18:293–297.

Weismann A (1883) On heredity. 1889 English translation. In: Poulton EB, Schönland S, Shipley AE (eds.). Essays upon heredity and kindred biological problems. Oxford University Press Warehouse, London. pp 69–105.

Single-Marker Assisted Introgression of Crown Rust Resistance in an Italian Ryegrass Breeding Program

25

R. Kölliker, V. Knorst, B. Studer, and B. Boller

Abstract

Ecotype populations of Italian ryegrass (*Lolium multiflorum* Lam.) provide a valuable gene pool for further improvement of this important forage grass species. Although they are often characterized by high yield potential and persistence, ecotype populations usually suffer from severe susceptibility to leaf diseases such as crown rust, caused by *Puccinia coronata*. In this study we tested the potential of single-marker assisted selection to introgress a resistance gene from a bi-parental mapping population into ecotype-based breeding material. A single SSR marker linked to a major QTL for crown rust resistance on LG2 of *L. multiflorum* was used to generate three breeding populations with contrasting intensities of marker assisted selection (homozygous for the marker allele, heterozygous, absence of the marker allele). In the framework of our breeding program, these populations were continuously genotyped and characterized for crown rust resistance over five generations. Observations on individual plants of the F_3 generation showed significant linkage of the marker allele with crown rust resistance but there was no significant difference between individuals homozygous or heterozygous for the marker allele, respectively. In plot trials with Syn_2 progeny, crown rust scores ranged from 1.44 to 2.00 (1=resistant, 9=susceptible) for populations based on positive selection for the marker allele while populations based on pure phenotypic selection had crown rust scores ranging from 3.11 to 4.56. Thus, our results demonstrate that genotypic selection based on a single marker can complement phenotypic selection and facilitate the introgression of resistance traits in breeding material.

R. Kölliker (✉) • V. Knorst
Molecular Ecology, Agroscope, Institute for Sustainability Sciences, Zurich, Switzerland
e-mail: roland.koelliker@agroscope.admin.ch

B. Studer
Forage Crop Genetics, Institute of Agricultural Sciences, Zurich, Switzerland

B. Boller
Fodder Crop Breeding, Agroscope, Institute for Sustainability Sciences, Zurich, Switzerland
e-mail: roland.koelliker@agroscope.admin.ch

I. Roldán-Ruiz et al. (eds.), *Breeding in a World of Scarcity: Proceedings of the 2015 Meeting of the Section "Forage Crops and Amenity Grasses" of Eucarpia*,
DOI 10.1007/978-3-319-28932-8_25

Keywords
Crown rust • Italian ryegrass • Marker assisted selection • QTL

Introduction

Italian ryegrass (*Lolium multiflorum* Lam.) is one of the most important forage grasses in temperature regions. It is particularly valued for its yield potential and its high forage quality. In central Europe, ecotype populations of Italian ryegrass can be found in natural or semi-natural grassland. They are usually well adapted to their respective habitats and can show a high yield potential but often suffer from severe susceptibility to diseases such as crown rust, caused by *Puccinia coronata* f.sp. *lolii* (Boller et al. 2009).

Due to its economic importance, resistance to crown rust is a prime breeding target in any ryegrass improvement program. Although phenotypic recurrent selection has resulted in cultivars with considerable levels of crown rust resistance, further progress is often hampered by the complex composition of cultivars and rapidly evolving pathogen populations (Kimbeng 1999). Consequently, considerable effort has been taken to understand this complex host-pathogen interaction, to elucidate the genetic control of resistance and to develop tools and strategies for improved resistance breeding (reviewed in Dracatos et al. 2010). Mainly based on bi-parental mapping populations, quantitative trait loci (QTL) explaining substantial proportions of phenotypic variance were discovered on most linkage groups of *L. perenne*. In *L. multiflorum*, broad spectrum QTL have been mainly identified on linkage groups (LG) one and two (Studer et al. 2007; Sim et al. 2007), while specific QTL conferring resistance to specific *P. coronata* isolates have been identified on LG two, four and seven (Dracatos et al. 2010).

Despite these advances, reports on utilization of markers linked to resistance genes and QTL in ryegrass breeding are scarce. The aim of this work was to test the suitability of single-marker assisted selection in a standard breeding program by (i) introgressing the QTL identified in a bi-parental mapping population into breeding lines and (ii) developing cultivar candidates with increased resistance to crown rust.

Materials and Methods

Plant Material and Selection Scheme

As a source for crown rust resistance, progeny of a bi-parental mapping population segregating for crown rust resistance were used (FALXtg03; Studer et al. 2007). In this population, a major QTL explaining up to 55.5 % of the phenotypic variance was identified on LG 1 in field experiments at different locations as well as in greenhouse evaluations and using a leaf segment test. As novel germplasm, we selected *L. multiflorum* ecotypes from a diverse collection characterized by high agronomic performance but poor crown rust resistance (Peter-Schmid et al. 2008). Seventeen plants from the mapping population and 17 from the ecotype collection were selected for 17 pair-crosses (Fig. 25.1). From the resulting 340 F1 progeny, 19 resistant plants with the marker allele Rx were selected and poly-crossed. From the resulting 240 F_2 plants, 82 plants that did not show rust symptoms in the breeding nursery were genotyped for the marker allele and phenotypic characteristics such as vigor and persistence. Thirty-eight plants were selected for three polycrosses with the parental marker allele composition Rx, RR and xx (Fig. 25.1). Single F_3 plants were phenotyped and genotyped and simultaneously, Syn_1 and Syn_2 populations were developed. In parallel, a similar scheme was used to develop breeding lines based on a different source of resistance and solely based on phenotypic characterization (Table 25.1).

Phenotypic and Genotypic Characterization

At all stages, resistance to crown rust was determined based on a visual scoring using a scale from 1 (resistant) to 9 (susceptible). Individual plants (up to F_3) were scored in the breeding nurseries at Agroscope, Zurich, CH. Resistance of the second

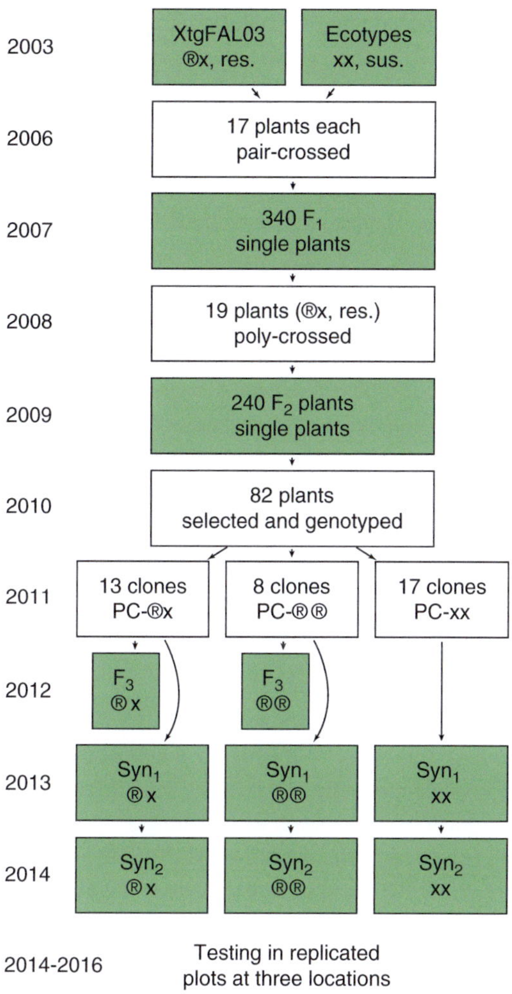

Fig. 25.1 Starting material and breeding scheme used for the development of cultivar candidates for *L. multiflorum* based on marker assisted and phenotypic selection (RR, Rx) or pure phenotypic selection (xx). Green boxes indicate the different generations produced and analyzed. Plants were genotyped using a single SSR linked to a crown rust resistance QTL (R-allele) and phenotypically characterized for crown rust resistance using a visual score from 1 (resistant) to 9 (susceptible)

generation synthetic populations (Syn_2) was assessed in replicated field trials at three locations.

The SSR marker NFFA012 (Saha et al. 2004), closely linked to the QTL in the mapping population, was used for genotyping of individual plants. Protocols for DNA extraction and marker genotyping were as described by Studer et al. (2007). Polycross parents homozygous for the resistance allele (RR), heterozygous for the resistance allele (Rx) and not carrying the resistance allele (xx) were selected to give rise to the respective synthetic progenies (Fig. 25.1).

Table 25.1 Mean rust scores of cultivar candidates based on different parental populations and selection strategies

Syn2 population	Selection strategy	Crown rust score
Ecotypes[1] × XtgFAL03[2]	MAS: RR	1.44 a[4]
Ecotypes × XtgFAL03	MAS: Rx	2.00 a
Ecotypes × XtgFAL03	MAS: xx	3.67 bc
Ecotypes × 'Tigris'[3]	Phenotypic	3.56 b
Ecotypes × 'Tigris'	Phenotypic	3.78 bc
Ecotypes × 'Tigris'	Phenotypic	3.11 b
Ecotypes	Phenotypic	4.56 c
Standard (Tigris)	Phenotypic	3.89 bc

MAS RR, Rx, xx indicate genotypic selection for the resistance allele R combined with phenotypic selection, phenotypic indicates phenotypic selection based on visual scores only. Syn_2 progeny were sown in replicated plot trials at three locations and rust scores were determined repeatedly during 2014 using a score from 1 (resistant) to 9 (susceptible)

[1]Ecotypes from diverse Swiss ecotype collection (Peter-Schmid et al. 2008)

[2]Progeny of a mapping population segregating for crown rust resistance (Studer et al. 2007)

[3]Breeding material derived from the Tigris (Agroscope, Switzerland) genepool

[4]Means followed by the same letter are not significantly different (Duncan's multiple range test; $P < 0.05$)

Statistical Analysis

Linkage of the marker allele with crown rust resistance was tested using a χ^2 test. Differences in rust resistance among the different populations were tested using analysis of variance and Duncan's multiple range test.

Results and Discussion

Characterization of the Resistance Allele

To characterize the occurrence of the allele of SSR marker NFFA012 which was linked to resistance in the mapping population FALXtg03

(R-allele), 48 vigorous and persistent individuals from seven ecotype populations cultivated as spaced plants as described by Peter-Schmid et al. (Peter-Schmid et al. 2008) were genotyped and characterized for crown rust resistance. Twenty individuals carried a marker allele of the same length as the R-allele, but only two of these were resistant to crown rust, while three of the 28 individuals not carrying the marker allele were also resistant. Thus, in ecotype populations no linkage was observed between the marker allele and crown rust resistance ($\chi^2=0.0064$, $p=0.93$).

On the other hand, significant linkage between the R-allele and crown rust resistance was observed after intercrossing individuals from ecotype populations and FALXtg03 ($\chi^2=20.76$, $p<0.0001$). In single plants of the F_3 generation which were the progeny of polycrosses between parental plants selected for crown rust resistance and the presence of the R-allele (Fig. 25.1), individuals homozygous or heterozygous for the R-allele showed a rust score of 1.60 or 1.76, respectively, while individuals from the Rx – polycross progeny (Fig. 25.1) not carrying any R-allele were significantly more susceptible with a rust score of 5.62 ($R^2=0.52$, $p<0.001$).

Introgression of Resistance Allele into Cultivar Candidates

The ultimate aim of the study was to investigate whether single-marker assisted selection can be used to enhance crown rust resistance in *L. multiflorum* candidate cultivars. To do so, Syn_2 populations of several candidate cultivars were grown in a replicated field plot trial at three locations and scored for crown rust resistance in 2014. Although disease pressure in the testing period was not very severe, distinct differences among the individual populations were observed (Table 25.1).

The synthetics created by marker assisted selection in addition to phenotypic selection showed significantly improved resistance when compared to populations based on phenotypic selection only (Table 25.1). No significant difference was observed for the population based on parents homozygous for the resistant allele when compared to the population based on heterozygous parents. The XtgFAL03 population based on parents where the R-allele was absent showed a similar level of resistance as the populations based on "Tigris" related germplasm indicating a similar value of both sources of resistance. On the other hand, ecotypes which were continuously selected for crown rust resistance but not intercrossed with a resistant plant showed significantly higher susceptibility (Table 25.1). The initial ecotype populations were evaluated in different trials and multi-annual data analysis was applied to be able to compare the scores of different trials. Compared to the values shown in Table 25.1, unimproved ecotype populations showed a rust score of 6.96. Thus, pure phenotypic selection resulted in a progress of 2.4 rust scores, while MAS resulted in a progress of 5.52.

Conclusions

In this study, single-marker assisted selection proved useful to introgress a novel source of resistance efficiently into breeding germplasm, and to fix it in a candidate variety by selecting homozygous polycross parents. The synthetics based on marker assisted and phenotypic selection were clearly superior regarding crown rust resistance when compared to synthetics based on phenotypic selection only.

References

Boller B, Peter-Schmid M, Tresch E, Tanner P, Schubiger F (2009) Ecotypes of Italian ryegrass from Swiss permanent grassland outperform current recommended cultivars. Euphytica 170:53–65

Dracatos PM, Cogan NOI, Keane PJ, Smith KF, Forster JW (2010) Biology and genetics of crown rust disease in ryegrasses. Crop Sci 50 (5):1605–1624

Kimbeng CA (1999) Genetic basis of crown rust resistance in perennial ryegrass, breeding strategies, and genetic variation among pathogen populations: a review. Aust J Exp Agric 39 (3):361–378

Peter-Schmid MKI, Kölliker R, Boller B (2008) Value of permanent grassland habitats for in-situ conservation of *Festuca pratensis* Huds. and *Lolium multiflorum* Lam. Euphytica 164 (1):239–253

Saha MC, Mian MAR, Eujayl I, Zwonitzer JC, Wang L, May GD (2004) Tall fescue EST-SSR markers with transferability across several grass species. Theor Appl Genet 109 (4):783–791

Sim S, Diesburg K, Casler M, Jung G (2007) Mapping and comparative analysis of QTL for crown rust resistance in an Italian x perennial ryegrass population. Phytopathology 97 (6):767–776

Studer B, Boller B, Bauer E, Posselt UK, Widmer F, Kölliker R (2007) Consistent detection of QTLs for crown rust resistance in Italian ryegrass (*Lolium multiflorum* Lam.) across environments and phenotyping methods. Theor Appl Genet 115 (1):9–17

Part III

Scarcity of Land

Current Status of Feed Quality Breeding and Testing in North America

26

J.G. Robins, B.S. Bushman, and K.B. Jensen

Abstract

Although selection for improved feed quality is an important objective for many North American forage breeding programs, especially for those based in dairy regions, it is generally of secondary importance to improved forage mass, persistence, and stress tolerance. The advent of more efficient and high-throughput phenotyping and genotyping procedures will decrease the cost of feed quality analysis and increase the efficiency of trait improvement. At the FRRL, our main focus is forage mass and abiotic stress tolerance in limited precipitation regions. We also seek to develop forage cultivars with increased feed quality. Current breeding populations and molecular biological tools allow us to better dissect the genetics of these traits and incorporate them into improved cultivars.

Keywords

Carbohydrates • Digestibility • Fiber • Forage • Selection

Introduction

Feed quality is the potential of feed to produce a desired animal response (Ball et al. 2001). Although this definition is straightforward, attempts to improve feed quality through breeding and selection are often not straightforward at all. Feed quality breeding objectives must be balanced with other breeding objectives, such as forage mass, pest resistance and abiotic stress tolerance, among others. Additionally, it can be difficult to determine which specific components of feed quality are most important for the intended use of the end feed. Complications to breeding objectives include negative phenotypic and/or genotypic correlations among agronomic and feed quality traits, complicated inheritance patterns in polyploidy organisms, expense of processing samples for laboratory analyses, development of appropriate equations for estimation with near-infrared reflectance spectroscopy (NIRS), extended time period required to

J.G. Robins (✉) • B.S. Bushman • K.B. Jensen
Forage and Range Research, USDA-ARS, Utah State University, Logan, UT 84322, USA
e-mail: joseph.robins@ars.usda.gov

I. Roldán-Ruiz et al. (eds.), *Breeding in a World of Scarcity: Proceedings of the 2015 Meeting of the Section "Forage Crops and Amenity Grasses" of Eucarpia*,
DOI 10.1007/978-3-319-28932-8_26

complete feed quality analyses, and changing feed quality expectations from animal producers. A successful feed quality breeding program incorporates the genetic potential of the feed quality of the targeted forage species, the feed quality needs of the targeted animal production system, the regional climate, and the relative importance of various forage quality measures, such as the importance of high protein compared to cell wall components.

An important consideration in the implementation of a feed quality improvement program is the target production environment of the improved cultivars. This is important because production of forages often occurs on marginal lands, or lands that possess attributes that limit row crop productivity and economics (Casler 2014). Limitations may include precipitation, soil chemical or physical properties, site topography/geography, and climate, among others. A discussion of the scarcity of land resources is therefore appropriate in the context of forage crops. Forage breeders must address the complications of developing productive and nutritious cultivars under site and management conditions that would not sustain other crop production methods (Casler 2014).

North America follows this trend. Despite the large tracts of arable land in Canada and the United States, much of that arable land is used to produce row crops rather than forages. Another pressure on forage-producing land is the recent rise in farmland and commodity prices, which resulted in replacement of forage production with more profitable row crops, such as maize and canola (Henderson 2008). Breeders in North America therefore focus on developing forages for marginal lands, including vast tracts of grassland/rangeland that receive no external inputs.

The major forage producing regions of the United States and Canada separate into five climate zones. These are (1) the hot, humid southeast United States, (2) the hot, arid southwest United States, (3) the cool, humid northwest United States and southwest Canada, (4) the cool, semi-arid north central United States and south central Canada, and (5) the cool, humid central to northeast United States and southeast Canada. The prevailing climate in each region dictates the potential forage species and corresponding feed quality, although some forage-producing areas in western North America are currently irrigated. Feed quality is further influenced by the prevailing livestock production system in each region.

To assess the feed quality objectives for each of these five regions of North America, we asked for input from forage breeders in each region. Based on their input, we developed a list of breeding priorities for North American forage breeders. Additionally, we discuss an application of the climate zone breeding priorities in the research program of the United States Department of Agriculture-Agricultural Research Service Forage and Range Research Laboratory (FRRL). The FRRL is the only forage breeding program in the Intermountain region of North America. It is responsible for developing improved forages for irrigated and rangeland production in semi-arid cool-season climates.

North American Forage Breeding for Improved Feed Quality

Relevance

Selection for improved feed quality has a long history in forage breeding programs (Casler 2001). Relatively small increases in feed quality result in substantial increases in animal performance (Anderson et al. 1988). But feed quality improvement is not always a high priority among North American forage breeding programs. Individual forage breeding programs place varying levels of emphasis on feed quality improvement, based on the needs of their target production environment and the prevailing livestock production operations in those environments, together with other considerations.

Although improved feed quality is highly relevant to animal production, the forage breeding programs most likely to emphasize this in North America are private alfalfa breeding programs and forage breeding programs in areas where dairy production is a major agricultural activity (i.e., the Great Lakes region and northeast United States/southeast Canada). Other forage breeding programs often select for improved feed quality, but consider it of secondary importance to traits such as forage mass, persistence, and tolerance to abiotic and biotic stresses. These programs often use feed quality as a means to better distinguish breeding material during the selection process. Forage mass alone distinguishes elite breeding material from poor breeding material, but this criterium often fails to distinguish differences among elite materials. Feeding quality parameters provide additional metrics for forage breeders during the selection process. Still other programs take little notice of feed quality. For example, forages produced in the southeast United States are mainly targeted for grazing beef cattle. Under conditions of high heat, humidity and grazing stress, persistence is a more important forage breeding objective than quality (Bouton et al. 1993). Likewise, a major forage base for beef cattle and sheep grazing operations in the semi-arid central and western regions of North America are the non-irrigated grasslands/rangelands (Adams et al. 1996). Under these semi-arid conditions, forage productivity peaks in the weeks following maximum precipitation. Under these conditions, the available forage provides little more than maintenance diets, making forage availability and persistence paramount to quality (Robins et al. 2013; Misar et al. 2015). For forage breeding programs focused on providing consistent forage resource in harsh environments, emphasis on forage quality is unlikely. Further in this paper we discuss the potential of new species to increase feed quality on semi-arid dryland pastures (see below).

Feed Quality Evaluation and Selection

Among those programs that emphasize improved feed quality, decreased neutral detergent fibre (NDF) and increased protein, digestibility, and neutral detergent fiber digestibility (NDFD) are common objectives. For the dairy industry, decreased neutral detergent fibre (NDF) and increased neutral detergent fibre digestibility (NDFD) are particularly important. There is also growing interest to develop forages that contain high concentrations of water soluble carbohydrates (WSC). Less important traits include acid detergent fibre, total digestible nutrients, and lignin, each of which have value for specific uses. Forage breeders want to characterize as many parameters as possible. Improvements in handheld instrumentation may facilitate this capability (Li et al. 2012).

The most important forage species in North America for improving feed quality are alfalfa, tall fescue, timothy, red clover, orchardgrass/cocksfoot and bromegrass (smooth and meadow). Breeding for improved feed quality in other species is not a major objective because they are used for livestock production that requires lower feed quality values, such as beef cattle grazing, or are used in harsh environments, where persistence is critical.

Forage breeders rely on near-infrared spectroscopy (NIRS) for evaluating feed quality (Norris et al. 1976). Previously, each forage program developed its own equations for NIRS prediction. However, the recent development of the NIRS Forage and Feed Testing Consortium points to increased collaboration on the development of relevant NIRS prediction equations for North American forage analysis (http://nirsconsortium.org). Many of the North American private and public forage breeding programs participate in this consortium and hopefully greater standardization of analyses and results will result.

To avoid confounding between maturity and management on feed quality, it is common to exclude first harvest results from feed quality analyses. Nevertheless, this decision does not come without problems, because greater genetic variation is detectable for feed quality during the spring. The tested plant material should also come from replicated sward plots under target environment production conditions. In terms of feed quality, good correspondence is seen in sward vs. space-planted plots (Waldron et al. 2008).

Feed Quality Testing During Candidate Cultivar Evaluation

Evaluation of feed quality during candidate cultivar testing is subject to debate. Proponents of such evaluation wish to know the actual animal performance before variety registration, despite the prohibitive cost and time required for such studies. Feed quality is a suitable and less expensive proxy for animal performance. No statute requires these criteria, however, which leaves feed quality data to the discretion of the breeder. Another issue is whether to heavily weight the importance of the pertinent feed quality parameters when they are used as justification for releasing an improved cultivar. Suggested weightings for feed quality range from 25 to 50 % for candidate cultivar evaluations. Evaluations must account for maturity differences and ensure uniform fertility. There is also a need to define threshold levels of feed quality to protect against declines. The use of economic weights would be ideal, but they are difficult to derive.

All of these are only wishes and suggestions, however. Neither the United States nor Canada have a Value for Cultivation and Use (VCU) system to ensure uniformity in the testing and evaluation process of candidate cultivars. Various universities conduct their own variety trials, and efforts are made to coordinate those trials, but individual breeders are free to choose the trials in which their candidate cultivars participate, and which traits will be measured. For reasons of cost and complexity, most of these coordinated trials evaluate only forage mass and persistence. In many cases, the forage breeders choose to evaluate the candidate cultivars in their own trials.

Drawbacks to Feed Quality Improvement

Does increased feed value comes at a cost of other traits including biomass yield, cold tolerance, and pest resistance? A number of studies identified inverse relationships between these key agronomic traits and feed quality. Nevertheless, a number of programs from across the world successfully breed for both feed value and biomass yield. Many of the studies finding inverse relationship between feed quality and agronomics, such as the correlations with lignin, did not include the correlated traits in their selection program. We must therefore meaningfully evaluate and select for all relevant traits. There is insufficient scientific knowledge to make broad generalizations, particularly across species and environments.

Cost is another drawback of prioritizing feed quality in breeding programs. Selection for improved feed quality is expensive, even with the use of NIRS. While NIRS provides a more rapid and efficient feed quality evaluation, the instrumentation is expensive and the processing of samples is time-consuming. Costs include sampling, drying, grinding/processing, chemical analyses, and associated labor. Despite the recognized importance of ongoing investments into forage breeding (BCRC 2014), funding for this research is diminishing. Lower volumes of seed sales as compared to row crops also limit the return on money invested in research by private forage breeders. Besides the financial cost, the forage quality analyses often require months to process and analyze. This delay precludes rapid selection decisions and leads to a lower emphasis on feed quality. Breeders would likely be more interested in breeding for feed quality if more rapid, accurate, and inexpensive measures were available.

Further Needs for Feed Quality Improvement

Forage breeding in North America is underfunded. As an indirect food crop, forages are not a seamless fit in many competitive grant funding programs in the U.S., and the extra rationalization necessary to submit such grants inhibits their success in highly competitive funding schemes. The cost associated with measurably improving feed quality is prohibitive for most North American forage breeding programs. Many North American breeders feel their efforts are exhausted trying to improve biomass yield and biotic and abiotic stress tolerances. Overall we need faster, more efficient, and less expensive phenotyping methodologies. NIRS is promising, but the labor-intensive sample preparation remains a major impediment to progress.

An Example of Breeding in a Semi-Arid Environment

Rangeland Forage Breeding

The FRRL has a history of developing improved perennial forage cultivars in semi-arid cool-season environments dating back to the "Dustbowl" era in the 1930s. The FRRL breeding program developed many of the common perennial grass species used for rangeland revegetation and forage production on western United States rangelands. This list includes cultivars Hycrest (Asay et al. 1985a) and Hycrest II crested wheatgrass (Jensen et al. 2009a), Vavilov (Asay et al. 1995) and Vavilov II (Jensen et al. 2009a) Siberian wheatgrass, and Bozoisky-Select (Asay et al. 1985b) and Bozoisky II Russian wildrye (Jensen et al. 2006). These cultivars are invaluable under harsh western North American rangeland conditions, particularly when annual precipitation levels are below 250–300 mm and irrigation is unavailable. They establish rapidly, persist, and provide a consistent forage base for livestock producers in these areas. Feed quality was not a consideration during the development of these cultivars. As mentioned above, persistence and forage mass trump feed quality in these harsh target environments. Despite documented genetic variation within these species for feed quality (Jensen et al. 2002), forage improvement is not prioritized for low precipitation areas.

While feed quality will not influence forage production decisions on low precipitation rangeland areas, recent findings suggest that options do exist for improving forage quality on sites receiving more than 300 mm annual precipitation. Ongoing breeding work on meadow bromegrass and intermediate wheatgrass suggest that these species are more drought-tolerant than originally thought (Robins et al. 2013). These species provide better feed quality than the traditionally used wheatgrasses and wildryes on the semiarid sites (Jensen et al. 2001). The meadow bromegrass cultivar 'Arsenal' is the latest release from the FRRL breeding program. It provides excellent forage mass and feed quality under semiarid conditions (Jensen et al. 2015).

Irrigated Forage Breeding

The FRRL forage breeding program also develops improved forage cultivars for irrigated production systems. In the future, grazing may be limited on western rangelands, thus the importance of forage productivity on scarcely-irrigated lands will increase. The FRRL's irrigated forage breeding program includes orchardgrass (the primary focus of our feed quality research), tall fescue, and timothy. Despite the importance of feed quality as a breeding objective, it is less important than forage mass and tolerance to drought and winter injury. Breeding objectives include increasing digestibility and decreasing NDF, but the major focus now is to increase the expression of water-soluble carbohydrates. To do so, we have identified a variety of populations from various sources and coupled these populations (Robins et al. 2015a, 2015b) with molecular biological tools.

Our initial objectives were to develop breeding populations appropriate for applied breeding objectives. We characterized half-sib families and semi-hybrids developed from nine populations or orchardgrass germplasm. We found substantial genetic variation within these populations for forage mass, protein, NDF, digestibility, and water soluble carbohydrates (Robins et al. 2015a). Broad-sense heritability ranged from approximately 0.40–0.70 for the feed quality traits. We also identified mid- and high-parent heterosis for each of these traits. Values of high-parent heterosis ranged from 9 to 30 $g \cdot kg^{-1}$ for the various feed quality traits. Based on these results, we made selections and developed advanced breeding populations that we are now evaluating in multi-location trials with continuing recurrent selection. Our current focus is to develop late maturing cultivars with high WSC. The later maturing cultivars will possess higher feed quality. We also anticipate that the water soluble carbohydrates will increase animal performance while simultaneously improving drought tolerance and winter injury (Sanada et al. 2007). Ongoing association mapping and genotype-by-sequencing projects are providing candidate genes and functional mutations for marker-assisted selection (Xie et al. 2012; Bushman, unpublished data). We hope to use the functional markers to enhance our ability to move toward phenotype stability in an efficient timespan.

Inheritance of water soluble carbohydrates does not happen in a straightforward manner. In collaboration with Aberystwyth University (UK), we evaluated the water soluble carbohydrate concentration of high carbohydrate perennial ryegrass cultivars under varying irrigation levels in the United States and at sites in Wales and Scotland (Robins and Lovatt 2015). We found substantial genotype-by-environment interaction for the expression of water soluble carbohydrates across the locations and years. There were limited changes in ranking of the included cultivars at the United Kingdom sites and the higher irrigation Utah sites. However, we found changes in ranking when comparing the low irrigation Utah sites to the other sites. The Aberystwyth cultivars are generally well-adapted to high production environments. Expression of the water-soluble carbohydrate trait was nearly double at the United Kingdom sites when compared to the Utah locations. The reasons for this discrepancy are not completely obvious, but we hypothesize that the hotter and sunnier Utah summer conditions limit the ability of the cool-season perennial ryegrass to maximize photosynthetic efficiency.

To more efficiently measure feed quality, we now utilize handheld NIRS technology in the field. We are in the process of testing the technology, comparing it to laboratory-based NIRS, and developing appropriate prediction equations. Our hope is that the handheld NIRS will replace the laboratory-based NIRS for our forage breeding program, making the selection process faster and less expensive.

References

Adams, DC, Clark RT, Klopfenstein TJ, Volesky, JD (1996) Matching the cow with forage resources. Rangelans 18:57–62.

Anderson B, Ward JK, Vogel KP, Ward MG, Gorz HJ, Haskins FA (1988) Forage quality and performance of yearlings grazing switchgrass strains selected for differing digestibility. Journal of Animal Science 66:2239–2244.

Asay KH, Dewey DR, Gomm FB, Johnson DA, Carlson JR (1985a) Registration of 'Hycrest' crested wheatgrass. Crop Science 25:368–369.

Asay KH, Dewey DR, Gomm FB, Johnson DA, Carlson JR (1985b) Registration of 'Bozoisky-Select' Russian wildrye. Crop Science 25:575.

Asay KH, Johnson DA, Jensen KB, Chatterton NJ, Horton WH, Hansen WT, Young SA (1995) Registration of 'Vavilov' Siberian wheatgrass. Crop Science 35:1510.

Ball DM, Collins M, Lacefield GD, et al. (2001) Understanding forage quality. American Farm Bureau Federation Publication 1-01, Park Ridge, IL, USA.

Beef Cattle Research Council (2014) Priority area review: forage and grassland productivity. pp. 1–48.

Bouton, JH. Gales RN, Belesky DP, Owsley (1993) Yield and persistence of tall fescue in the Southeastern Coastal Plain after removal of its endophyte. Agronomy Journal 85:52–55.

Casler MD (2001) Breeding forage crops for increased nutritional value. Advances in Agronomy 71:51–107.

Casler, M (2014) Forage breeding programs aimed at increasing productivity of marginal lands. J. Anim. Sci Vol. 92, E-Suppl. 2/J. Dairy Sci. Vol. 97, E-Suppl. 1:155.

Henderson, J (2008) Will farmland values keep booming? Economic Review 93:81:104.

Jensen KB, Asay KH, Johnson DA, Larson SR, Waldron BL, Palazzo, AJ (2006) Registration of 'Bozoisky-II' Russian wildrye. Crop Science 46:986.

Jensen KB, Johnson DA, Asay KH, Olson KC (2002) Seasonal-accumulated growth and forage quality of range grasses for fall and winter grazing. Canadian Journal of Plant Science 82:329–336.

Jensen KB, Asay KH, Waldron BL (2001) Dry matter production or orchardgrass and perennial ryegrass at five irrigation levels. Crop Science 41:479–487.

Jensen KG, Larson SR, Waldron BL, Robins JG (2009a) 'Hycrest II', a new crested wheatgrass cultivar with improved seedling establishment. Journal of Plant Registrations 3:57–60.

Jensen KG, Singh D, Bushman BS, Robins JG (2015) Registration of 'Arsenal' meadow bromegrass. Journal of Plant Registrations doi:10.3198/jpr2015.03.0018crc.

Li F, Mistele B, Hu Y, Yue X, Yue S, Miao Y, Chen X, Cui Z, Meng Q, Schmidhalter U (2012) Remotely estimating aerial N status of phonologically differing winter wheat cultivars grown in contrasting climatic and geographic zones in China and Germany. Field Crops Research 138:21–32.

Misar CG, Xu L, Gates RN, Boe A, Johnson P (2015) Stand persistence and forage yield of 11 alfalfa (Medicago sativa) populations in semiarid rangeland. Rangeland Ecology and Management 68:79–85.

Norris KH, Barnes RF, Moore JE, Shenk JS (1976) Predicting forage quality by infrared reflectance spectroscopy. Journal of Animal Science 43:889–897.

Robins JG, Bushman BS, Escribano S, Jensen KB (2015a) Heterosis for protein, digestibility, fiber, and water soluble carbohydrates in nine sources of orchardgrass germplasm. Euphytica 204:503–511.

Robins JG, Bushman BS, Jensen KB, Escribano S, Blaser G (2015b) Genetic variation for dry matter yield, forage quality, and seed traits among the half-sib progeny of nine orchardgrass germplasm populations. Crop Science 55:275–283.

Robins JG, Jensen KB, Jones TA, Waldron BL, Peel MD, Rigby CW, Vogel KP, Mitchell RB, Palazzo AJ Cary TJ (2013) Stand establishment and persistence of perennial cool-season grasses in the Intermountain West and the Central and Northern Great Plains. Rangeland Ecology & Management 66:181–190.

Robins JG, Lovatt JA (2015) Cultivar by environment effects of perennial ryegrass cultivars selected for high water soluble carbohydrates managed under differing precipitation levels. Euphytica doi: 10.1007/s10681-015-1607–1609.

Sanada Y, Takai T, Yamada T (2007) Ecotypic variation of water-soluble carbohydrate concentration and winter hardiness in cocksfoot (Dactylis glomerata L.). Euphytica 153:267–280.

Waldron BL, Robins JG, Peel MD, Jensen KB (2008) Predicted efficiency of spaced-plant selection to indirectly improve tall fescue sward yield and quality. Crop Science 48:443–449.

Xie W, Robins JG, Bushman BS (2012) A genetic linkage map of tetraploid orchardgrass (Dactylis glomerata L.) and quantitative trait loci for heading date. Genome 55:360–369.

Important Feeding Value Parameters in Ruminant Nutrition

27

J.L. De Boever and S. De Campeneere

Abstract

The main priority in breeding forages is attaining high yields and resist disease. Highly productive animals, scarcity of land and natural resources and increasingly strict environmental protections put more and more emphasis on the quality of the feed. Interest in the implementation of feed evaluation in grass/legume breeding and variety testing is growing. Therefore, it is important to know which feed quality parameters have to be taken into account when formulating a ration for ruminants, particularly for dairy cows. The main quality parameters of forages, i.e., energy content, the protein value and voluntary intake, are discussed using own experimental data from fresh grass, grass and grass clover silages as well as published data. Additionally, the importance of structure value, water soluble carbohydrates, fats, trace elements and vitamins is discussed.

Keywords

Grass • Clover • Energy value • Protein value • Voluntary intake • Structure value • Digestibility • WSC • Fat

Introduction

Grasses and legumes have great potential as forage for ruminants, because they can be grown on farm and are low-cost, protein-rich feeds. Grazing these crops is positive for the health and well-being of the animals. Another advantage is that animal manure can (partly) be recycled on the land. The main negative points of these forages are the variable nutritive value when grazed, the general low efficiency of protein utilization by the ruminant, particularly after ensiling, and the relatively high methane emissions. When forages are incorporated in the ration, several parameters have to be taken into account like the energy value, the protein value, voluntary intake, the structure value, the contents of sugars, fat, minerals, vitamins and possibly anti-nutritional factors. The aim of this paper is to examine in more detail

J.L. De Boever (✉) • S. De Campeneere
ILVO (Institute for Agriculture and Fisheries Research), Animal Sciences Unit,
Melle 9090, Belgium
e-mail: johan.deboever@ilvo.vlaanderen.be

I. Roldán-Ruiz et al. (eds.), *Breeding in a World of Scarcity: Proceedings of the 2015 Meeting of the Section "Forage Crops and Amenity Grasses" of Eucarpia*,
DOI 10.1007/978-3-319-28932-8_27

how these parameters vary within forages and what the factors of influence are. This discussion is based on experiments at the Institute for Agricultural and Fisheries Research (ILVO) with mainly grasses, grass silages and grass-clover silages as well as data from literature.

Feed Quality Parameters

Energy Value

One of the most important functions of a feed is energy provision. Dairy cows need energy for maintenance, milk production and growth and reproduction. Energy is often the limiting factor for high producing cows at the onset of lactation: an energy shortage depresses milk production and milk protein content. Moreover, a negative energy balance (more energy expenditure than energy intake) at the start of lactation may result in ketonemia and lower fertility. The energy content of feeds for ruminants is expressed in either metabolizable energy (ME) or net energy (NE). To obtain ME, energy losses through fermentation gasses in the rumen (CO_2, CH_4), through feces and urine have to be subtracted from the gross energy (heat of combustion) of the feed. To obtain NE, the heat losses during metabolism also have to be taken into account. The ME of ruminant feeds is mostly determined in digestibility trials with sheep. ILVO has been performing digestion trials for 40 years. Based on the contents and digestibility of crude protein (CP), crude fibre (CF), crude fat and the other carbohydrates (calculated as 100 – CP – CF – fat – crude ash) the net energy value was calculated according to the Dutch energy evaluation system (Van Es 1978). In this study, the data from six digestion trials with fresh grass, 66 trials with grass silage and 17 trials with grass-clover silage are discussed. In Table 27.1 the mean and standard deviation (SD) of the chemical composition (g/kg DM), of the *in vivo* digestibility of the organic matter (DC_{OM}) and of the net energy lactation (NEL) is presented. The NEL is expressed on an OM basis, to avoid the confounding effect of the variable crude ash content.

On average in our data set, half of the dry matter of fresh and ensiled grass consisted of cell walls; grass-clover silages had a somewhat lower percentage. In fresh grass, hemicellulose had a greater contribution to the cell walls than cellulose, whereas the reverse was observed for silages from both grasses and grass-clover. Fowler et al. (2003) mentioned that during ensiling of grasses hemicellulose content decreased by 3–11 %, the cellulose content remained unchanged, and lignin

Table 27.1 Chemical composition (g/kg DM), *in vivo* digestibility and net energy lactation of fresh grass, grass silages and grass clover silages examined at ILVO

	Fresh grass n=6	Grass silage n=66	Grass clover silage[a] n=17
Dry matter (g/kg)	160±15	323±145	424±75
NDF (total cell walls)	565±27	509±62	449±75
Hemicellulose (NDF-ADF)	309±20 (55 %)[b]	213±35 (42 %)[b]	151±46 (34 %)[b]
Cellulose (ADF-ADL)	220±29 (39 %)[b]	268±32 (53 %)[b]	256±40 (57 %)[b]
ADLignin	37±15 (6 %)[b]	28±10 (6 %)[b]	42±19 (9 %)[b]
Crude protein	209±60	144±32	163±30
Water Soluble Carbohydrates	89±35	22±34	39±24
Lactic ac. + VFA + Alcohols	–	97±47	65±23
Crude fat	41±10	42±9	29±8
Crude ash	108±12	162±72	131±32
In vivo DC_{OM} (%)	80.4±6.1	74.2±5.8	70.5±5.4
NEL_{OM} (MJ/kg DM)	7.35±0.76	6.83±0.71	6.19±0.67

[a]9 grass-white clover silages (clover portion between 20 and 60 %) and 8 grass-red clover silages (30–90 %)
[b]Between brackets: as percentage of NDF

content increased by 23 %. The increase of lignin by ensiling was not observed in our data. Crude protein amounted to about 20 % of DM in fresh grass and was on average 6 % points higher than in ensiled grass, despite the lower cell wall content in the latter. This can partly be explained by volatilization of ammonia from silages during oven drying. Further, the fresh grass contained almost 10 % water soluble carbohydrates (WSC). The sugars were almost completely fermented during ensiling, giving rise to mainly lactic acid, and to a lesser extent acetic acid and alcohols. The crude fat content was on average similar for fresh grass and silage. Crude ash content amounted to 11 % in fresh grass, but was some 5 % points higher in grass silage, which is explained by contamination with soil. Fresh grass was on average more digestible than silage and grass silage appeared to be more digestible than grass clover silage. This ranking order was also reflected in the net energy value.

To examine the feed characteristics which affect the energy value, correlations with most important nutrients were calculated for the three groups of forages (Table 27.2).

The energy value was highly correlated (r >0.94) with the *in vivo* digestibility of the organic matter for the three groups of forages. This strong relationship can mainly be explained through the digestibility of the cell walls, particularly for fresh and ensiled grass. From the cell wall parameters and components, only ADLignin showed a strong negative correlation with the energy value for the three groups of forages. For grass-clover silage, crude fat content also showed a high positive correlation with the NEL content.

Table 27.2 Correlations of chemical parameters and *in vivo* digestibility of OM and of crude fiber with NEL_{OM}

	Fresh grass	Grass silage	Grass clover silage
NDF	−0.82	−0.44	−0.65
ADF	−0.83	−0.27	−0.87
ADLignin	−0.92	−0.71	−0.83
Hemicellulose	ns	−0.49	ns
Cellulose	ns	ns	−0.71
WSC	ns	ns	ns
Crude protein	ns	ns	ns
Crude fat	ns	0.67	0.93
In vivo DC_{OM}	0.98	0.98	0.99
In vivo DC_{CF}	0.98	0.90	0.75

Protein Value

Proteins, or more specifically amino acids, are important building blocks for the body, particularly for the muscle tissue (meat) as well as for the bones and the hair (keratin). For lactating cows they are important for the formation of milk protein. Until 1991 the Digestible Crude Protein system (DCP: the difference between CP ingested with the feed and CP excreted with the feces) was used to express the protein value of ruminant feeds as well as the protein requirements of the animals. Protein that can be effectively utilized is only a part of the ingested CP, which is digested in and subsequently absorbed from the small intestine (SI) as amino acids. The DCP system is a poor predictor of the amount of true protein absorbed from the SI, because it does not indicate to which degree the CP in a feedstuff is degraded in the rumen, nor does it take into account microbial protein synthesis in the rumen. These are important limitations because CP that is degraded in the rumen and lost as ammonia and urea, cannot be utilized by the animal. Therefore, in many countries more sophisticated systems describing the digestion and metabolism of N in a more detailed way than the DCP system were introduced in practice. In Belgium, we apply the Dutch protein evaluation system (Tamminga et al. 1994), which was recently updated by Van Duinkerken et al. (2011).

In this system two values are considered: the true protein digested in the intestines (PDI) and the degraded protein balance (DPB). These values are described in more detail below to give better insight into the factors of influence.

The PDI is calculated according to the formula: **PDI = DRBP + DMP − EPL** with

(a) DRBP: digestible rumen bypass protein calculated as DRBP = CP × %RBP × %dRBP and %RBP: the percent of feed protein bypassing rumen degradation, estimated from incubation of feed in nylon bags incubated in the rumen of cannulated animals

%dRBP: the digestibility of that bypass protein in the intestines, estimated from incubation of rumen bypass feed residue kept in nylon bags that pass through the digestive tract

(b) DMP: digestible microbial protein calculated as DMP = MPE × 0.75 × 0.85

with MPE: the microbial protein synthesized in the rumen based on available energy with a variable growth efficiency depending on the substrate: 174 g MP per kg sugars, 138 g MP per kg cell walls, 99 g MP per kg crude protein

0.75 = the ratio of amino acid N to total N

0.85 = the true digestion of AAN in the intestines

(c) EPL: endogenous protein loss (due to the digestive process) calculated as 75 g per kg undigestible dry matter

The DPB shows the (im) balance between microbial protein synthesis potentially possible from available rumen degradable CP (MPN) and that potentially possible from the energy extracted during anaerobic fermentation in the rumen (MPE). It is calculated using the formula: **DPB = MPN − MPE**

When positive, the DPB value gives the loss of N from the rumen; when negative, microbial protein synthesis may be impaired due to a shortage of N in the rumen. The RBP value in a ration should at least be zero, but in practice a surplus of 150–200 g per day is recommended (De Brabander et al. 2011).

In this study the data from ten batches of fresh grass, 25 grass silages and 14 grass clover silages are discussed (Table 27.3).

From Table 27.3 it is clear that crude protein content, which is mostly used by grass breeders to evaluate protein value, is much higher than the content of digestible protein, by an average factor of 2.3 for fresh grass and by a factor of 2.8–2.9 for the silages. In fact, CP only affects the first component of PDI (i.e., the rumen bypass protein) and this effect is rather weak due to the negative relation between CP content and the rumen bypass percentage. Further, it appears that for the studied forages the contribution of microbial protein is larger than that of feed protein bypassing the rumen. The lower PDI content of grass silage as compared with fresh grass can partly be explained by the lower CP content, but mainly by the lower rumen bypass percentage as well as by the lower intestinal digestibility. The rumen bypass percentage is lower for grass silage than for fresh grass, because of extensive proteolysis in the silo, particularly when grass is not or less wilted. The percentage of RBP of grass clover silage is clearly higher for grass silage. The higher rumen bypass of white clover protein may be explained by the presence of saponins (Das et al. 2012), whereas that of red clover can be

Table 27.3 Protein value of fresh grasses, grass silages and grass clover silages examined at ILVO

	Fresh grass n = 10	Grass silage n = 25	Grass-clover silage[a] n = 17
Crude protein (g/kg DM)	170 ± 33	156 ± 31	174 ± 33
PDI (g/kg DM)	75 ± 6	56 ± 9	60 ± 9
DRBP (g/kg DM)	43 ± 9	31 ± 6	42 ± 7
DMP (g/kg DM)	50 ± 9	44 ± 5	43 ± 5
EPL (g/kg DM)	18 ± 9	20 ± 3	25 ± 4
%RBP	30.6 ± 3.5	25.9 ± 4.2	34.1 ± 5.4
%dRBP	84.9 ± 2.5	79.0 ± 5.0	72.1 ± 3.8
DPB (g/kg DM)	41 ± 32	48 ± 28	48 ± 28
MPN (g/kg DM)	119 ± 28	117 ± 27	115 ± 29
MPE (g/kg DM)	78 ± 5	69 ± 8	67 ± 8

[a]9 grass-white clover silages (clover portion between 20 and 60 %) and 8 grass-red clover silages (30–90 %)

ascribed to polyphenol oxidase (Lee 2014). In contrast, intestinal digestibility of the bypass protein is on average lower for grass clover than for grass silage. All three groups of forages are characterized by a surplus of available N in the rumen for microbial protein synthesis as compared with available energy, resulting in a strongly positive RPB value. The RPB value expresses the balance on a daily basis, which may hide large hourly fluctuations, particularly if roughages and concentrates are fed separately and twice daily. This means that shortly after feeding grass silage, the discrepancy between MPN and MPE is even more pronounced.

Voluntary Intake

Efficient feeding of dairy cows relies on a good estimate of their feed intake. Particularly knowledge of roughage intake is important, as it allows to calculate the amount and nature of the concentrates which should be supplemented in order to meet the animal requirements and to avoid excessive nutrients, which are expensive and overload the environment. Moreover, one should aim to maximize the intake of roughages, because they are cheaper than concentrates. The regulation of feed intake is very complex and concerns physical, physiological as well as thermostatic control (De Brabander et al. 2011). Further, the voluntary intake (VI) depends on animal related, feed related and environmental factors. The most important animal factors are breed, genetic predisposition, body weight, age, production level, lactation stage and body condition (fat). The most important environmental factor is temperature. Feed factors such as the quality of the ration, the rumen fermentation, the structure value, the DM content, the feeding strategy and water availability all may play a role. According to Balch and Campling (1969) the VI of roughages is determined by the rate of disappearance from the rumen, which in its turn is a function of the degradation rate by the combined action of microbial fermentation and mechanical breakdown through eating, rumination and rumen contractions. At ILVO between 1970 and 1985, De Brabander et al. (1985) carried out many intake trials mainly with Holstein and Belgian White-Red cows. In 23 trials with wilted grass silage they obtained a mean VI of 12.2 kg DM per day, varying between 9.8 and 14.2 kg. The best chemical parameter to predict VI was crude fiber (CF) content; per percent of higher CF content in the DM, VI decreased with 0.24 kg DM. Voluntary intake of forages clearly decreases with later growth stage of the plant. From these data also appeared a significant increase of VI with 0.17 kg DM per 1 % point increase in OM digestibility. A similar finding was done by Thomas (1980) in a review of dairy cow experiments with well-preserved grass silages, who calculated that per 1 % higher DOMD, DM intake increased with 0.15 kg. Oba and Allen (1999) statistically evaluated dairy cow performances from 13 sets of forage comparisons in literature and found that a 1-point increase in NDF digestibility *in vitro* or *in situ* was associated with a 0.17 kg increase in DMI. Further, they stated that *in vitro* or *in situ* NDFD might be a better indicator of DMI than *in vivo* NDFD, because forages with high *in vitro* or *in situ* NDFD might have shorter rumen retention times, allowing greater DMI at the expense of NDF digestibility *in vivo*. In that respect, they mentioned that ruminal retention time of NDF from perennial grasses is generally longer than for legume NDF, despite being less lignified, resulting in a lower DMI of the former.

Concerning the effect of grass quality on enteric methane emission, Bannink et al. (2010) made a theoretical study of microbial fermentation processes in the rumen and large intestine of cows. At a feed intake level of 18 kg DM/day, including 90 % grass product and 10 % concentrates, the simulated amount of CH_4 emitted per kg feed DM was higher with low N fertilization compared to high N fertilization, irrespective of the unit of expression (g CH_4 per day, per kg DM or per kg milk). A late-cut grass silage resulted in higher (with high N fertilization) or equal (with low N fertilization) CH_4 emission per kg DM; but in both cases more CH_4 with late-cut grass silages when expressed per kg milk.

Structure Value

For an efficient ruminant production, an optimal functioning of the rumen is required, which is characterized by strong and frequent rumen contractions, an average pH of rumen liquor around 6.4 and never lower than 6, a ratio of acetic to propionic acid of at least 3, high digestibility of cell walls, high passage rate of nutrients, the presence of a structural layer and a maximal microbial protein synthesis (De Brabander et al. 2011). ILVO developed a structure evaluation system based on chewing activity measurements with cows and dairy cow experiments in which the critical roughage portion in the ration was investigated (De Brabander et al. 1999). In this system a standard cow producing 25 kg milk with 4.4 % fat requires a structure value (SV) in the ration of at least 1 per kg DM. The SV of grassland products clearly increases with the growth stage; for grass silage the following relationship was obtained: SV=0.15+0.0060 NDF (g/kg DM). It is clear that grass silage as such, even harvested in a very young stage or chopped finely, always will provide enough physical structure. Also for solely fresh grass, lack of physical structure will hardly occur, as the minimum SV value from six observations amounted to 1.85 per kg DM. A risk may occur with high-sugar grasses, because sugars are fermented very rapidly in the rumen, resulting in a low pH. However, Taweel (2004) observed no effect on rumen pH, nor on degradation or clearance rate of NDF with cows fed or grazing perennial ryegrass cultivars differing in WSC; only the composition of volatile fatty acids (VFA) was altered in favour of propionate at the expense of acetate when feeding high sugar grass. The author explained the small effects of WSC by the relative slow ingestion of fresh grass and the subsequent slow release of sugars from the cell contents in the rumen.

Water Soluble Carbohydrates (WSC)

Contrary to the expectations that a high sugar content in grass improves its sweetness and thus palatability – resulting in a higher DMI – Taweel (2004) found no differences in intake between cows fed or grazing perennial ryegrasses differing in WSC. Apart from the small effects on VFA composition (see above), ruminal ammonia concentration and milk urea content were reduced as a result of feeding high WSC grass, which may also be explained by the lower N content of the high sugar grass. Further, milk yield and milk composition were not significantly affected. Considering that in grass products, rumen microbes lack adequate supply of energy (carbon) for growth relative to the large amount of available N, Miller et al. (2001) proposed that increasing the sugar content of grasses could rectify the imbalance, thereby increasing efficiency of N utilization by microbes, decreasing N loss from the animal and increasing the supply of protein to the ruminant. However, when reviewing published results, Ellis et al. (2011) found considerable variation in the magnitude of the responses. To explain this variation, they used a dynamic mechanistic model to predict observed N and milk yield results from the literature and from simulated data. The greatest benefit in terms of N utilization and urine N levels were seen when the WSC content of grass increased at the expense of CP and less pronounced by a trade for NDF. Simulated milk yield decreased slightly when WSC increased at the expense of CP but increased strongly when WSC increased at the expense of NDF. In a subsequent study concerning the effects on methane emission, Ellis et al. (2012) found a higher CH_4 production per kg milk when WSC increased at the expense of N, whereas CH_4 production per kg milk was lower when WSC replaced NDF.

Another possible advantage of grass with more WSC is that the resulting silage is characterized by a lower pH and ammonia fraction (Downing and Gamroth 2007). On the other hand, a high residual sugar content in the silage may increase the risk for heating and mold growth when opening the silo.

Fats

Fat is required to provide the animal with essential fatty acids (linoleic, linolenic acid) and because it contains fat-soluble vitamins. Moreover, fat has a high caloric value and is often added to the ration to increase energy content. However, too much fat, particularly when it is unsaturated, may be harmful for rumen function as it may depress cell wall degradation and hence feed intake. The total concentration of fat in the ration should be limited to maximum 6 % on DM, knowing that an unsupplemented ration usually contains about 3–3.5 % fat. Apart from the fat content, the nature of the fat provided in the diet affects the fatty acid composition of milk and meat. Enriching ruminant products with polyunsaturated fatty acids (PUFAs) receives a great deal of attention.

The lipid fraction in leaves of herbs and grasses ranges from 30 to 100 g/kg DM, much of which is contributed by chloroplast lipids (Bauchart et al. 1984). Sources of variation in lipid content are plant species, growth stage (leaf/stem ratio), temperature and light intensity. Approximately 95 % of the triglycerides consists of linolenic, linoleic and palmitic acid (Hawke 1973). However, lipids in plants are not static entities, but are continuously subject to turnover. Thus, immediately after grazing or cutting, lipolysis occurs by plant lipases. Dewhurst and King (1998) found that wilting prior to ensiling reduced the content of total fatty acids by almost 30 %, with a reduction of up to 40 % for linolenic acid. In the rumen between 75 and 90 % of linoleic (C18:2 cis-9, cis-12) and between 85 and 95 % of α-linolenic acid (C18:3 cis-9, cis-12, cis-15) are subjected to microbial bio-hydrogenation (Elgersma et al. 2006). Furthermore, these authors state that a varying portion of dietary PUFA, ranging from less than 2 % to over 20 %, is recovered in the milk as vaccenic acid, which has a fairly constant ratio of 2–2.5 to rumenic acid (CLA). Both are products of bio-hydrogenation. CLA has been associated with anticarcinogenic properties in rats and possibly in humans.

Minerals, Trace Elements and Vitamins

Minerals, trace elements and vitamins have diverse functions in the organism. De Brabander et al. (2011) considered the provision by feeds of these minor components in relation to the requirements of dairy cows and concluded that grass products provide mostly sufficient calcium and phosphorus. Further, they often contain too much potassium, which may thin the feces. There may be lack of magnesium and sodium, but this can be corrected by fertilization. Some trace elements like cupper, iodium, selenium and cobalt may be marginal. Grass is rich in β-carotene and also contains a lot of vitamin E. The latter is known as anti-oxidant, which is desirable considering the high content of unsaturated fat in grass.

Conclusions

Forage with a better digestibility has several positive effects on the feeding value, by increasing the energy content and voluntary intake and by decreasing endogenous protein losses. This multiplicative effect is also reflected in better animal performances. Thomas (1980) found an increase of 0.29 kg milk per 1 % point increase in DOMD of well-preserved grass silages. Oba and Allen (1999) calculated that an increase of forage NDFD with 1 % point resulted in 0.25 kg more fat-corrected milk. Keady et al. (2013) reviewed experiments with grass silage for different ruminant systems in Ireland and concluded that digestibility is the most important factor influencing the feeding value of grass silage and consequently of the performance of animals offered grass-silage based diets. They calculated that each 1 % point increase in DMD resulted in:

- Daily milk yield of dairy cows: +0.33 kg
- Daily carcass gain of beef cattle: +23.8 g
- Daily carcass gain of finishing lambs: +9.3 g
- Lamb birth weight: +52.3 g
- Ewe weight post lambing: +1.3 kg

In addition to high digestibility, a high content of WSC, particularly when this replaces part of the cell walls, is interesting for a better N efficiency, higher milk production as well as lower methane production. Finally, grass products with more unsaturated fat and measures to reduce bio-hydrogenation could improve the health-promoting value of milk and meat.

References

Balch CC, Campling RC (1969) Voluntary intake of food. In Handbuch der Tierernahrung. pp. 554–579.

Bannink A, Smits MCJ, Kebreab E, Mills JAN, Ellis JL, Klop A, France J, Dijkstra J (2010) Simulating the effects of grassland management and grass ensiling on methane emission from lactating cows. Journal of Agricultural Science (Cambridge) 148: 55–72.

Bauchart D, Vérité R, Rémond B (1984) Long-chain fatty acid digestion in lactating cows fed fresh grass from spring to autumn. Canadian Journal of Animal Science 64 (Suppl.): 330–331.

Das TK, Banerjee D, Chakraborty D, Pakhira MC, Shrivastava B, Kuhad RC (2012) Saponin: Role in Animal system. Veterinary World 5(4): 248–254.

De Brabander DL, Andries JI, Boucque ChV, Buysse FX (1985) Prediction of voluntary intake of maize silage and preserved grassland products by dairy cattle. Book of Abstracts of the 36th Annual Meeting of EAAP, Kallithea-Haldiodiki (Greece), NC 4.13, pp. 324–325.

De Brabander DL, De Boever JL, Vanacker JM, Boucqué ChV, Botterman SM (1999). Evaluation of physical structure in dairy cattle nutrition. In PC Garnsworthy, J Wiseman, eds, Recent Advances in Animal Nutrition 1999, Nottingham University Press, pp. 111–145.

De Brabander D, De Campeneere S, Ryckaert I, Anthonissen A (2011) Brochure Melkveevoeding. ILVO Mededeling 101, Beleidsdomein Landbouw en Visserij, Vlaamse Overheid, 112 p.

Dewhurst RJ, King PJ (1998) Effects of extended wilting, shading and chemical additives on the fatty acids in laboratory grass silages. Grass and Forage Science 53 (3): 219–224.

Downing T, Gamroth M (2007) Nonstructural carbohydrates in cool-season grasses. Oregon State University Extension Special report 1079, 13 p.

Elgersma A, Tamminga S, Ellen G (2006) Modifying milk composition through forage. Animal Feed Science and Technology 131: 207–225.

Ellis JL, Dijkstra J, Bannink A, Parsons AJ, Rasmussen S, Edwards GR, Kebreab E, France J (2011) The effect of high-sugar grass on predicted nitrogen excretion and milk yield simulated using a dynamic model. Journal of Dairy Science 94: 3105–3118.

Ellis JL, Dijkstra J, France J, Parsons AJ, Edwards GR, Rasmussen S, Kebreab E, Bannink A (2012) Effect of high-sugar grasses on methane emissions simulated using a dynamic model. Journal of Dairy Science 95: 272–285.

Fowler PA, McLauchlin AR, Hall LM (2003) The potential industrial uses of forage grasses including Miscanthus. Paper from the BioComposites Centre, University of Wales, Bangor (UK), 37 p.

Hawke JC (1973) Lipids. In GW Butler, RW Bailey, eds, Chemistry and Biochemistry of Herbage. Academic Press, London (UK), pp. 213–263.

Keady TWJ, Hanrahan JP, Marley CL, Scollan NB (2013) Production and utilization of ensiled forages by beef cattle, dairy cows, pregnant ewes and finishing lambs – A review. Agricultural and Food Science 22 (1): 70–92.

Miller LA, Moorby DR, Davies DR, Humphreys MO, Scollan ND, MacRae JC, Theodorou MK (2001) Increased concentration of water-soluble carbohydrates in perennial ryegrass (Lolium perenne L.): Milk production from late-lactation dairy cows. Grass and Forage Science 56: 383–394.

Lee MRF (2014) Forage polyphenol oxidase and ruminant livestock nutrition. Frontiers in Plant Science 5, article 694, 9 p.

Oba M, Allen MS (1999) Evaluation of the importance of the digestibility of neutral detergent fiber from forage: effects on dry matter intake and milk yield of dairy cows. Journal of Dairy Science 82 (3): 589–596.

Tamminga S, Van Straalen WM, Subnel APJ, Meijer RGM, Steg A, Wever CJG, Blok MC (1994) The Dutch protein evaluation system: the DVE/OEB-system. Livestock Production Science 40: 139–155.

Taweel HZH (2004) Perennial ryegrass for dairy cows: grazing behaviour, intake, rumen function and performance. Ph.D. Thesis, Wageningen University (The Netherlands), 156 p.

Thomas C (1980) Conserved forages In WH Broster, CL Johnson, JC Taylor, eds, Feeding Strategies for Dairy Cows, Agricultural Research Council, London (UK), pp. 8.1–8.14.

Van Duinkerken G, Blok MC, Bannink A, Cone JW, Dijkstra J, Van Vuuren AM, Tamminga S (2011) Update of the Dutch protein evaluation system for ruminants: the DVE/OEB2010 system. Journal of Agricultural Science 149: 351–367.

Van Es AJH (1978) Feed evaluation for ruminants. I. The system in use from May 1977 onwards in The Netherlands. Livestock Production Science 5: 331–345.

Phalaris arundinacea L.: Evaluation of Accessions and Cultivars in Terms of Shattering Reduction and Seed Yield Improvement

28

K. Förster, C. Wittmann, and U. Feuerstein

Abstract

The focus of this investigation was an assessment of the seed shattering potential of 32 gene bank accessions and cultivars of *Phalaris arundinacea* L. (reed canary grass) (GBIS/I, GRIN). For five plants per accession, seed losses (natural and technically provoked shattering) and harvestable seed yield were determined. The ratio of the seed yield to the total grown seed mass varies from 0.2 % (total seed shattering) to 94.2 % (high shattering resistance). Shattering resistance is a significant feature of the individual accessions. In addition to this trait, the usable seed yield is determined by the magnitude of seed setting per plant. Significant differences in seed setting were also observed in the investigated accessions.

Keywords

Reed canary grass • Shattering reduction • Seed yield

K. Förster (✉) • C. Wittmann
Crop Physiology, Institute of Agricultural and Nutritional Sciences, Martin Luther University Halle-Wittenberg,
Halle (S.) D-06099, Germany
e-mail: karin.foerster@landw.uni-halle.de

U. Feuerstein
Deutsche Saatveredelung AG (DSV),
Steimker Weg 7, Asendorf D-27330, Germany

Introduction

Use of biomass for energy production has led to an increased interest in reed canary grass, a perennial rhizomatous grass species (Lewandowski et al. 2003). Reed canary grass (RCG) has a high biomass yielding potential. It is better adapted to wet and dry growing conditions, and it is more tolerant of

I. Roldán-Ruiz et al. (eds.), *Breeding in a World of Scarcity: Proceedings of the 2015 Meeting of the Section "Forage Crops and Amenity Grasses" of Eucarpia*,
DOI 10.1007/978-3-319-28932-8_28

flooding events than other grass species. One major drawback is that RCG seeds are prone to shattering, even when they are not yet ripe for combine harvesting. For biomass production using RCG, the challenge of increased seed yield must first be addressed. The aim of this study was to evaluate seed yielding potential and shattering behavior of Gene bank accessions and cultivars under the conditions of the dry region of central Germany.

Materials and Methods

Plant Material and Experimental Design

A total of 32 reed canary grass accessions (Table 28.1) were evaluated in two field trials carried out at the experimental research station of the University Halle-Wittenberg in Etzdorf (Saxony-Anhalt, Germany; 51°26'05 N, 11°45'18 E). The site is characterized by soil type Chernosem, a mean annual precipitation of 459 mm, and a mean annual temperature of 9 °C. Fertilizer was applied at the beginning of the growing season at a rate of 80 kg N/ha and in autumn at a rate of 30 kg N/ha. In 2012, the young plants were irrigated.

Two sets of accessions were grown on separate nurseries (Experiment 1, Experiment 2). Plots consisted of five plants of each accession planted successively in rows. Plant distance between rows was 0.90 m; within the row 1.00 m.

Experiment 1: Ten of 22 gene bank accessions (GBIS/I 2015; Table 28.1, no. 1–10) met the criteria of adaptation to different European regions and seed availability. In 2011, plants were cultivated and vernalized in a plastic tunnel at Asendorf (Lower Saxony, Germany; 52°46'03 N, 9°01'36 E). In spring 2012, young plants were space-planted in one row of a RCG nursery at Etzdorf. Some of the border rows on either side were also sometimes planted with RCG.

Experiment 2: Twenty-two of 112 accessions (Table 28.1, no. 11–32) taken from gene banks (GBIS/I 2015; GRIN 2015) were selected for the experiment. They met the criteria of varying shattering behavior, morphological characters, and different geographical origin (GRIN 2015). In autumn 2012, seedlings were planted in two rows of a nursery according to the experimental design of experiment 1.

Table 28.1 Accession number, cultivar name, and origin for 32 reed canary grass samples evaluated 2012 – 2014

No.	Accession number[a]/cultivar	Country of origin	No.	Accession number[a]/cultivar	Country of origin
1	GRA 349	Unknown	17	PI 345662/Donskoi 18	Former Soviet Union
2	GRA 357	Unknown	18	PI 357645/Grove	Canada
3	GRA 687/Motycka	Poland	19	PI 422030/Ioreed	USA
4	GRA 2694	Poland	20	PI 422031/Auburn	USA
5	GRA 3396	Germany	21	PI 433725	Germany
6	GRA 3397	Germany	22	PI 531088/Palaton	(USA)
7	GRA 3398	Germany[b]	23	PI 531089/Venture	USA
8	GRA 3399	Germany[b]	24	PI 578790/Arkansas Upland	USA
9	GRA 3404	Germany	25	PI 578791/SYN 4 Ioreed	USA
10	GRA 3406	Germany	26	PI 578792/Superior	USA
11	GRA 3554/Nagykaloi	Unknown	27	PI 578794/Vantage	USA
12	GRA 3555/common reed	Unknown	28	PI 578795/Cana	USA
13	Lipaula	Germany	29	PI 578796/Rise	USA
14	PI 255887	Poland	30	PI 578797/MN-76	USA
15	PI 272122/Motycka	Poland	31	PI 587092/Bellevue	Canada
16	PI 272123/Nakielska	Poland	32	PI 587193/Lakeside LA	Hungary

[a]GRA from GBIS/I 2015; PI from GRIN 2015
[b]Former German Democratic Republic

Harvest Parameters

In 2013 and 2014, for determination of shattering behavior, about ten panicles per plant were covered with perforated bags (Crispac bag, 305 × 600 mm, Baumann, Saatzucht, Waldenburg, Germany). A cotton ball placed under the edge of the bag prevented seed loss. Only pollinated panicles were isolated (during developmental stages full flowering to end of flowering). The bagged panicles were tied to poles. Approx. 50 cm long culms were cut when panicles had become fully ripe and the first seeds had been shattered. Harvested plant material was stored under normal indoor conditions. The test of shattering resistance was performed 4–6 months after harvest. For each sample (bag) three components were estimated: (1) mass of naturally shattered seeds, (2) mass of seeds shattered after finishing a laboratory test under controlled shattering provocation conditions and (3: seed yield) mass of non-shattered seeds. The seed samples were cleaned by the General Seed Blower (Hoffman Manufacturing Comp., Albany, OR, USA). Shattering resistance was calculated as percent by mass of seed yield (3) in relation to the total seed set (sum of shattering losses (1), technical losses (2) and seed yield (3)). To compare shattering resistance values of the accessions across 2 years the shattering resistance was calculated for each accession in relation to the annual mean of an accession subset (see Figs. 28.2 and 28.4).

Statistical Analysis

To determine accession effects, a Kruskal-Wallis one-way analysis of variance was performed. Dunnett's test with adjustment by Bonferroni was used as post hoc test to compare each of the accessions with accession No. 1 of experiment 1 or accession No. 11 of experiment 2, respectively, as control. The accessions used as controls corresponded to the wild type and allowed the recognition of accessions with an improved shattering resistance. To compare seed yield of the accession to the respective control, a log10 transformation of the left share values was necessary. The software package R 3.1.2 was used.

Results and Discussion

Experiment 1

In 2013, accessions of experiment 1 planted in spring 2012 were very well developed. The number of culms varied from 2 to 320 per plant. Fig. 28.1 shows the great seed losses in ripening (shattering losses) and post-harvest treatment

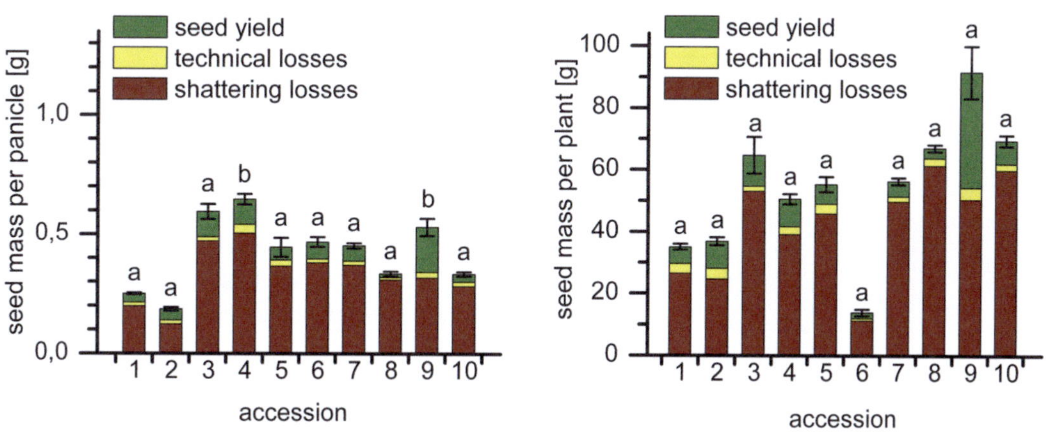

Fig. 28.1 Experiment 1: Seed yield components in 2013. Data shown as seed mass per panicle (on the *left*) and seed mass per plant (on the *right*). Values are means (N=46, SE of seed yield). Each seed yield is compared to that of accession 1 ($p<0.05$). Different letters: means are significantly different. 1, 2, and 4–10: accessions, 3: cultivar 'Motycka'

(technical losses). The influence of accession origin on the effective seed yield per panicle was significantly (p=0.004) different. Accessions 4 and 9 had a significantly higher seed yield per panicle than accession 1. The absolute values were lower than that of a breeding collection after mutagenic treatment (Förster et al. 2014). The seed yield per plant of the individual accessions was also different between accessions (p=0.055).

Shattering resistance (Fig. 28.2) proved to be a characteristic value to differentiate the studied accessions under the very different growing conditions in 2013 and 2014. In spring 2014, growth of RCG accessions 1, 2, 4, 6, and 7 started late. The plants were not able to develop mature panicles. Therefore, comparison of shattering across 2 years could be assessed only for a subset of accessions (3, 5, 8–10). Comparable averaged values for accessions in the subset were determined with the 16 % (2013) and 34 % (2014) annual means, respectively. In 2013, values of all investigated accessions are calculated with respect to the annual mean of the subset.

Although the underlying annual means differ by factor two, the estimated shattering resistance values of the accessions in 2013 and 2014 are located on similar characteristic places in Fig. 28.2. In both years, the median of relative shattering resistance of accessions 3 and 9 was 50 % higher than the annual accession mean

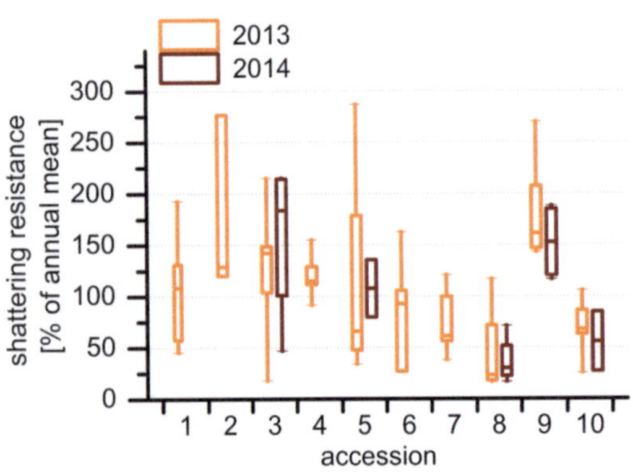

Fig. 28.2 Experiment 1: Shattering resistance of ten accessions in relation to the annual mean of a subset consisting of accession 3, 5, 8–10 (N=18). 2013: annual mean: 16 %; 2014: annual mean: 34 %. In 2013, all data (N=46) are shown using the subset annual mean

(100 %). Accession 8 showed typical wild type behavior. In 2013, single plants in accessions 2, 5 and 9 showed excellent seed retention. Only accession 9 showed a good performance of the traits seed yield per panicle, seed yield per plant and shattering resistance in both years (Figs. 28.1 and 28.2). This makes it an interesting candidate for seed yield improvement.

Experiment 2

In 2013, the RCG accessions of experiment 2, planted in autumn 2012, developed slowly and started heading in late summer. For this reason, results for single plants in 2014 were reported, for which 2–185 culms were recorded at the beginning of flowering. The distribution of seed yield components is shown in Fig. 28.3. Cultivars or accessions were significantly different from accession 11 ('Nagykaloi') at $p<0.05$. The six cultivars 'Auburn', 'Palaton', 'Venture', 'Superior', 'Cana' and 'Rise' were characterized by significantly higher seed yields per panicle than 'Nagykaloi'. In the case of seed yield per plant, only 'Palaton' (22), 'Superior' (26), and 'Cana' (28) were significantly different from number 11. The mean seed yield per plant of accession 9 (experiment 1; 37.4 g) was exceeded by 'Cana' (28) with 58.4 g. 'Palaton' (35.8 g) and 'Superior' (32.8 g) had a slightly lower yield than accession 9. Seed yields per plant of accessions 14–18, 23 and 30 were very low.

In contrast to the accessions from European regions (experiment 1, Fig. 28.1), it is notable that in experiment 2 (accessions or cultivars from different geographical origin) one aspect of the proportion between the three seed yield components had changed: the seed yields of six cultivars exceed their shattering losses (Fig. 28.3 on the top). This indicates a high potential for seed production of these accessions. If the trait seed yield per plant is taken into consideration, the number of accessions with an improved seed yield potential is then reduced to five (Fig. 28.3 on the bottom: 20, 22, 26, 28 and 29).

Fig. 28.4 shows the results of shattering resistance of 2 years in relation to the annual

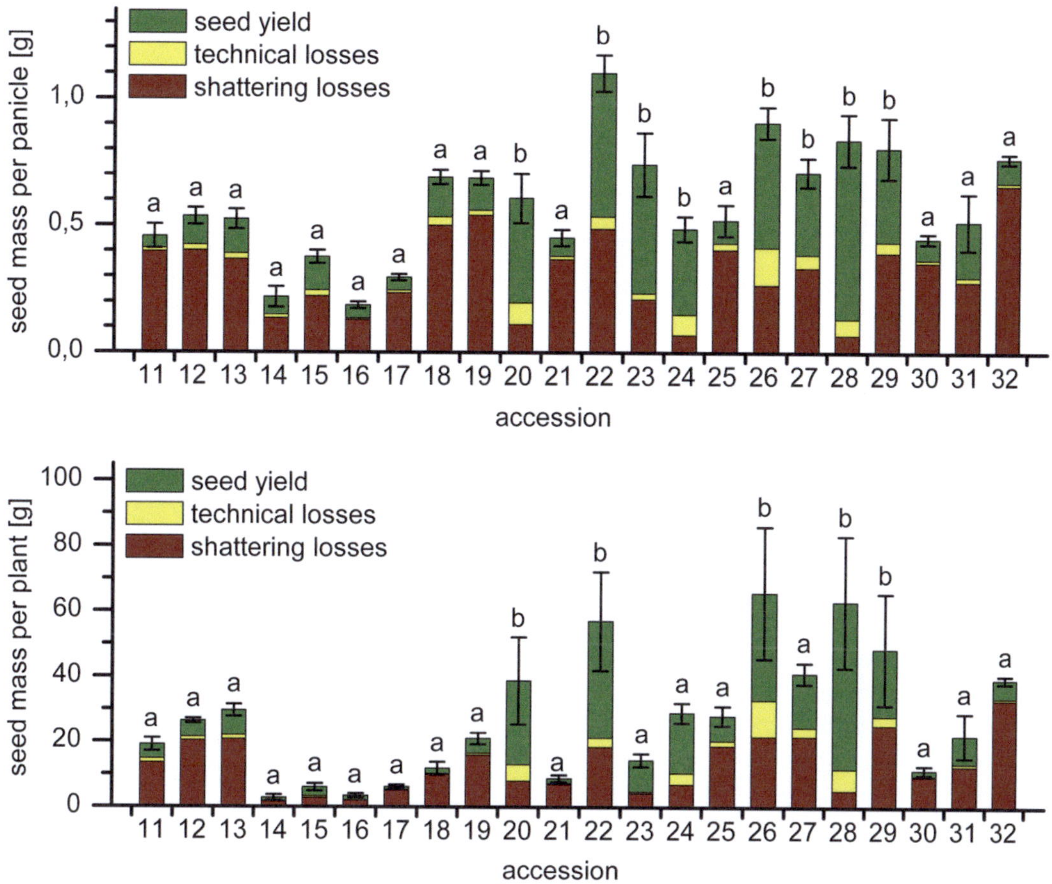

Fig. 28.3 Experiment 2: Seed yield components in 2014. Data shown as seed mass per panicle (on the *top*) and seed mass per plant (on the *bottom*). Values are means (N=70, SE of seed yield). Each seed yield is compared to that of accession 11 ($p<0.05$). Different letters: means are significantly different. RCG accession/cultivar – No. 11: 'Nagykaloi', 12: origin unknown, 13: 'Lipaula', 14: origin Poland, 15: 'Motycka', 16: 'Nakielska', 17: 'Donskoi 18', 18: 'Grove', 19: 'Ioreed', 20: 'Auburn', 21: origin Germany, 22: 'Palaton', 23: 'Venture', 24: 'Arkansas Upland', 25: 'SYN 4 Ioreed', 26: 'Superior', 27: 'Vantage', 28: 'Cana', 29: 'Rise', 30: 'MN-76', 31: 'Bellevue', 32: 'Lakeside LA'

mean. Independent of the different plant development in 2013 and 2014, the annual mean 2013 (37 %) is similar to that of 2014 (39 %). Again, the estimated shattering resistances in Fig. 28.4 are located in characteristic places in each year. In both years, cultivars 'Palaton' (22), 'Venture' (23) and 'Cana' (28) showed more than 50 % higher shattering resistance compared to the annual mean (100 %). In two cases (cultivar 22 and 28) yield per plant and shattering resistance are positively correlated. Sahramaa et al. (2004) found also differences between cultivars and wild germplasm in seed yield and shattering.

Conclusion

In a 2-year study, shattering resistance proved to be a significant feature for the differentiation of gene bank accessions and cultivars of *Phalaris arundinacea* L. in two experiments. Applied in the form of a relative parameter in relation to an annual mean, it compensated for the very different growing conditions in 2013 and 2014. It was the basis for the selection of three accessions (experiment 1: 9, experiment 2: 22, 23) which showed shattering resistance values 50 % higher than the annual mean (100 %) in both years.

The seed yield per panicle also allows differentiation of the accessions. Ten accessions

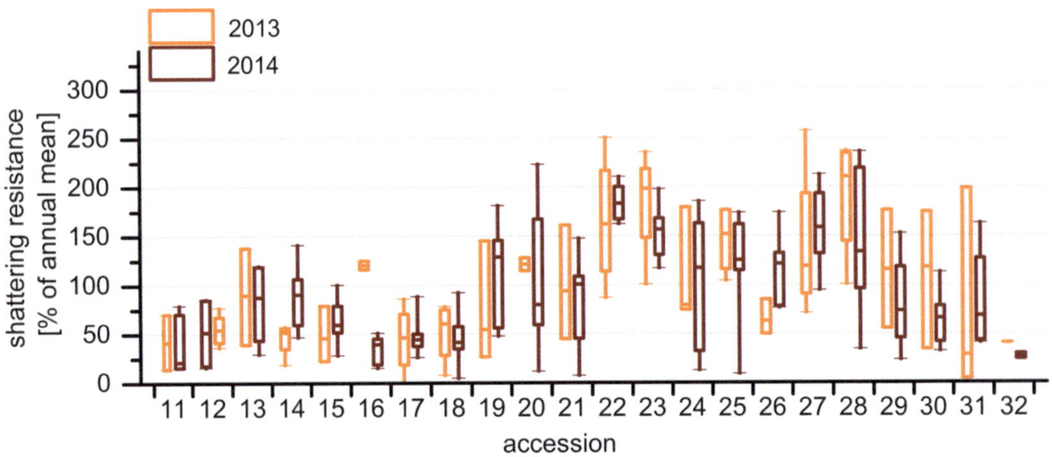

Fig. 28.4 Experiment 2: Shattering resistance of 22 accessions in relation to the annual mean of a subset consisting of accessions harvested in both years (N=70). 2013: annual mean: 37 %; 2014: annual mean: 39 %. In 2014, all data (N=105) are shown using the subset annual mean

and cultivars were selected. The space-planted accessions developed greatly different numbers of panicles per plant in one year compared to the other. The seed yield per plant thus varied strongly between years. Only one of the best five accessions met the seed yield-quality criteria improved shattering resistance, high yield per panicle and high yield per plant. Further selections should be focused on the combination of these traits.

Acknowledgments This project is supported by the Federal Ministry of Food, Agriculture and Consumer Protection on the basis of a decision by the German Bundestag.

References

Förster K, Pentschew S, Braune J, Feuerstein U (2014) Phalaris arundinacea L.: Variations in seed shattering and related traits in a breeding collection. Sokolovic D et al. (eds.), Quantitative traits breeding for multifunctional grasslands and turf. Springer Science + Business Media, Dordrecht. pp 339–343.

GBIS/I (2015) Gene bank of the IPK Gatersleben. https://gbis.ipk-gatersleben.de/-GBIS_I/ergebnisliste.jsf;jsessionid=MWDDVCTdP5Vc2qM6TX5FTDgP0Q9vg2nJ516v1hkvtbPNvyLppy7G!-46513717!1430393725253.

GRIN (2015) USDA, ARS, National Genetic Resources Program. Germplasm Resources Information Network – (GRIN) [Online Database]. National Germplasm Resources Laboratory, Beltsville, Maryland. http://www.ars-grin.gov/cgi-bin/npgs/acc/search.pl?-accid=Phalaris%¬20¬arundinacea.

Lewandowski I, Scurlock JMO, Lindvall E, Christou M (2003) The development and current status of perennial rhizomatous grasses as energy crops in the US and Europe. Biomass and Bioenergy 25, pp 335–361.

Sahramaa M, Hömmö L, Jauhiainen L (2004) Variation in seed production traits of reed canarygrass germplasm. Crop Sci. 44, pp 988–996.

Comparing Juvenile Development of Perennial Ryegrass, Meadow Fescue and Different Hybrid Generations

29

T. Lunenberg, S. Hartmann, and W. Friedt

Abstract

Meadow fescue (*Festuca pratensis* Huds., Fp) is well adapted to various biotic and abiotic stresses and has a high nutritional value. Unfortunately, meadow fescue is not sufficiently competitive in mixed stand with the major forage grass perennial ryegrass (*Lolium perenne* L., Lp) since the intensity of grassland cultivation has increased. Main disadvantages of Fp are its slow juvenile development and low formation of side shoots. To broaden the genetic diversity of meadow fescue, individuals of Fp and Lp were hybridized and the F_1 (*Festulolium*) was backcrossed (BC) to Fp. The BC seeds were used in an experiment on juvenile development. In the first experiment, a method to measure the juvenile development by assessing three major traits was examined on 13 Lp and 7 Fp varieties. No clear and significant differences for the three traits were found between the two species as intraspecies variation was high, but in general Lp varieties emerged faster and tillering started earlier. In a second experiment the method was applied to seeds of 151 F_1 and BC_1 *Festulolium* hybrids. Hybrid plants (potential BC_1 and BC_2 plants) showing phenotypic similarity to Fp were selected. The best BC_1 and BC_2 plants were characterized by a satisfactory seed quality and a fast juvenile development. They represent valuable basic material for further forage grass breeding.

Keywords

Juvenile development • Meadow fescue • Perennial ryegrass • *Festulolium*

T. Lunenberg (✉) • S. Hartmann
Institute for Crop Science and Plant Breeding; Forage Plant Research Breeding and Grassland Systems, Bavarian State Research Center for Agriculture, Am Gereuth 4, Freising 85354, Germany
e-mail: tatjanalunenberg@gmail.com; stephan.hartmann@lfl.bayern.de

W. Friedt
Department of Plant Breeding, IFZ, Justus-Liebig University of Giessen, Heinrich-Buff-Ring 26-32, Giessen 35392, Germany

Introduction

Species of the genus *Lolium* are the most abundant forage grasses in Europe due to their high yield potential, nutritive quality and tolerance to frequent defoliation. *Lolium* species are consequently also the most intensively bred grasses. Compared to

I. Roldán-Ruiz et al. (eds.), *Breeding in a World of Scarcity: Proceedings of the 2015 Meeting of the Section "Forage Crops and Amenity Grasses" of Eucarpia*,
DOI 10.1007/978-3-319-28932-8_29

Lolium, representatives of the genus *Festuca* are more tolerant to biotic and abiotic stresses such as drought or frost, but they have disappeared from pastures in southern Germany because of the slower juvenile development and the increased cutting frequency in the last decades (Kölliker et al. 1999; Heinz and Kuhn 2008). In Bavaria, 33.7 % of the agricultural area was used as permanent pasture in 2013 (Bayerisches Staatsministerium für Ernährung, Landwirtschaft und Forsten 2014), but not all habitats are fully suitable for continued growing of *Lolium*. For this reason a combination of both species aims to combine good stress tolerance with the positive characteristics of *Lolium*. It is possible to successfully hybridize *Lolium* and *Festuca* species (Jenkin 1933; Ghesquière et al. 2010); the resulting hybrids are called *Festulolium*. Two breeding schemes for *Festulolium* can be found: (1) using amphiploid material to combine the whole genomes of both species, and (2) introgression breeding, where favorable genes of one species are transferred to the other species (Thomas et al. 2003). Intergeneric recombinations occur in all regions of the hybrid genomes (Canter et al. 1999). The Bavarian Research Institute for Agriculture aimed to develop a method for assessing juvenile development and to introgress the trait from diploid perennial ryegrass (*L. perenne*, Lp) into diploid meadow fescue (*Festuca pratensis*, Fp).

Materials and Methods

Plant Material and Experiment

In a preliminary "variety experiment" 13 diploid *Lolium perenne* (Lp) varieties and 7 *Festuca pratensis* (Fp) varieties were phenotyped for the trait juvenile development. The experiment was designed as randomized complete block design with three replicates. Each entry consisted of 22 seeds per replicate. Seeds were sown in potting soil in the greenhouse and grown at a minimum temperature of 16/18 °C (night/day) and a maximum of 19/20 °C (night/day).

F_1 hybrids between Lp and Fp were manually produced by emasculation of single florets and hand pollination in the greenhouse. To produce BC_1 or BC_2 plants, F_1 hybrids and BC_1 individuals were planted in the field as single plants alternating with six Fp genotypes of the varieties Pradel, Preval, Kolumbus and Cosmolit and were isolated to prevent cross-

Table 29.1 Time in days to emergence, emergence to tillering and emergence to level 4 (n = number of analyzed plants)

Time to emergence			Time from emergence to tillering			Time from emergence to level 4		
Grandmother plant	Mean [d] #	n	Grandmother plant	Mean [d] #	n	Grandmother plant	Mean [d] #	n
Aberavon	7.2[a]	6	Cosmolit	22.5[a]	6	Aberavon	20.8[a]	5
Cosmolit	6.9[ab]	7	BC_1 FEL 37*	20.0[ab]	30	Cosmolit	18.0[ab]	7
WSC PX4***	6.4[ab]	129	WD PX4_1**	19.0[abc]	3	Barata	17.0[abc]	9
Barata	6.3[ab]	9	Aberavon	17.0[bc]	4	Ivana Klon	17.0[abc]	11
Ivana Klon	6.0[ab]	12	Arabella	19.0[bc]	79	WD PX4_1**	16.0[abc]	3
Iduna**	6.0[ab]	43	Barata	18.6[bc]	7	Arabella	16.0[bc]	86
BC_1 FEL 37*	5.9[ab]	31	F_1 Respect	18.3[bc]	4	F_1 Respect	14.0[bc]	4
Arabella	5.6[bc]	86	Iduna**	18.7[bc]	32	F_1 FEL 37*	15.0[bc]	31
Niata	5.0[bcd]	2	Ivana Klon	18.9[bc]	7	Iduna**	15.5[bc]	41
F_1 Lipresso	4.8[bcd]	4	Niata	15.0[bc]	2	WSC PX4***	15.5[bc]	126
F_1 Respect	4.0[cd]	4	WSC PX4***	18.0[bc]	120	F_1 Lipresso	12.0[c]	4
WD PX4_1**	3.0[d]	3	F_1 Lipresso	13.3[c]	4	Niata	11.0[c]	2

#Means, different letters show significant differences (LSD test, $P<0.05$)
*Unknown grandparents
***Meadow fescue material grown in the breeding nursery
**Perennial ryegrass material grown in the breeding nursery

pollination by *Lolium* plants grown in the vicinity. The seeds of 151 F_1 and BC_1 plants were harvested in 2013. Denominations of the produced *Festulolii* (see Table 29.1) indicate the mother plant variety of the F_1 hybrids. For recording juvenile development of BC_1 and BC_2 (BC experiment) seeds were spread in germination trays and cultivated under analogous conditions. The experiment was run as a block design with three blocks and without replicates of the mother plants.

Measuring Juvenile Development

Juvenile development for both the variety and the BC experiment was assessed by recording different traits for up to 60 days after sowing. For measuring leaf elongation, special frames covered with mesh were installed at four heights (level 1: 10 cm, level 2: 13 cm, level 3: 17 cm, level 4: 21 cm) and the time points when the plants reached these levels were recorded for each entry. Furthermore, the dates of emergence (BBCH09) and beginning of tillering (BBCH21) were recorded on a single plant basis within both experiments.

Statistical Analysis

ANOVA with subsequent LSD test was calculated with the statistical software R and the package agricolae (R Development Core Team 2014; Mendiburu 2014). The analysis of the BC experiment was performed using the model

$$y = \mu + variety\ of\ the\ grandmother\ plant + block + \varepsilon$$

The model

$$y = \mu + variety + block + \varepsilon$$

was used to analyze the variety experiment.

Results and Discussion

Differences Among Varieties to Define the Trait Juvenile Development

In general, Lp varieties emerged faster, started tillering earlier and had a faster length growth than Fp types. But the Fp variety Liherold proved to be competitive with most of the Lp varieties (Figs. 29.1 and 29.2). Thus the assumption that Lp varieties generally have a faster juvenile development could only be partially confirmed, as high variation was found in Lp. In accordance with previous experiments, seed quality has a strong influence on germination time. Degenerated plants started tillering quickly but had a very slow length growth. Therefore a combination of the traits length growth, emergence and tillering has to be considered to confidentially detect fast juvenile development. The procedure described here is considered to be a suitable tool to select for this trait.

Differences of Hybrids in Juvenile Development

The method described above was used to evaluate the juvenile development in the BC experiment. Only BC plants showing Fp characteristics (indicated by rolled youngest leaf) were rated as successful backcrosses and considered for further analysis.

The BC_1 and BC_2 plants were grouped according to the female plant of the F_1 hybrid, the so-called grandmother plant. Differences between the families could be detected for all three traits influencing juvenile development (Table 29.1). Especially interesting was the great difference between the offspring of fescue cultivars Cosmolit and WSC PX4. Although they had nearly the same germination time, the progenies of WSC PX4 grew faster and started tillering earlier. The progenies of F_1 Lipresso, Niata, WSC PX 4 and Iduna had a fast elongation growth and started tillering early. It can

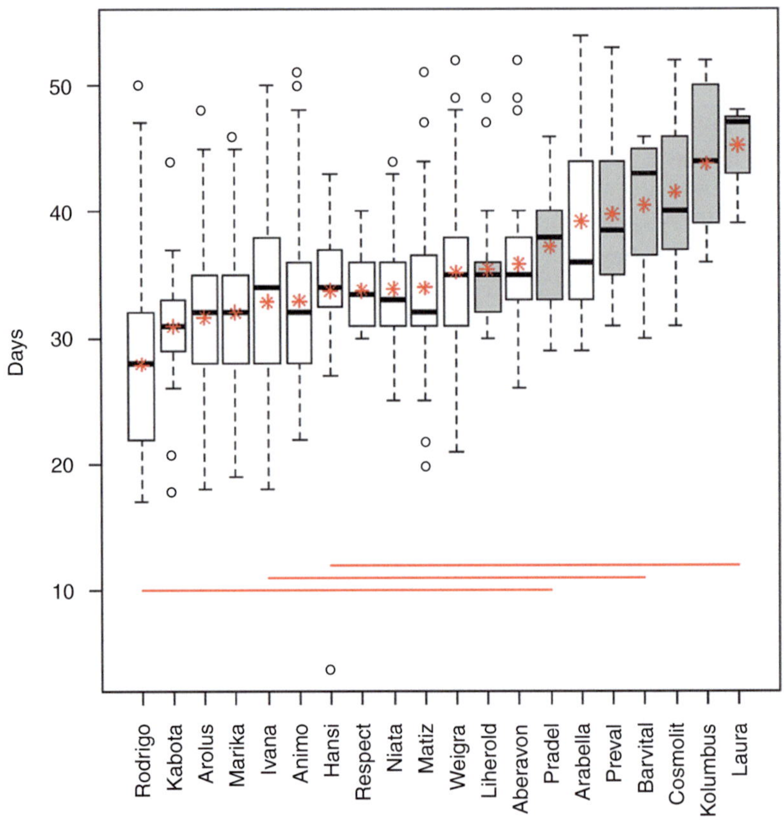

Fig. 29.1 Time to tillering. Varieties above the same *horizontal line* are not significantly different (LSD test, P<0.05). *Grey boxes* indicate meadow fescue varieties, *white boxes* perennial ryegrass varieties

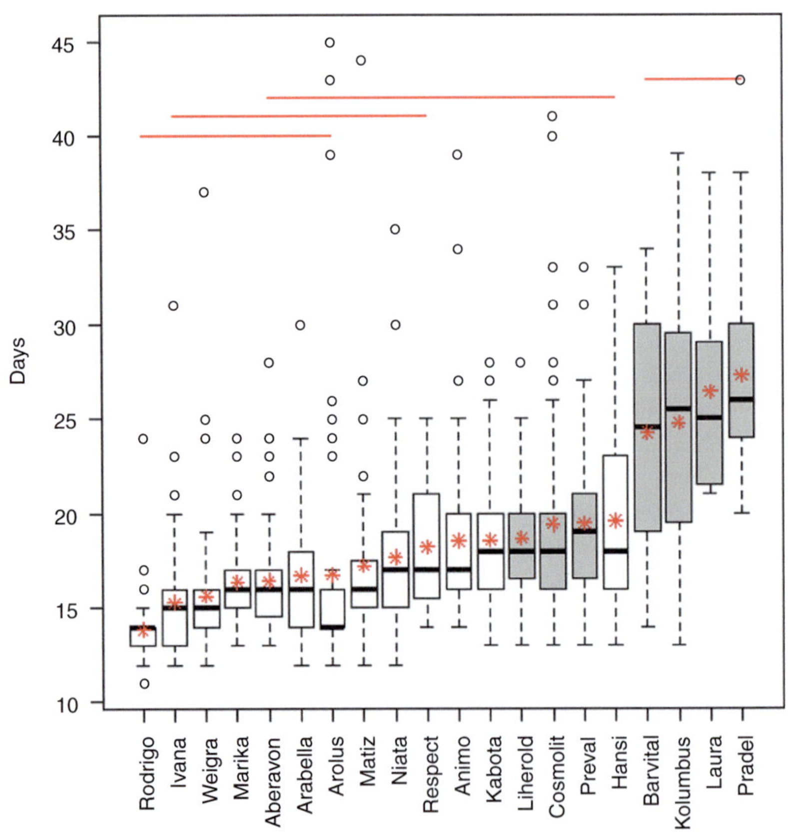

Fig. 29.2 Time from emergence to level 4. Varieties under the same *horizontal line* are not significantly different (LSD test, P<0.05). *Grey boxes* indicate meadow fescues varieties, *white boxes* perennial ryegrass varieties

therefore be stated that these progenies are characterized by a fast juvenile development.

Conclusion

As already known from previous experiments, the quality (i.e., age) of the seed sample and the air temperature had the strongest influence on the speed of germination. Furthermore, in all experiments, the traits germination time and tillering were not highly related or were not correlated at all. The diploid varieties of the species Lp and Fp are not always distinguishable according to their similar juvenile development. Some Fp varieties grew as fast as Lp varieties. The described method is a suitable tool to select for juvenile development. Note that the present experiments were performed with diploid material. Any statement about the development and differences of tetraploid material is not possible. In case of the present experiment with diploid hybrid seeds, the best BC_1 and BC_2 plants were characterized by a satisfactory seed quality and a fast juvenile development. Therefore, it represents valuable basic material for further forage grass breeding.

References

Bayerisches Staatsministerium für Ernährung, Landwirtschaft und Forsten (2014). Bayerischer Agrarbericht 2014 Fakten und Schlussfolgerungen. www.stmelf.bayern.de/mam/cms01/agrarpolitik/dateien/agrarbericht2014.pdf. Accessed 23 February 2015.

Canter, P.H., Pasakinskiene, I., Jones, R.N., Humphreys, M.W. and et al. (1999). Chromosome substitutions and recombination in the amphiploid Lolium perenne x Festuca pratensis cv Prior (2n = 4x = 28). Theoretical and applied genetics 98: 809–814.

Ghesquière, M., Humphreys, M., Zwierzykowski, Z., (2010). Festulolium Hybrids: Results, Limits and Propects. In: Huyghe, Christian, Sustainable Use of Genetic Diversity in Forage and Turf Breeding. Springer Science + Business Media B.V., Dordrecht Heidelberg London New York 495–507.

Heinz, S. & Kuhn, G. (2008). 20 Jahre Bodendauerbeobachtung in Bayern – Teil 2: Vegetation auf Äckern und im Grünland. Schriftenreiche der Bayerischen Landesanstalt für Landwirtschaft. ISSN 1611-4159.

Jenkin, T.J. (1933). Interspecific and intergeneric hybrids in herbage grasses: Initial crosses. Journal of genetics 28.2:205–264.

Kölliker, R., Stadelmann, F.J., Reidy, B. and Nösberger, J. (1999). Genetic variability of forage grass cultivars: A comparison of Festuca pratensis Huds., Lolium perenne L., and Dactylis glomerata L. Euphytica 106: 261–270.

Mendiburu, F. de (2014). agricolae: Statistical Procedures for Agricultural Research. R package version 1.2-0. http://CRAN.R-project.org/package=agricolae

R Core Team (2014). R: A language and environment for statistical computing. R Foundation for Statistical Computing, Vienna, Austria. http://www.R-project.org/.

Thomas, H.M., Morgan, W.G. and Humphreys, M.W. (2003). Designing grasses with a future – combining the attributes of Lolium and Festuca. Euphytica 133.1: 19–26.

Enhancing the Productivity in Forage Grasses on the European Scale using Interspecific Hybridization

30

M. Ghesquière, J. Baert, S. Barth, V. Černoch, D. Grogan, M.W. Humphreys, P. Murray, L. Østrem, D. Sokolović, E. Paszkowski, and Z. Zwierzykowski

Abstract

Ryegrass x fescue interspecific hybridization (so-called *Festulolium)* has great potential for forage diversification because of its many amphiploid and introgression-bred forms. *Festulolium* breeding efforts aim to improve adaptation to mainly winter cold and summer drought through higher abiotic stress tolerance. This goal must be met with a minimum of compromise regarding productivity in the first years of full exploitation. To compare field performances, a 21-entry trial of *Festulolium* and controls

M. Ghesquière (✉)
INRA/URP3F, Le Chêne,
RD 150, CS 80006,
Lusignan 86600, France
e-mail: marc.ghesquiere@lusignan.inra.fr

J. Baert
ILVO, Caritasstraat 21, Melle 9090, Belgium

S. Barth
Teagasc, Oak Park Crops Research Centre,
Carlow, Ireland

V. Černoch
DLF – TRIFOLIUM, Hladké Životice, s.r.o,
Fulnecka 95, Hladké Životice 742 47,
Czech Republic

D. Grogan
Department of Agriculture, Food and the Marine,
Crop Evaluation and Certification, Backweston Farm,
Leixlip Co. Kildare, Ireland

M.W. Humphreys
IBERS, Aberystwyth University,
Gogerddan Campus, Aberystwyth, Ceredigion, Wales
SY23 3EE, UK

P. Murray
Rothamsted Research, North Wyke Research Station,
Okehampton, Devon EX20 2SB, UK

L. Østrem
Bioforsk - Norwegian Institute for Agricultural and
Environmental Research, Fureneset,
Hellevik i Fjaler 6967, Norway

D. Sokolović
Institute for forage crops Kruševac,
Globoder Kruševac 37251, Serbia

E. Paszkowski
DANKO Plant Breeding Ltd., Branch in Szelejewo,
Szelejewo Drugie, Piaski 64-820, Poland

Z. Zwierzykowski
Institute of Plant Genetics, Polish Academy of
Sciences, Strzeszynska 34, Poznan 60-479, Poland

I. Roldán-Ruiz et al. (eds.), *Breeding in a World of Scarcity: Proceedings of the 2015 Meeting of the Section "Forage Crops and Amenity Grasses" of Eucarpia*,
DOI 10.1007/978-3-319-28932-8_30

of pure species has been ongoing since 2012 in eight European countries. The yield data collected in 2012 and 2013 in nine locations showed that the *Festulolium* cultivars performed on average quite well compared to pure species controls. In amphiploids, the annual yield appeared to be mainly driven by the *Lolium* sp. parent combined with *F. pratensis (Fp)*; the *Lm* x *Fp* amphiploids performed thus far better on average than the *Lp* x *Fp* equivalents. The only amphiploid from *F. glaucescens* included in the study had an intermediate response over locations, which was closer to tall fescue than to *F. pratensis*. Interestingly, a broad variability for yield among the *Lm* x *Fp* cultivars appeared to be associated with climate interactions when cold, wet locations were contrasted with more temperate locations with early spring growth.

Keywords
Fescue • Ryegrass • Festulolium • Interspecific hybridization • Amphiploid • Introgression • Climate adaptation • Forage yield

Introduction

Plant species adapted to severe climatic conditions often have traits that lower their productivity when the incidence or severity of abiotic stress increases: in other words, there is often a trade-off between productivity and climatic adaptation. In nature, the best fitness strategy is avoidance, including cessation of growth when confronted with stress conditions. At the same time, in order to guarantee their competitiveness within plant communities, perennials also have to optimize their growing capacity. Climatic adaptations are also not always needed, as periods of stress can vary considerably over a number of years.

In this context, the *Festuca/Lolium* complex of grass species provides promising opportunities to limit as far as possible trade-offs between productivity and adaptation. The complex includes many species diversely adapted to climate conditions through plant size, phenology, and response of leaf growth to temperature and to soil water deficit. So far, two major amphiploid forms have been bred by using cultivated species, mainly by crossing *F. pratensis* (*Fp*) with either *L. multiflorum* (*Lm*) or *L. perenne* (*Lp*) and vice versa (Ghesquière et al. 2010). But further amphiploids were also produced by using wild relatives in the broad-leaved fescues such as *F. glaucescens* (*Fg*), the 4x-progenitor of tall fescue, and more recently, *F. mairei*, Atlas fescue from North Africa (M.W. Humphreys, pers. comm.). As back-crossing amphiploid hybrids into any *Lolium* or *Festuca* population is possible at any step of a breeding program, many introgression forms have been developed in parallel. The present *Festulolium* cvs, all synthetic varieties, represent thus a totally new redeployment of the genetic variability in the complex that has only been distributed at a species scale so far. Agronomic assessment of *Festulolium* has been often limited by seed availability or has been restricted to particular locations, controlled environments or the use of clones in spaced-plant field trials. For the first time, we report yield data for a relative large set of *Festulolium* cvs (15) obtained in a multi-location trial involving nine locations across Western Europe.

Materials and Methods

A set of 21 cultivars (15 *Festulolium* and six pure species controls) were included in the study. *F. pratensis*, cv Fure, *F. arundinacea*, cv Kora, *L. multiflorum*, cvs Podium (2x) and Caballo (4x) and *L. perenne*, cvs AberMagic (2x) and AberBite (4x) were used as pure species controls. The

Festulolium cultivars were drawn from the EU list of cultivars. The cultivars are thus representative of the breeding efforts undertaken by several European institutions and illustrate the potential of interspecific hybridization within the complex. Twelve of the chosen *Festulolium* are amphiploids (8 *Lm* x *Fp*, 3 *Lp* x *Fp* and 1 *Lm* x *Fg*) and three are hybrids derived from introgression breeding into either 2x or 4x *L. multiflorum*, from *Lm* x *Fa* or *Lm* x *Fp* primary hybrids. The genome composition of 10 of these *Festulolium* varieties has been described by Kopecký et al. (2006)), who confirmed that the varieties selected are representative of current usable interspecific variability, both in terms of *Festuca:Lolium* genome balance and interspecific recombination rate.

A network of nine trials of the same genotypes was launched in 2012 in eight countries: Belgium (BE, ILVO), Czech Republic (CZ, DLF), France (FR, INRA), Ireland (IR, Crop Evaluation and Certification), Norway (NO, Bioforsk/Graminor), Poland (PO, DANKO), Serbia (SRB, IKBKS) and UK (IBERS and North Wyke Research Station). All of the trials were sown according to a complete block design with 3 replicates, in spring or autumn 2012 and 2013. A total of 41 cuts were taken: eight cuts in the year of establishment (Year 0), with 1–3 cuts in BE, CZ, PO and SRB, followed by 33 cuts in the first year of full exploitation (Year 1), with 3–5 cuts in all locations except in BE (no data available for Year 1 from this location by the time of manuscript preparation).

A mixed model of analysis of variance (ANOVA) was used to estimate the least square means of the 21 genotypes with 'cultivar', 'location' and their interaction as fixed effects and 'cut (within location)' and all other interactions as random effects. Student's t-statistics were computed to test the contrast between parent species and to describe the genotype x location interactions relative to the parent species from which the respective *Festulolium* variety was derived.

To illustrate the variety x cut (location) interactions, unstandardized Principal Component Analyses were performed at each location and the scores of the six parental species control cvs on the two first components were plotted. Summarizing the variation of yield among *Festulolium* cvs relative to parent species was finally illustrated by mean score value across all locations.

Results and Discussion

The mean yield per cut in year 1 varied considerably over locations from 2.7 T DM/ha in North Wycke and SRB to 6.9 T DM/ha in PO. The analysis of variance showed that the main cultivar effect was highly significant, although slightly less supress in year 0 than in year 1 due to lower number of cuts and therefore lower power of the test ($F=7.34$, $P<.0001$; $F=11.85$, $P<.0001$, respectively). However, in both years, the cultivar x location interaction was much lower than the cultivar effect ($F=1.68$, $P<.0156$; $F=1.95$, $P<.0001$, respectively); yield thus correlated relatively well among cultivars between the 2 years ($r=0.66$, $P<0.0010$).

Both years gave a similar profile for the interspecific contrasts, largely in favor of *L. multiflorum*, and even increasing from year 0 to year 1 against *L. perenne* (Table 30.1). Conversely, the mean contrast between *L. perenne* and *F. pratensis* was not significant in year 1. The contrasts between parent species were only rarely significant within location due to large variance of yield between cuts. The *Lm-Fa* contrast was positive and significant in IBERS and FR, and negative (but not significant) in NO, CZ and PO, indicating that tall fescue yielded higher than Italian ryegrass controls in these latter locations.

The two first components of the Principal Component Analyses accounted for almost all the variance of total yield within location, from 82 % (CZ) to 98 % (BE). As the first component is highly correlated with total yield, the ranking of parent species along this component reflects the contrasts of yield in each location (Table 30.1). However, the second component, which accounted for a maximum of 35 % of the variance (in CZ), illustrates how the cultivar ranking may reverse from one cut to the next in a given location, and from one location to another (Fig. 30.1).

Table 30.1 Contrast of yield/cut between the *Lolium* and *Festuca* cultivars

Year-0				
Contrast	*Lm – Lp*	*Lm – Fp*	*Lm – Fa*	*Lp – Fp*
All locations	$4.48^{<0.001}$	$7.27^{<0.001}$	$2.80^{0.003}$	$3.61^{<0.001}$
CZ	0.15^{ns}	1.24^{ns}	-0.78^{ns}	1.12^{ns}
SRB	0.95^{ns}	0.87^{ns}	1.37^{ns}	0.09^{ns}
BE	$2.02^{0.023}$	–	1.46^{ns}	–
PO	$2.10^{0.019}$	$3.26^{<0.001}$	1.21^{ns}	0.09^{ns}
Year-0				
Contrast	*Lm – Lp*	*Lm – Fp*	*Lm – Fa*	*Lp – Fp*
All locations	$7.09^{<0.001}$	$7.38^{<0.001}$	$2.67^{0.004}$	1.59^{ns}
IBERS	$2.29^{0.011}$	$2.66^{0.004}$	$2.23^{0.013}$	0.79^{ns}
FR	$1.90^{0.029}$	$2.07^{0.020}$	$1.87^{0.031}$	0.52^{ns}
North Wyke	0.79^{ns}	0.85^{ns}	0.82^{ns}	0.21^{ns}
IR	-0.21^{ns}	0.69^{ns}	0.27^{ns}	0.86^{ns}
SRB	0.75^{ns}	0.43^{ns}	0.25^{ns}	-0.19^{ns}
NO	-0.21^{ns}	0.13^{ns}	-0.17^{ns}	0.30^{ns}
CZ	0.35^{ns}	0.18^{ns}	-0.52^{ns}	-0.10^{ns}
PO	$3.08^{0.001}$	$2.33^{0.010}$	-0.85^{ns}	-0.18^{ns}

The Least Square Means used for the contrasts come from a 21-cultivar trial replicated across nine locations using three to six cuts within location. The location acronyms refer to the international country code with IBERS and North Wyke, both located in the UK. *Lm* and *Lp* are the mean of one 2x and one 4x cultivar from each *Lolium* species. *P* value of the t-tests indicated in exponent with ns as not significant when $P > 0.05$

The diagram representing the cultivar mean scores on the first two components shows that the *Festulolium* cvs overall performed as well as the best parent of their initial hybrid combination (*Lm* x *Fp*) or even better (*Lp* x *Fp*) (Fig. 30.2). However, the ranking between combinations seems to be directly determined by the potential of the *Lolium* parent species: the *Lp* x *Fp* cvs yielded much less than the *Lm* x *Fp* cvs. The only *Lm* x *Fg* cv available ranked intermediately, in the vicinity of *Fa*, consistent with its genealogy and with the growth traits of *Fg*, which are more similar to *Fa* than to *Fp*. The second component exemplifies the cultivar x cut (location) interactions displayed by the individual Principal Component Analyses above, likely due to variability of heading date and cutting management in spring at each location. Thus, the extended variability among *Lm* x *Fp* amphiploid and introgressive *Festulolium* cvs compared to tall fescue can be explained principally by the earlier heading date of tall fescue as compared to Italian ryegrass. However, the broad contrast among ploidy levels within *Lolium* spp., in both cases favor the 4x controls (AberBite (*Lp*)–Caballo (*Lm*)) rather than the 2x controls (AberMagic (*Lp*)–Podium (*Lm*)). This suggests that phenology would not be the only factor affecting yield variability among *Festulolium* cvs. Although later heading, the results of the *Lm* x *Fg* cv Lueur, similar to tall fescue cv Kora - its fescue species control–supports this conclusion. The spread of the varieties along the second component illustrates how present *Festulolium* breeding has the potential for wide and positive differentiation from any initial interspecific hybrid combination over a wide range of climates and management conditions.

Conclusion

These first results from a multi-site pan-European field trial revealed a large variability of yield between and within *Festulolium* cultivars during their first 2 years of growth, with considerable local interactions. Studying

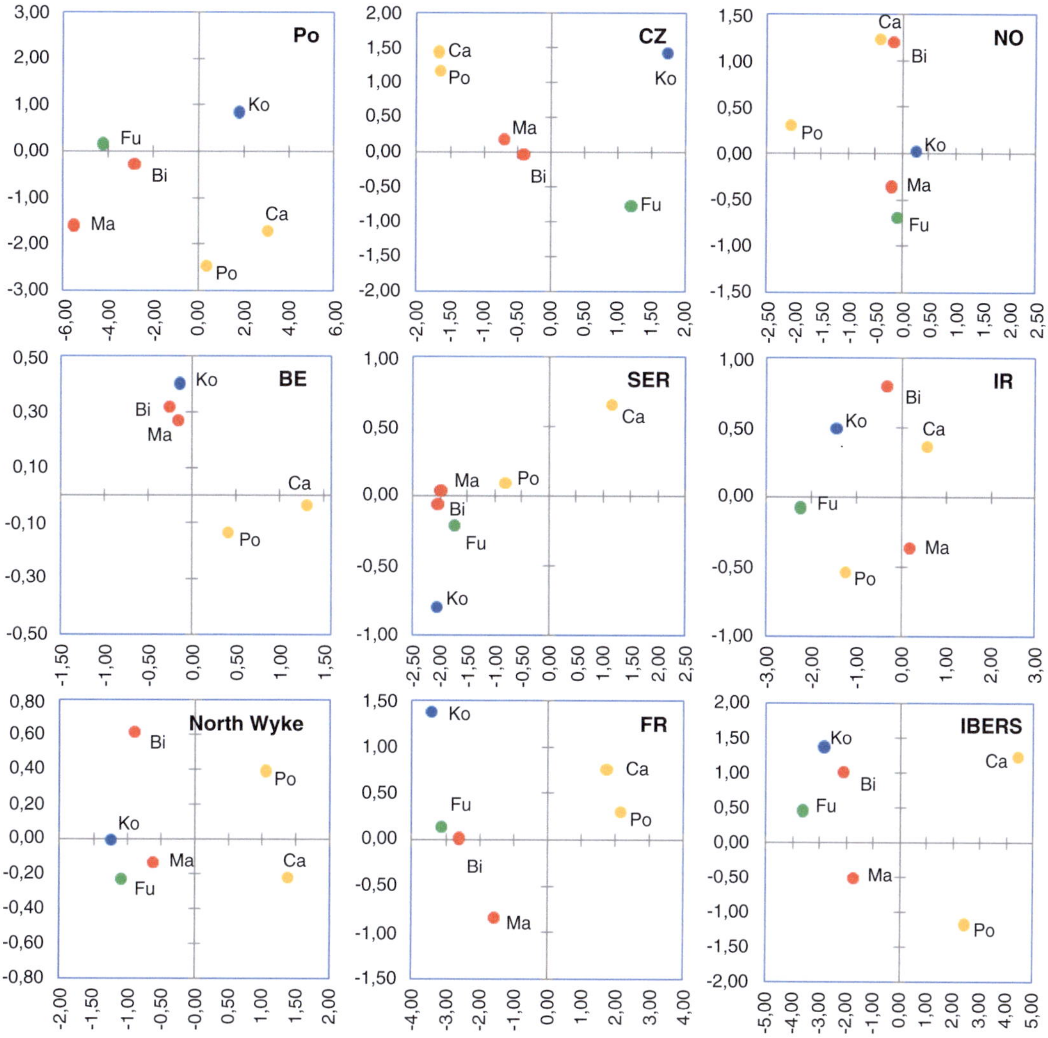

Fig. 30.1 Scores of the *Lolium* and *Festuca* cultivars. Scores were obtained from unstandardized Principal Component Analysis of yield (T DM/ha) in a 21-genotype trial replicated across nine locations using three to six cuts within each location. Note that the scale on the first component (X-axis) and on the second component (Y-axis) is not the same across locations. The location acronyms refer to the international country code with IBERS and North Wyke, both in the UK. Ko: *Fa* cv Kora; Fu: *Fp* cv Fure; Po: *Lm*-2x cv Podium; Ca: *Lm*-4x cv Caballo; Ma: *Lp*-2x cv AberMagic; Bi: *Lp*-4x cv AberBite

how this might change during subsequent years is our future objective. This will help to better understand climate adaptation in the grasses across a large geographic scale, how to breed for it, and to which extent *Festuca* x *Lolium* hybridization may contribute to ensure sustainable grassland agriculture in Europe in the future. The next challenge will be to determine which relevant optimal inter- and intraspecific genome combinations are able to ensure consistent high production when confronted by the climate variations and stresses increasingly encountered throughout Europe.

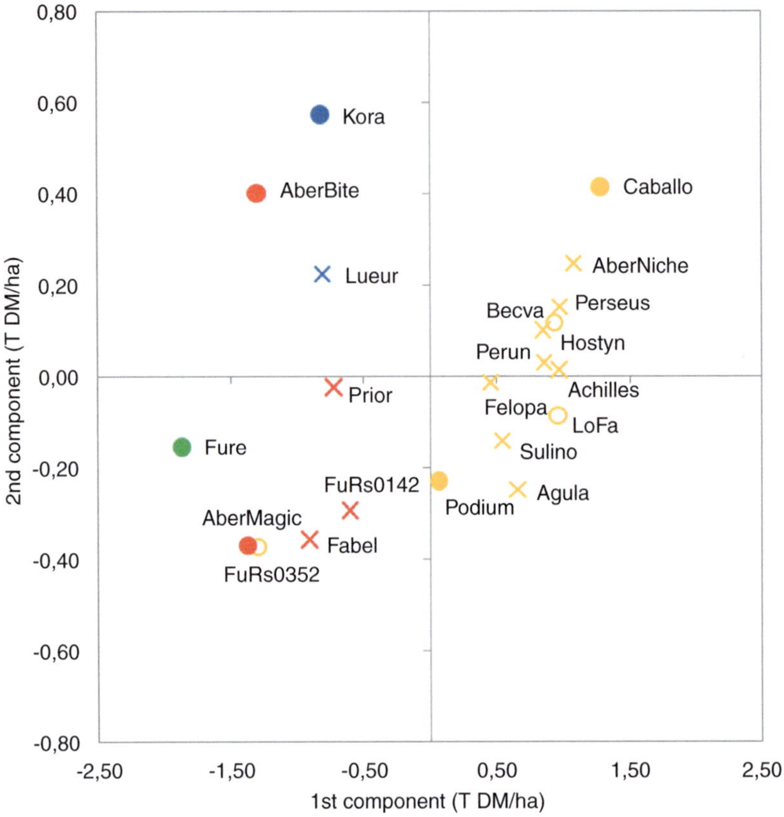

Fig. 30.2 Mean DM yield among 15 *Festulolium* cvs and six parent species controls. The coordinates are the mean scores on the two first components of unstandardized Principal Component Analysis on yield of three to six successive cuts in each location over 1–2 years. The solid circle symbols plot the parent species controls: *Lp* (*red*), *Fp* (*green*), *Fa* (*blue*) and *Lm* (*yellow*); the cross symbols plot the amphiploid cvs : *Lm* x *Fp* (*yellow*), *Lp* x *Fp* (red), *Lm x Fg* (*blue*); the open circle symbols *yellow plot* the cvs coming from introgression of *Fa* or *Fp* into *Lm*

References

Ghesquière, M., Humphreys, M. W., Zwierzykowski, Z. (2010). Festulolium. In: Beat Boller, Ulrich K. Posselt, Fabio Veronesi, dir., Fodder crops and amenity grasses. Handbook of Plant Breeding, 5. New York, USA : Springer. (p. 293–316).DOI : 10.1007/978-1-4419-0760-8_12

Kopecký, D., Loureiro, J., Zwierzykowski, Z., Ghesquiere, M., Doležel, J. (2006). Genome constitution and evolution in Lolium x Festuca hybrid cultivars (*Festulolium*). Theoretical and Applied Genetics, 113 (4):, 731–742. DOI : 10.1007/s00122-006-0341-z

Comparison of Lucerne Genotypes and Varieties in Pure and Mixed Stands

31

B. Julier, G. Louarn, F. Gastal, F. Surault, J.-P. Sampoux, and A. Maamouri

Abstract

Grasslands sown with grass/legume mixtures combine agricultural and environmental advantages, but genetic progress is still needed. Current breeding and evaluation of alfalfa varieties is conducted in monoculture, which may not allow for the creation of varieties that perform well in mixture with grasses (such as tall fescue or cocksfoot). We report on recent results on lucerne–grass mixtures which reveal that the ranking of lucerne genotypes or varieties for biomass production is substantially different in mixture and in monoculture. Biomass production of a mixture mainly resulted in additive effects of the varieties of each species, without statistically significant interaction. We suggest that the performance in mixture for lucerne variety breeding and evaluation should be taken into account.

Keywords

Mixture • Nitrogen • Yield • Lucerne • Grass

Introduction

The agronomic, economic and environmental advantages of mixtures as compared to monocultures are now well-established (Luscher et al. 2014). The choice of species is quite well defined, as least for binary mixtures. Lucerne (*Medicago sativa*) is often used as monoculture but may be used in mixture with tall fescue (*Festuca arundinacea*) or cocksfoot (*Dactylis glomerata*) in order to balance protein and energy contents of the forage, stabilize yield production over cuts and limit weed development (Chamblee and Colins 1988; Spandl et al. 1999). However,

B. Julier (✉)
Unité de Recherche Pluridisciplinaire Prairies et Plantes Fourragères, Institut National de la Recherche Agronomique (INRA), Lusignan, France
e-mail: bernadette.julier@lusignan.inra.fr

G. Louarn • F. Gastal • F. Surault
J.P. Sampoux • A. Maamouri
UR4 Unité de Recherche Pluridisciplinaire Prairies et Plantes Fourragères, INRA, Lusignan F-86600, France

I. Roldán-Ruiz et al. (eds.), *Breeding in a World of Scarcity: Proceedings of the 2015 Meeting of the Section "Forage Crops and Amenity Grasses" of Eucarpia*,
DOI 10.1007/978-3-319-28932-8_31

lucerne varieties are bred and evaluated for registration according to agronomic value in monoculture. A crucial question is whether their agronomic value in monoculture is correlated to that in mixed stands. Comparing the agronomic value of varieties when they are grown in monoculture and in mixture is challenging because of multiplicative effects of the levels of the factors (grass variety × legume variety) but also because it has to integrate numerous traits (yield, quality, quantity of fixed N_2, etc.). This question can be divided into two aspects: (1) is the ranking of varieties in monoculture the same as in mixture? (2) in mixtures, does the ranking of the varieties of one species depend on the variety of the companion species? Our recent experimental results answer these questions.

Ranking of Lucerne Genotypes or Varieties Is Somewhat Different in Monoculture and in Mixture

A design with three treatments for lucerne was established: (1) dense monoculture, and mixed stands of lucerne either with (2) a forage-type tall fescue or (3) a turf-type tall fescue (Fig. 31.1). The genotypes were cloned to produce three repetitions. Each plot contained seven plants. In a first experiment, 46 lucerne clones taken from various varieties were studied. Forage yield was measured in four cuts per year during 2 years. In a second experiment, 200 genotypes originating from the progeny of a biparental cross were studied.

On the set of 46 genotypes, a strong correlation ($r=0.80$, $P<0.001$) was found between lucerne biomass in mixture and in monoculture (Fig. 31.2a). However, statistical interactions between genotypes and treatments (monoculture or mixture) were significant. Focusing on the most productive genotypes (at least 20 g per plant in monoculture of Fig. 31.2a), the correlation between the two treatments was no longer significant. On the set of 200 genotypes originated from a cross, the correlation between biomass in monoculture and biomass in mixture was much lower than with the 46 genotypes, especially when considering the most productive genotypes (Fig. 31.2b). These results were obtained with clones to test a large genetic variation (Julier et al. 2000) but the plots were very small. Nevertheless, they indicated a tendency that the most productive genotypes were the same in both treatments, and that some genotypes were relatively more productive in one treatment or another.

A field experiment with eight lucerne varieties was sown either in monoculture or in mixture with cocksfoot (three varieties) in four locations. Preliminary results indicated that the correlation between yield of the monoculture and yield of the mixture was not very high (Fig. 31.3).

These results indicated that biomass production of mixtures was generally positively correlated to biomass production of monocultures. However, some genotypes or varieties have

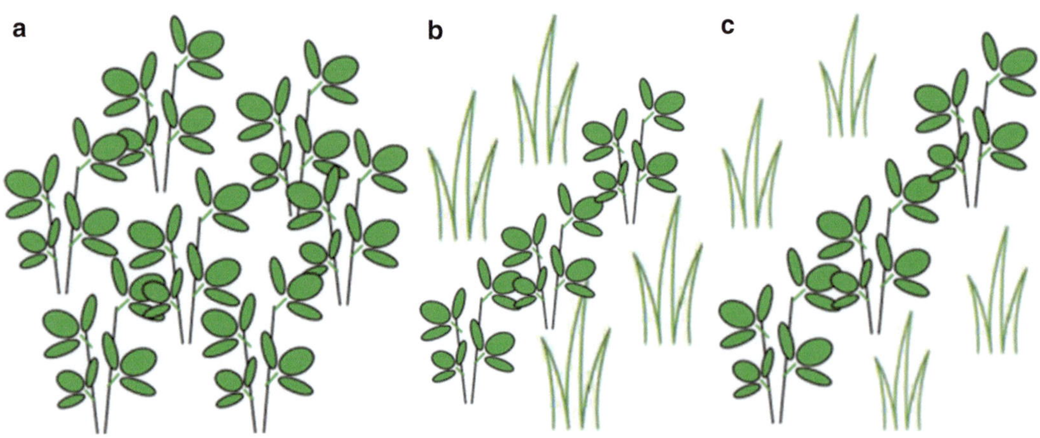

Fig. 31.1 Experimental design with 4 treatments (**a**) lucerne dense monocultures, (**b**) mixture of lucerne and tall fescue of forage type, (**c**) mixture of lucerne and tall fescue of turf type. Each treatment was studied with 46 lucerne genotypes

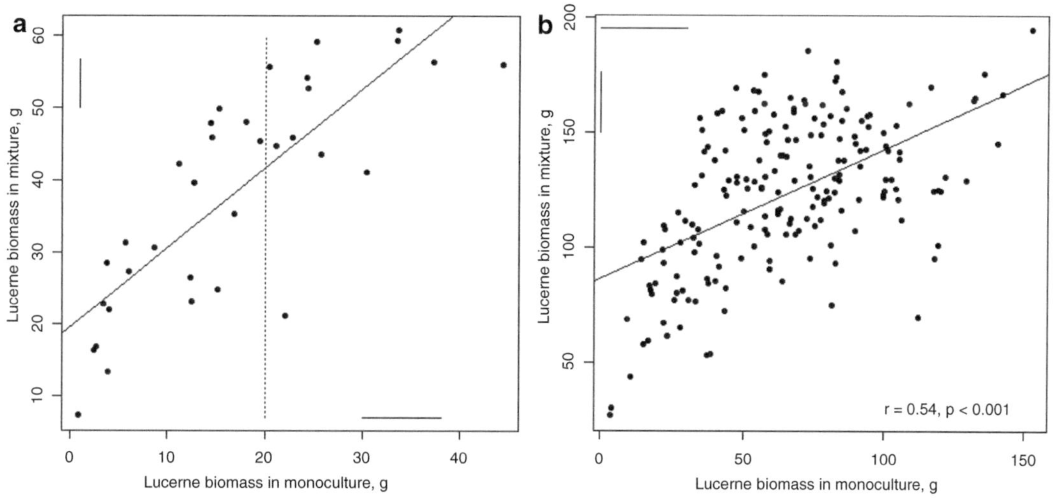

Fig. 31.2 Correlation between lucerne biomass in monoculture and in mixture in g/plant (**a**) average of 8 cuts in 2011 and 2012 for 46 genotypes, (**b**) data collected in the first cut of 2012, for 200 genotypes. Bars indicate residual standard error

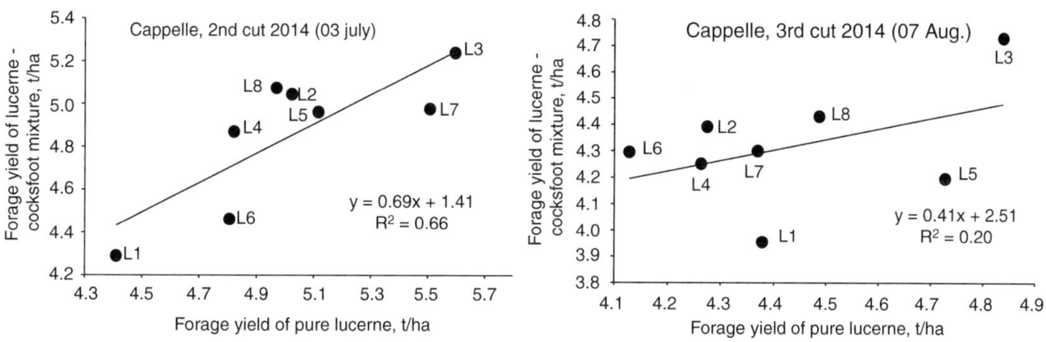

Fig. 31.3 Relationship between yield in monoculture and yield in mixture of eight lucerne varieties (L1 to L8) (average of three cocksfoot varieties) in two cuts in one location (Cappelle-en-Pévèle)

relatively higher production either in mixture or in monoculture. Similar results were obtained in other locations or when considering total annual forage yield. These findings should be completed with other agronomic traits such as the forage quality and the proportion of each species.

In Mixture, the Ranking of Varieties Is not Affected by the Choice of Companion Variety

In the above-described design with 46 genotypes, lucerne biomass production was studied when the companion genotype of tall fescue was a forage or a turf type. The statistical effect of lucerne genotype was significant, but there was generally no effect of fescue genotype and no interaction between lucerne genotype and fescue genotype (Fig. 31.4) (Maamouri et al. 2015). Similarly, in the design sown with eight lucerne and three cocksfoot varieties, no statistical interaction between lucerne variety and cocksfoot was significant for the biomass production of the mixtures. In both trials, these results are based on a large lucerne genetic variation but lucerne was generally dominant over the companion grass. A similar result was described on white clover with no interaction between clover variety and perennial ryegrass variety (Annicchiarico 2003). This shows that the effect of general aptitude to mixture (Gallais 1970) is much more important

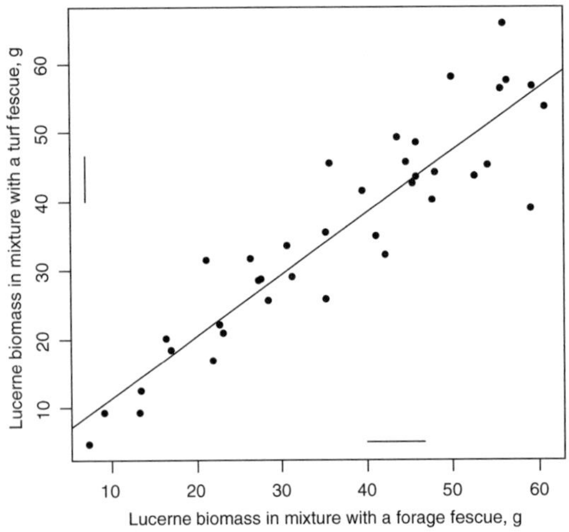

Fig. 31.4 Correlation between biomass production of lucerne genotypes grown in mixture with tall fescue of forage or turf types. Average data over four cuts in 2011 and four cuts in 2012. Each point is for one lucerne genotype. *Bars* are for residual standard errors

that the effect of specific aptitude to mixture. Thus, lucerne variety selection or evaluation for a use in a binary mixture might simply require testing the material in mixture with a single variety of the companion species.

Conclusion

Because the majority of temporary grasslands are now sown with a mixture of species, the value in mixture of varieties has become a key agronomic trait. We have shown that the value of a lucerne variety in mixture can be only partially deduced from its value in monoculture. Breeding schemes as well as variety testing should be designed to test the performance of the variety in mixed stands.

Acknowledgments Région Poitou-Charentes (project Expoleg-AV 2011–2015), Ministry of Agriculture (project Aglaé, Casdar C-2011-07, 2012–2015) and the Departments BAP and EA of INRA supported this work. We also thank field teams of INRA URP3F and Ferlus and private companies of ACVF grass and lucerne sections.

References

Annicchiarico P (2003) Breeding white clover for increased ability to compete with associated grasses. Journal of Agricultural Science 140: 255–266.

Chamblee DS, Colins M (1988) Relationships with other species in a mixture. In: AA Hanson, DK Barnes, RR Hill (Eds), Alfalfa and alfalfa improvement. ASA, Madison, WI. pp 439–461.

Gallais A (1970) Modèle pour l'analyse des relations d'associations binaires. Biométrie-Praximétrie 11: 51–80.

Julier B, Huyghe C, Ecalle C (2000) Within- and among-cultivar genetic variation in alfalfa: forage quality, morphology, and yield. Crop Science 40: 365–369.

Luscher A, Mueller-Harvey I, Soussana JF, Rees RM, Peyraud JL (2014) Potential of legume-based grassland-livestock systems in Europe: a review. Grass Forage Science 69: 206–228.

Maamouri A, Louarn G, Gastal F, Béguier V, Julier B (2015) In mixtures, lucerne genotype affects morphology, biomass production and nitrogen content of both lucerne and tall fescue. Crop and Pasture Science 66: 192–204.

Spandl E, Kells JJ, Hesterman OB (1999) Weed invasion in new stands of alfalfa seeded with perennial forage grasses and an oat companion crop. Crop Science 39: 1120–1124.

32 Optimizing Species and Variety Choice in Legume–Cereal Mixtures as Forage Crops in a Dry Mediterranean Region

I. Thami Alami, L. Pecetti, A. Souihka, and P. Annicchiarico

Abstract

During two seasons, this study assessed the forage yield of legume–cereal mixed crops including common vetch, Narbon vetch or pea grown together with oat or triticale, and their respective pure stands; verified the effect of mixture complexity; and assessed differences between a semi-dwarf and a tall pea type in pure stand and in mixtures. Crop mixtures including common vetch and oat generally gave highest dry-matter yield. The best mixtures over-yielded the pure stands of their component species. In a participative farmer assessment, legume pure stands and legume-based mixtures were more appreciated than cereal monocultures. On average, binary mixtures out-yielded the pure stands by 22 %. Any advantage of complex mixtures (four components) over simple mixtures appeared to be influenced by legume species and climatic conditions. Some superiority of the semi-dwarf pea type over the tall type was only observed in the first (wetter) season, possibly due to its better resistance to lodging.

Keywords

Competitive ability • Forage legume • Intercropping • Oat • Pea • Triticale • Vetch

Introduction

Overexploitation of forage resources, insufficiency of high-protein feedstuffs and climate change are jeopardizing livestock systems in dry regions of the Mediterranean basin. Cultivation of different functional groups of plants may contribute to sustainable intensification through better resource complementarity (Cardinale et al. 2007). Intercropping annual legumes with cereals tends

I. Thami Alami • A. Souihka
Institut National de la Recherche Agronomique, Rabat, Morocco

L. Pecetti (✉) • P. Annicchiarico
Council for Agricultural Research and Economics, Centre for Fodder Crops and Dairy Productions (CRA-FLC), Lodi, Italy
e-mail: luciano.pecetti@entecra.it

I. Roldán-Ruiz et al. (eds.), *Breeding in a World of Scarcity: Proceedings of the 2015 Meeting of the Section "Forage Crops and Amenity Grasses" of Eucarpia*,
DOI 10.1007/978-3-319-28932-8_32

to enhance the crop forage yield and quality and improves the efficiency of soil resource exploitation (Hauggaard-Nielsen and Jensen 2001). Mixtures of annual forage legumes with winter cereals are widespread in severe drought-prone Mediterranean regions (Lithourgidis et al. 2006). In such mixtures, the cereal companion improves light interception, provides structural support and facilitates the mechanical harvest, while the legume produces high-protein herbage, thereby improving forage quality (Roberts et al. 1989). Additional benefits of such mixtures can be greater (complementary) water and nutrient uptake, lower (or no) nitrogen fertilization, enhanced weed suppression, and better soil conservation (Ranells and Wagger 1997).

The choice of the best components in a mixture is a prerequisite for its success, as the companions should ensure a synergistic effect on forage yield. Different species, as well as different cultivars (or plant types), may affect the performance of the mixture by influencing the competition dynamics between companions (Hauggaard-Nielsen and Jensen 2001; Annicchiarico et al. 2012). The mixture composition may also be optimized in terms of diversity by varying the number and the proportion of components (Picasso et al. 2011).

Oat (*Avena sativa* L.) and triticale (× *Triticosecale* Wittm.) have often been reported as suitable cereals for mixtures with annual legumes (Lithourgidis et al. 2006). Common vetch (*Vicia sativa* L.), Narbon vetch (*V. narbonensis* L.) and field pea (*Pisum sativum* L.) have shown potential as annual legume crops for low-rainfall environments (Abd El Moneim 1993; Siddique et al. 1999). Competitive disadvantage for the legume has been observed in pea–cereal mixed cropping. Selecting pea varieties with better ability to compete for light and nutrients has been suggested to alleviate this hindrance (Hauggaard-Nielsen and Jensen 2001). Differences in plant stature can be found in pea germplasm due to the presence/absence of dwarfing mutations.

This study, carried out in a harsh Mediterranean environment of northwestern Morocco, assessed the forage yield of legume–cereal mixed crops and of their pure stands, verified the effect of mixture complexity, and compared the overall performance of a semi-dwarf vs. a tall pea type.

Material and Methods

Sixteen forage crops were grown during two growing seasons (2012–2013 and 2013–2014) in Marchouch, Morocco (33°33′ N, 6°41′ W) in plots of 3 m×4 m replicated four times in a randomized complete block design. The crops included four legume and two cereal pure stands, eight legume-cereal binary mixtures, and two complex (4-component) mixtures (Table 32.1). A semi-dwarf pea type (*cvs.* Attika in 2012–2013 and Kaspa in 2013–2014) and a tall pea type (line 2/38b/7, bred at CRA-FLC, sown in both years) were evaluated. Common vetch *cv.* Barril, Narbon vetch *cv.* Bozdag, oat, *cvs* Bionda in 2012–2013 and Genziana in 2013–2014, and triticale *cv.* Amarillo were the other species used in pure and mixed stands. Complex mixtures included the two cereals with either the two pea types or the two vetch species. Seeding rates for pure stands were 70 germinating seeds m^{-2} for the two pea types and the Narbon vetch, 140 seeds m^{-2} for the common vetch, and 280 seeds m^{-2} for the two cereals. These rates were halved in the sowing of binary mixtures, and reduced to one-fourth in the sowing of complex mixtures. In pre-sowing fertilization, 30 kg ha^{-1} N were applied to cereal pure stands and 15 kg ha^{-1} N to legume pure stands and mixtures, together with 45 kg P_2O_5 ha^{-1} to all plots. Nitrogen fertilization was repeated at the end of winter. Broadcast sowing was in autumn (November 23 and December 12) following a cereal crop (barley and durum wheat in the first and second year, respectively). The trial was carried out under rainfed conditions. The two seasons of evaluation were very different, with 571 mm and 296 mm annual (September to August) rainfall, being much wetter and drier than average, respectively (long term mean: 403 mm). During the crop cycle (November to April) rainfall amounted to 435 mm and 286 mm, respectively. The aerial biomass was harvested in the 2 years on April 19 and April 17, respectively,

Table 32.1 Total forage dry matter (DM) yield, legume proportion, weed proportion out of total aerial biomass, and mean score attributed by farmers, for a set of legume-cereal forage crops and their pure stands evaluated for two seasons in northwestern Morocco (the farmer score was only recorded in the second year)

Crop[a]	Total forage DM yield (t ha^{-1})		Legume % on total forage DM yield[b]		Weeds % on total biomass		Farmer score (1 min, 5 max)
	2012–2013	2013–2014	2012–2013	2013–2014	2012–2013	2013–2014	2013–2014
P_1O	4.23	10.71	54.6	36.4	47.1	1.6	3.9
P_2O	3.08	11.22	45.8	40.1	51.9	0.1	4.5
VO	7.17	11.04	60.1	17.9	30.8	1.0	4.1
NO	2.74	10.10	20.5	30.0	62.1	3.6	4.0
P_1T	3.96	9.19	52.0	45.8	47.1	2.4	4.0
P_2T	3.20	8.60	66.5	46.9	56.4	1.7	4.5
VT	6.04	7.77	61.3	38.4	43.7	1.7	4.0
NT	3.30	8.30	12.0	54.6	58.7	2.1	4.0
P_1P_2OT	3.38	9.67	33.1	43.3	51.3	3.0	3.5
VNOT	7.00	9.31	66.9	36.0	34.1	0.3	4.7
P_1	3.07	6.80	(100)	(100)	41.7	4.6	4.7
P_2	1.72	6.88	(100)	(100)	53.2	1.6	5.0
V	6.09	5.27	(100)	(100)	11.4	4.6	4.7
N	1.00	5.61	(100)	(100)	78.2	12.1	4.2
O	4.04	11.74	(0)	(0)	53.0	3.7	3.6
T	5.48	9.94	(0)	(0)	43.0	2.0	2.7
LSD $_{(P=0.05)}$	1.69	1.61	21.5	11.7	16.5	5.8	–

[a]P_1 semi-dwarf field pea, P_2: tall field pea, *N* Narbon vetch, *V* common vetch, *O* oat, *T* triticale
[b]Comparison among mixed crops, excluding pure stand crops (in parentheses)

when cereals were at late heading/early milky stage. In that period, legume grains were at waxy stage. A 1-m^2 sampling area in the middle of each plot was harvested and used to separate sown species from any weeds. After oven-drying at 65 °C, the proportion of weeds out of total biomass, the dry-matter (DM) yield of sown species, and the proportion of legume(s) and cereal(s) in mixtures were determined.

Shortly prior to harvest in the second season, a group of local farmers visited the trial and gave an evaluation of the different crops expressed on a scale from 1 = very poor to 5 = excellent, taking into account their potential for forage quantity and quality, the ease of cropping and the value as feed.

An analysis of variance (ANOVA) was carried out for DM yield, proportion of legume species and proportion of weeds, testing the factors 'crop' and 'season' and their interaction. ANOVAs with orthogonal contrasts compared binary mixtures *vs* pure stands (across species), binary *vs* complex mixtures within pea and vetches, the two pea types in pure stand and mixed stand, and the respective interactions with the season.

Results and Discussion

Differences among crops and among seasons, and 'crop × season' interactions were significant ($p < 0.001$) for all characters (data not reported). Crop mixtures including common vetch (V) and oat (O) tended to highest DM yield across seasons (Table 32.1). However, the common vetch pure stand tended to be out-yielded by the two pea lines in the second, drier season. Oat became more competitive respective to the legumes in the mixtures in the second season. In general, the mixtures including oat yielded better than the corresponding mixtures including triticale (Table 32.1). The yield advantage of the former over the latter (+19.7 % across seasons) was fully consistent with the difference reported by

Lithourgidis et al. (2006)) for oat and triticale mixtures with common vetch.

Binary mixtures with either pea type (P_1 or P_2) were noteworthy for their generally well balanced legume proportion (Table 32.1).

Unlike other studies (Lithourgidis et al. 2006, and the literature cited there), the best mixtures yielded better than not only the legume pure stand but also the pure stand of the cereal in question (Table 32.1). In the dry 2013–2014 season, cereal monocultures were less appreciated by farmers, whereas common vetch and pea pure stands received highest farmer scores (Table 32.1). These findings make a clear point towards the cultivation of adapted forage legume crops in drought-prone livestock systems of this region.

Contrast interactions with the season of growth were generally not significant at $p < 0.05$ (with the exception of DM yield of binary *vs* complex vetch mixtures). For that reason contrast results are reported across seasons in Table 32.2. Overall, binary mixtures yielded better than the pure stands by 22 %, with an equal proportion of weeds (Table 32.2). Increasing the complexity of the mixture had slightly different effects on the forage production depending on the mixture. While binary and complex mixtures including pea had very similar DM yields and weed proportions in both seasons, the complex mixture with the two vetch species had higher yield than the mean of binary vetch mixtures only in the first season (7.00 t ha^{-1} *vs* 4.81 t ha^{-1}, respectively, $p < 0.05$), thereby accounting for the abovementioned significant contrast interaction. Any advantage of complex mixtures over simple mixtures thus appeared to be influenced by legume species and climatic conditions.

According to contrasts, the semi-dwarf and the tall pea types did not differ for DM yield, legume proportion and weed proportion, either in pure stand or in binary mixture conditions (data not reported). Only in pure stand in the first season, the semi-dwarf pea type showed a trend towards higher yield and lower weed incidence (Table 32.1), likely due to its better lodging resistance than the tall type in the presence

Table 32.2 Comparison by analysis of variance and contrast for total forage dry-matter (DM) yield and weed proportion on total biomass between: (a) pure stand and binary mixtures, across species; (b) binary and complex (4-component) mixtures

Crop	Total forage DM yield (t ha^{-1})	Weeds % on total biomass
(a)		
Pure stands	5.64	25.8
Binary mixtures	6.92	25.7
Contrast probability	<0.05	ns
(b)		
Field pea:		
Binary mixtures	6.78	26.0
Complex mixture	6.53	27.2
Contrast probability	ns	ns
Vetches:		
Binary mixtures	7.06	25.5
Complex mixture	8.15	17.2
Contrast probability	ns	ns

ns not significant

of high rainfall (Annicchiarico et al. 2012). On the other hand, although the type × growing condition interaction was not significant at $p < 0.05$, the tall type had slightly lower decrease in legume DM yield in binary mixtures compared to the pure stand (2.97 t ha^{-1} *vs* 4.30 t ha^{-1}, −31 % across seasons) than the semi-dwarf type (3.07 t ha^{-1} *vs* 4.94 t ha^{-1}, −38 %) possibly owing to better competitive ability of the former type towards the cereal companions (Annicchiarico et al. 2012). Selecting varieties with high plant stature within the semi-dwarf type (exploiting the existing variation for this trait) could be an interesting compromise to combine competitive ability and lodging tolerance.

Acknowledgments This study was carried out in the framework of the FP7-ArimNet project "Resilient, water- and energy-efficient forage and feed crops for Mediterranean agricultural systems (REFORMA)". The pea material was developed by CRA-FLC with the support of the project "Plant Genetic Resources – FAO Treaty" funded by the Italian Ministry of Agriculture and Forestry Policy

References

Abd El Moneim AM (1993) Agronomic potential of three vetches (*Vicia* spp.) under rainfed conditions. Journal of Agronomy and Crop Science 170: 113–120.

Annicchiarico P, Ruda P, Sulas C, Pitzalis M, Salis M, Romani M, Carroni AM (2012) Optimal plant type of pea for mixed cropping with cereals. S. Barth, D. Milbourne (eds.), Breeding Strategies for Sustainable Forage and Turf Grass Improvement. Springer Science+Business Media, Dordrecht. pp 341–346.

Cardinale BJ, Wright JP, Cadotte MW, Carroll IT, Hector A, Srivastava DS, Loreau M, Weis JJ (2007). Impacts of plant diversity on biomass production increase through time because of species complementarity. Proceedings of the National Academy of Science of the United States of America 104: 18123–18128.

Hauggaard-Nielsen H, Jensen ES (2001) Evaluating pea and barley cultivars for complementarity in intercropping at different levels of soil N availability. Field Crops Research 72: 185–196.

Lithourgidis AS, Vasilakoglu IB, Dhima KV, Dordas CA, Yiakoulaki MD (2006) Forage yield and quality of common vetch mixtures with oat and triticale in two seeding ratios. Field Crops Research 99: 106–113.

Picasso VD, Brummer EC, Liebman M, Dixon PM, Wilsey BJ (2011) Diverse perennial crop mixtures sustain higher productivity over time based on ecological complementarity. Renewable Agriculture and Food Systems 26: 317–327.

Ranells NN, Wagger MG (1997) Grass–legume bicultures as winter annual cover crops. Agronomy Journal 89: 659–665.

Roberts CA, Moore KJ, Johnson KD (1989) Forage quality and yield of wheat-common vetch at different stages of maturity and common vetch seeding rate. Agronomy Journal 81: 57–60.

Siddique KHM, Loss SP, Regan KL, Jettner RL (1999) Adaptation and seed yield of cool season grain legumes in Mediterranean environments of south-western Australia. Australian Journal of Agricultural Research 50: 375–388.

Yield of Forage Pea-Cereal Intercropping Using Three Seed Ratios at Two Maturity Stages

33

S. Vasiljević, D. Milić, Đ. Karagić, V. Mihailović, A. Mikić, D. Živanov, B. Milošević, and S. Katanski

Abstract

The main purpose of this study was to evaluate productivity and profitability of different forage winter pea-cereal mixtures at two maturity stages: (1) in the middle of pea flowering, and (2) emergence of the first pea pods. A field experiment was conducted during the 2012–2013 and 2013–2014 growing seasons. The yield of winter pea (*Pisum arvense* L.), oat (*Avena sativa* L.) and triticale (×*Triticosecale*) monocultures as well as mixtures of winter pea with both cereals, in three seed weight ratios (50:50, 75:25 and 85:15), was investigated. To determine the efficiency of intercropping, the land equivalent ratio (LER) at the second maturity stage was calculated. The results showed that mixtures of winter pea with both cereals at the seed ratio of 85:15 had a significant advantage from intercropping which was attributed to better land use efficiency.

Keywords

Winter pea • Oat • Triticale • Intercropping • Forage yield • Land Equivalent Ratio

Introduction

Intercropping of winter annual forage legumes with cereals is extensively applied in southern Europe for the development of sustainable forage production systems, with limited external inputs in dry farming conditions during spring and summer. The major agronomic advantages of legume-cereal intercropping are higher productivity and profitability, increased forage quality, better lodging resistance and higher ability to adapt to climatic and biotic stresses (Mariotti et al. 2006; Ćupina B et al. 2011; Živanov et al. 2014). Several factors can affect the growth of the species used in intercropping, including cultivar selection, seeding ratios and competition between mixture components (Droushiotis 1989; Roberts et al. 1989;

S. Vasiljević (✉) • D. Milić • Đ. Karagić •
V. Mihailović • A. Mikić • D. Živanov • B. Milošević
S. Katanski
Department of Forage Crops, Institute of Field and Vegetable Crops, Novi Sad, Serbia
e-mail: sanja.vasiljevic02@gmail.com

I. Roldán-Ruiz et al. (eds.), *Breeding in a World of Scarcity: Proceedings of the 2015 Meeting of the Section "Forage Crops and Amenity Grasses" of Eucarpia*,
DOI 10.1007/978-3-319-28932-8_33

Papastylianou 1990; Caballero et al. 1995; Carr et al. 2004; Dhima et al. 2007). A major problem with legume-cereal intercropping for forage is determining the optimal harvest time because the growth cycle of consociated species is often not synchronized (Mariotti et al. 2006).

The efficiency of an intercropping system as compared to sole cropping can be evaluated by the LER index, defined as the relative land area required when growing a sole crop to produce the yield achieved in intercropping (Mead and Willey 1980).

Winter forage pea is a valuable component in the diet of all types of ruminants, because it has a high content of crude protein and calcium, while the crude fiber content is very similar to alfalfa. The differences in chemical characteristics between alfalfa and winter forage pea haylage are minimal, which is of great importance for the uniformity of feed and avoiding stress in animals. Traditionally in the Republic of Serbia, mixtures of winter pea with winter cereals (triticale and oat) are used extensively for forage production as the first crops in a double cropping system with maize and sorgho for silage as second crops.

The main aim of this study was to determine the effects of different proportions of winter forage pea (cv. *NS Pionir*) with either oat (cv *NS Jadar)* or triticale (cv *Odisej*) on forage yield and land use efficiency at two maturity stages: (1) in the middle of pea flowering and (2) emergence of the first pea pods.

Material and Methods

The experiments were carried out at the Institute of Field and Vegetable Crops in Novi Sad, Serbia, during the 2012–2013 and 2013–2014 growing seasons on a field characterized as a chernozem soil. The previous crop was spring vetch. Climatic data during the growing seasons are given in Table 33.1.

Winter pea (*Pisum arvense* L.), oat (*Avena sativa* L.) and triticale (× *Triticosecale*) monocultures as well as mixtures of winter pea with both cereals, in three seed weight ratios (50:50. 75:25 and 85:15) were planted with a special sowing machine. The row spacing was 12.5 cm and the seeds were mixed and sown together in both years at the end of October. The experimental design was a randomized complete block with nine treatments (three monocultures and six mixtures of winter pea with cereals); replicated four times. Plot size was 5 m × 20 m.

For determination of the dry matter yield. 1 m^2 was harvested in the center of each plot at two maturity stages, generally corresponding to: (1) the middle of pea flowering and (2) emergence of the first pea pods. Samples taken from the plots were chopped to approximately 1 cm particle size and dried in a drying oven at 60 °C for 72 h to determine dry matter content.

The land equivalent ratio (LER) was calculated as:

$$\mathrm{LER} = (\mathrm{Ypm} / \mathrm{Ypp}) + (\mathrm{Ycm} / \mathrm{Ycp})$$

Table 33.1 Precipitation and mean monthly temperature for growing season (2012–2014) and long term average (1963–2013)

Month	Precipitation (mm)			Temperature (°C)		
	2012–2013	2013–2014	1963–2013	2012–2013	2013–2014	1963–2013
X	49	67	45	13.7	14.7	11.7
XI	36	41	52	10.0	9.03	6.0
XII	55	2	46	1.0	2.0	1.7
I	60	24	37	3.0	4.3	–0.3
II	49	10	33	4.3	7.0	1.7
III	68	50	39	6.0	10.4	6.3
IV	30	49	47	13.7	13.4	11.7
V	118	203	60	18.4	16.4	17.0

where Ypp and Ycp are yields of forage pea and cereals in pure stands, respectively, and Ypm and Ycm are yields of forage pea and cereals in mixtures.

If LER is greater than one, this means that intercropping favors the yield of the species and vice versa (Ofori and Stern. 1987; Caballero et al. 1995).

Data were statistically analyzed using the statistical software STATISTICA 12.

Results and Discussion

Extremely hot weather with a high amount of precipitation during the autumn period in both years was favorable for sowing and establishment of the winter crops until the end of October (Table 33.1). Large amounts of moisture during the winter and a rainy spring, especially in March and May in 2013, had a favorable impact on the growth and development of winter forage pea whose share increased linearly during the growing season in all mixtures with cereals.

Large amounts of rainfall during 2014 in the first part of the growing season had a favorable effect on achieving high yields of winter pea as well of cereals. Oats appeared much more aggressive due to intensive tillering, which was reflected in lower share of pea in the mixtures.

The highest dry matter yield in 2013 was obtained at the stage of the first pea pods with a sole crop of triticale (11.1 t ha^{-1}) followed by pea-triticale 50 %: 50 % (10.42 t ha^{-1}) (Table 33.2). At the same maturity stage in 2014 clearly higher dry matter yield was obtained with mixtures of winter pea with both oat (12.0 t ha^{-1}) and triticale (12.87 t ha^{-1}) at the seeding ratio 85:15.

Land Equivalent Ratio (LER index) was calculated for each intercropping at the stage of the first pea pods (Table 33.3). The LER values were greater than one in both of years for the pea-cereal mixtures (85:15) indicating that in these systems, there was an advantage of intercropping for exploiting the resources of the environment.

Conclusion

The results of this experiment suggested that sowing mixtures of winter forage pea with both cereals (oat and triticale) at the 85:15 seeding ratio had a yield advantage of intercropping at the stage of the first pea pods for exploiting the resources of the environment compared with the other intercropping systems used in this study.

Acknowledgment This study has been supported by the Ministry of Education, Science and Technological Development of the Republic of Serbia (the project TR – 31024)

Table 33.2 Dry matter yield of winter pea and cereals as pure stands and intercrops at two maturity stages and during two growing seasons

	Dry matter yield (t ha^{-1})			
	Middle of pea flowering		First pea pods	
Treatment	16 May 2013	20 May 2014	29 May 2013	02 June 2014
Pure crops of oat (150 kg ha^{-1})	7.22	7.03	7.85	12.8
Winter pea-oat mixture (50:50 %)	6.83	7.50	7.28	10.38
Winter pea-oat mixture (75:25 %)	5.15	6.70	7.1	9.98
Winter pea-oat mixture (85:15 %)	5.80	10.80	8.0	12.0
Pure crops of pea (140 kg ha^{-1})	3.85	3.90	7.27	7.25
Pure crops of triticale (250 kg ha^{-1})	7.70	8.12	11.1	9.9
Winter pea-triticale mixture (50:50 %)	5.75	6.22	10.42	8.45
Winter pea-triticale mixture (75:25 %)	6.25	7.40	8.23	11.42
Winter pea-triticale mixture (85:15 %)	5.95	10.57	9.23	12.87
LSD 0.05	1.56	1.86	2.29	2.87
0.01	2.11	2.52	3.11	3.88

Table 33.3 Land equivalent ratio (LER) for mixtures of winter forage pea with oat and triticale at the stage of the first pea pods during two seasons

Winter pea-cereal mixture	Land equivalent ratio (LER) at the stage of the first pea pods	
	29.05.2013	02.06.2014
Winter pea-oat mixture (50:50 %)	0.925	0.815
Winter pea-oat mixture (75:25 %)	1.075	0.837
Winter pea-oat mixture (85:15 %)	1.400	1.100
Winter pea-triticale mixture (50:50 %)	0.925	0.922
Winter pea-triticale mixture (75:25 %)	0.900	1.225
Winter pea-triticale mixture (85:15 %)	1.000	1.225
LSD 0.05 0.01	0.275 0.380	0.470 0.650

References

Caballero R, Goicoechea EL, Hernaiz PJ (1995) Forage yields and quality of common vetch and oat sown at varying seeding ratios and seeding rates of common vetch. Field Crops Research 41: 135–140.

Carr PM, Horsley RD, Poland WW (2004) Barley, oat and cereal-pea mixtures as dryland forages in the Northern Great Plains. Agronomy Journal 96: 677–684.

Ćupina B, Mikić A, Stoddard FL, Krstić Đ, Justes E, Bedoussac L, Fustec J, Pejić B (2011) Mutual Legume Intercropping for Forage Production in Temperate Regions. Sustainable Agriculture Reviews. Genetics, Biofuels and Local Farming Systems (ed. E. Lichtfouse), Springer 7: 347–365.

Dhima KV, Lithourgidis AS, Vasilakoglou IB and Dordas CA (2007) Competition indices of common vetch and cereal intercrops in two seeding ratios. Field Crops Research, 100: 249–256.

Droushiotis DN (1989) Mixtures of annual legumes and small-grained cereals for forage production under low rainfall. Journal of Agricultural Science 113: 249–253.

Mariotti M, Masoni A, Ercoli L, Arduini I (2006) Forage potential of winter cereal/legume intercrops in organic farming. Italian Journal of Agronomy / Rivista di Agronomia 3: 403–412.

Mead R, Willey RW (1980) The concept of a land equivalent ratio and advantages in yields for intercropping. Experimental Agriculture 16: 217–228.

Ofori F, Stern WR (1987) Cereal–legume intercropping systems. Advances in Agronomy 41: 41–90.

Papastylianou I (1990) Response of pure stands and mixtures of cereals and legumes to nitrogen fertilization and residual effects on subsequent barley. Journal of Agricultural Science 115: 15–22.

Roberts CA, Moore KJ, Johnson KD (1989) Forage quality and yield of wheat-vetch at different stages of maturity and vetch seeding rate. Agronomy Journal 81: 57–60.

Živanov D, Jeftić R, Tančić S, Vasiljević S, Maširević S (2014). Control of winter forage pea diseases by pea-oat intercropping under field conditions. Pesticides and Phytomedicine (Belgrade), 29 (2): 131–136, DOI:10.2298/PIF1402131Z.

Association Mapping of *LpCCR1* with Lignin Content and Cell Wall Digestibility of Perennial Ryegrass

34

F.R.D. van Parijs, T. Ruttink, G. Haesaert, I. Roldán-Ruiz, and H. Muylle

Abstract

In perennial ryegrass (*Lolium perenne*), a decrease of lignin content will improve ruminal digestibility and increase nutritional value. In order to identify molecular markers that can be used to breed for improved cell wall digestibility, an association mapping study was conducted with *LpCCR1*, a candidate gene known to be involved in monolignol biosynthesis. A high level of diversity was discovered in this gene, with LD decaying over small distances. An intronic SNP was significantly associated with cell wall digestibility and Klason lignin content of stem material, and is therefore a potential candidate for use in marker assisted breeding.

Keywords

Cell wall digestibility (NDFD) • Lignin • Perennial ryegrass (*Lolium perenne*) • Cinnamoyl-CoA reductase (CCR) • Association mapping

Introduction

In Europe, perennial ryegrass *(Lolium perenne)* is used extensively as fodder grass in dairy farming (Humphreys et al. 2010). Although the digestibility of blade and stem is already high in ryegrasses (Frame 1991), this is mainly due to a high water soluble carbohydrate (WSC) content. For current varieties, a large part of the cell wall is not digested due to the recalcitrance of lignin surrounding the cellulose microfibrils. Decreasing the lignin content will result in significant digestibility gains. By improving the accessibility of the cellulose to microbial enzymes that decompose it into simple sugars, more energy will be released from the same amount of grass. Moreover, because these sugars are released more gradually than WSC, rumen acidosis is prevented. It has been estimated that increasing the

F.R.D. van Parijs • T. Ruttink • I. Roldán-Ruiz
H. Muylle (✉)
Plant Sciences Unit, Institute for Agricultural and Fisheries Research (ILVO), Melle, Belgium
e-mail: hilde.muylle@ilvo.vlaanderen.be

G. Haesaert
Faculty Bioscience Engineering, Department of Applied Biosciences, Ghent University, Ghent, Belgium

I. Roldán-Ruiz et al. (eds.), *Breeding in a World of Scarcity: Proceedings of the 2015 Meeting of the Section "Forage Crops and Amenity Grasses" of Eucarpia*,
DOI 10.1007/978-3-319-28932-8_34

neutral detergent fiber digestibility (NDFD) by one unit results in an increase of dry matter intake by 170 g/day and milk yields by 250 g/day (Oba and Allen 1999).

Lignin is composed of monolignol units synthesized by the phenylpropanoid and monolignol biosynthesis pathways, which comprise at least 10 gene families (Van Parijs et al. 2015). Cinnamoyl-CoA reductase (CCR) is one such gene family. Transgenic silencing of one of its members, *LpCCR1*, has previously been shown to decrease lignin content (Tu et al. 2010). Here, we explore the natural genetic variation within this gene in a large collection of perennial ryegrass accessions and identify polymorphisms associated with total lignin content (TL) and NDFD.

Materials and Methods

Polymorphism Screening in LpCCR1

LpCCR1 (GenBank accession KM516130) was identified in a genome-wide screen for the CCR gene family as described by (Van Parijs et al. 2015). This 6906 bp region was resequenced in a total of 746 genotypes by probe capture enrichment followed by Illumina HiSeq sequencing (Ruttink et al. 2015). Eighty bit-pair PE reads were mapped onto the reference genome scaffolds using BWA-mem, and SNPs and indels were called using GATK v3.2–2. The genotype calls were quality filtered for a minimum PHRED score of 30 and minimum read depth of 6. Further, only positions that were called in at least 200 genotypes and had a minor allele count of 5 were withheld, giving a final count of 118 k polymorphisms. From this dataset, the polymorphisms of *LpCCR1* were extracted and used for calculating LD decay in R. This was done by fitting the logarithm of the distance between all pairs of SNPs onto the r^2 values between them, and determining the expected distance for $r^2=0.3$. In a next step, missing genotype calls were imputed for all 118 k positions using fastPHASE. Population genetics analyses were performed with PopGenome, using the imputed data for *LpCCR1*. Population structure was determined using fastSTRUCTURE with a simple prior on all non-imputed data (Raj et al. 2014), while kinship was determined using the VanRaden algorithm on all imputed data (VanRaden 2008).

Phenotyping for NDFD and Total Lignin Content

In 2013, 622 genotypes from the set of 746 described above were planted in 12-l pots in three clonal repetitions. The plant material comprises natural accessions, pre-breeding material and commercial varieties. The plants were harvested at heading date, which was determined as the date when three spikes emerged. The plant material was dried in a ventilated oven for 48 h at 70 °C. Subsequently, blade and stem were manually separated, and both fractions were milled through a 0.5-mm filter (Fritsch Pulverisette 19) and scanned using near-infrared spectroscropy (NIRS, FOSS XDS). Neutral detergent fiber digestibility (NDFD) and total lignin content (TL) were predicted by NIRS, based on closed calibration models containing either 70 blade or 90 stem samples. NDFD was determined as described by Goering and Van Soest (1970). Total lignin content was estimated by the Klason method based on the NREL/TP-510–42618 standard, but modified for use with the ANKOM Fiber Analyzer. The heading date effect was determined in R by linear regression on individual plants. Best linear unbiased predictions (BLUPs) per genotype and trait were calculated on heading date corrected values, using the lme4 R package.

Association Mapping

Association mapping was performed using the imputed SNP and indel data set. First, each polymorphism was associated with a trait using a mixed model that takes structure and kinship into account, as implemented in GAPIT (Lipka et al. 2012). Then the gene sequence was divided in blocks of 200 bp, and the SNP with the most significant effect was identified per region. The set of p-values of this subset of SNPs was used to adjust for the false discovery rate. This approach is appropriate, given the LD range estimated within this gene and identical associations identified for neighboring polymorphisms.

Results and Discussion

Variation for Lignin Content and Cell Wall Digestibility in Stem and Blade

Because farmers harvest perennial ryegrass close to heading date, individual plants were harvested at this maturity stage. Even though all plants were harvested at the same maturity, a heading date effect remained: later heading plants contain significantly more TL/NDF (0.3 units per week for blade, 0.2 for stem) and have a significantly lower NDFD (1.3 units per week for blade, 0.8 for stem). This effect is larger for blade than for stem, because the blade has been growing for a longer period at heading date when the plant heads later, while for stem, the period between the start of stem elongation and spike emergence is almost constant. Nonetheless, a small heading date effect was also detected for stem, which is probably due to later heading genotypes generally growing slower.

As heading date is a confounder for assessing the organ-specific lignin content or NDFD properly, its effect was partialled out. This was done by subtracting the median lignin content or NDFD for the corresponding heading date from each trait value (HD correction). For example, a positive value after HD correction means that the plant has a higher lignin content or NDFD than expected for plants heading on that day of the year. The genotypic values were then calculated as BLUPs over the HD corrected values of the biological replicates. The variation among BLUPs was slightly smaller for stem than for blade for TL/NDF and NDFD (Table 34.1).

Population Structure and Relation to Phenotypic Traits

FastSTRUCTURE identified four subpopulations among our 622 genotypes. The largest subpopulation (Q1) comprises genotypes from Central and Northern Europe and derivations. Q2 contains genotypes derived from a commercial breeding program. Genotypes from New Zealand and warmer European regions were assigned to Q3. Q4 comprises the two parents and 30 F_1 plants of a mapping population. These subpopulations, which differ in genetic background, also display significant differences in NDFD and TL/NDF for blade (Table 34.1). However, for stem, no significant differences were found between the subpopulations for either trait.

SNP Density, Diversity and LD Decay in LpCCR1

A total of 1143 polymorphisms were discovered in the *LpCCR1* gene, which includes a 1000 bp promoter region. On average, 17 polymorphisms occur

Table 34.1 Mean and SD over HD corrected BLUPs in blade and stem for the total set (All) and for each subpopulation (Q1-Q4)

	All (622)	Q1 (392)	Q2 (103)	Q3 (95)	Q4 (32)	P-value
Blade						
NDFD	80.0 ± 1.3	80.1 ± 1.3	80.5 ± 1.2	79.6 ± 1.3	79.6 ± 1.4	<0.0001
TL/NDF	6.5 ± 0.4	6.4 ± 0.4	6.4 ± 0.3	6.6 ± 0.4	6.6 ± 0.4	<0.0001
Stem						
NDFD	77.5 ± 1.1	77.5 ± 1.1	77.6 ± 0.8	77.5 ± 1.3	77.1 ± 0.9	0.138
TL/NDF	6.8 ± 0.3	6.8 ± 0.3	6.8 ± 0.3	6.8 ± 0.4	6.8 ± 0.3	0.145

The number of genotypes is given in parentheses. The p-value represents the significance of differences among subpopulations for each trait (ANOVA)

every 100 bp over all genotypes, but 67 % of these are rare (MAF<5 %). Excluding these, the SNP density reduces to 5 polymorphisms per 100 bp. The density of common polymorphisms (MAF >5 %) is markedly higher in introns (11 %) than in exons (4 %), while the promoter contains strikingly few polymorphisms (2 %). However, this may be due to a smaller sequencing depth (Fig. 34.1). Nucleotide diversity, measured by the π statistic, estimates the average number of nucleotide differences per site between two randomly chosen genotypes. Contrary to SNP densities, the diversity increases with the size of the subpopulation (Table 34.2). As expected, the pair cross (Q4) shows the least diversity. Introns are more diverse than coding regions in all subpopulations, while the promoter appears to be least diverse. Overall, 30 % of the total genetic variation is explained among subpopulations, while 70 % is explained within subpopulations.

LD decay was calculated for each subpopulation (Table 34.2, Fig. 34.1). The shallowest decay was found for the mapping population (Q4), where r^2 falls below 0.3 over a mean distance of 3378 bp. This is expected, as fewer cross-over events occur in a pair-cross compared to wild material. The steepest decay was found for the largest subpopulation (Q1), where r^2 falls below 0.3 over a mean distance of 294 bp. Genotyping a common polymorphism every 200 bp should therefore be sufficient for association mapping.

Association Mapping

No significant marker-trait associations were identified for blade (Fig. 34.2), even though NDFD and TL/NDF were more variable for blade than for stem. Two SNPs located in different

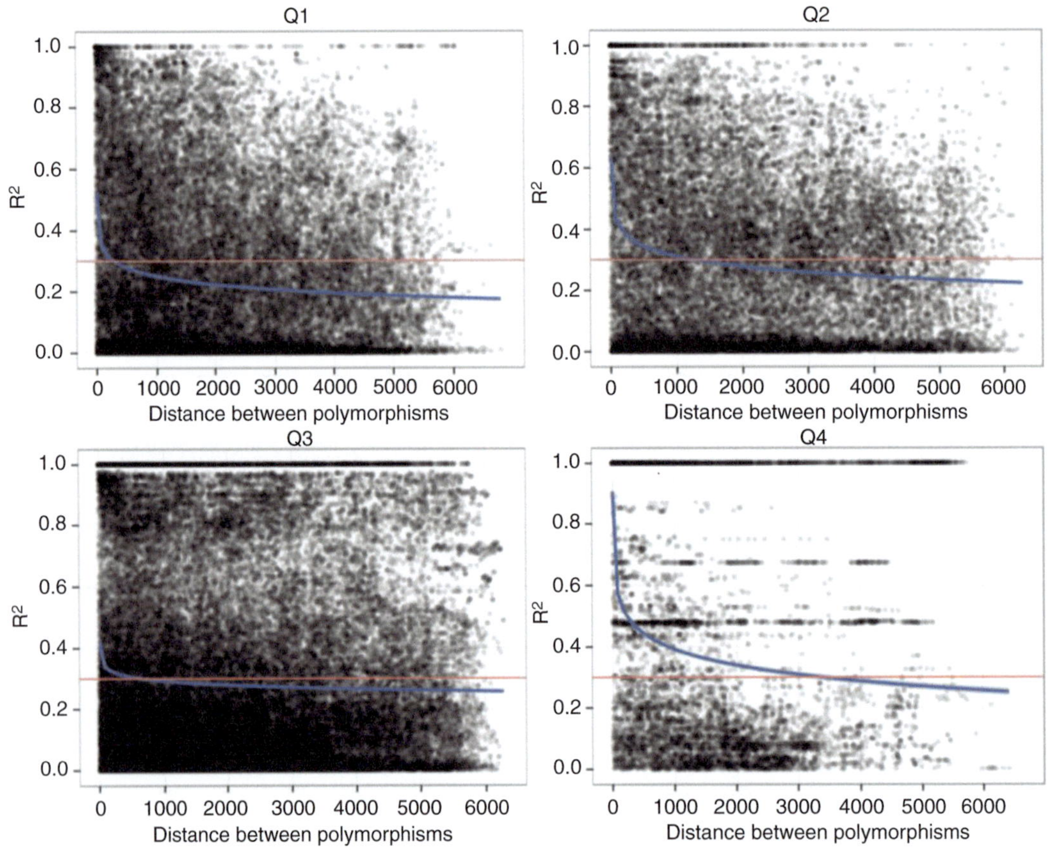

Fig. 34.1 LD decay plot for each subpopulation. *Each dot* represents LD (r^2) between a pair of SNPs, in function of the distance (in bp). The *blue line* shows the regression used to describe the relationship between the pairwise distance and r^2, the *horizontal line* shows the cutoff at $r^2=0.3$

Table 34.2 Number of common polymorphisms (MAF >5 %), SNP densities (percentages) and nucleotide diversities π (average number of pairwise variants per 1000 bp) in each subpopulation and gene region

	Genotypes	Gene (6906 bp)			CDS (1089 bp)		Intron (4234 bp)		Promoter (1000 bp)	
		Variants	π	LD decay	Variants	π	Variants	π	Variants	π
All	746	373 (5 %)	5.0	113 bp	44 (4 %)	2.3	277 (11 %)	6.8	22 (2 %)	0.4
Q1	392	320 (5 %)	7.7	294 bp	32 (3 %)	1.5	245 (10 %)	11.0	17 (2 %)	0.5
Q2	103	248 (4 %)	5.0	1270 bp	37 (3 %)	2.3	176 (7 %)	6.8	11 (1 %)	0.5
Q3	95	429 (6 %)	3.9	663 bp	46 (4 %)	2.3	315 (13 %)	4.9	20 (2 %)	0.9
Q4	32	165 (2 %)	3.6	3378 bp	24 (2 %)	2.1	124 (5 %)	4.9	8 (1 %)	0.0

For the entire gene, also LD decay is given ($r^2 < 0.3$)

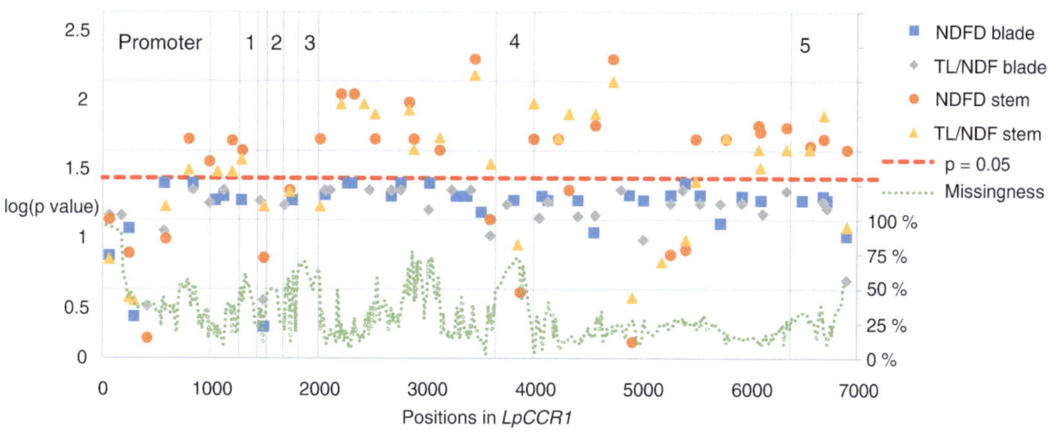

Fig. 34.2 Manhattan plot showing FDR adjusted –log(p) values for the additive effects of *LpCCR1* polymorphisms on TL/NDF and NDFD of blade and stem. The exons are indicated by numbers 1–5. All associations above –log (p) = 1.3 (p = 0.05) are significant, as indicated by the *horizontal dashed line*. Missingness before imputation (% of genotype calls missing) is indicated by the dotted line

introns (positions 3455 and 4734) were associated with NDFD of stem with a high significance, i.e., with an FDR-adjusted p-value of 0.007. These SNPs were also significantly associated with TL of stem. They have a minor allele frequency of 11 % and 27 %, respectively. Detailed analysis revealed that the effect at position 3455 is recessive, as only the minor homozygous class is more digestible compared to the other classes. For position 4734, the effect is truly additive. Although both SNPs are separated by 1279 bp, their dependence is rather high ($r^2 = 0.63$). In order to assess their conditional effects, both SNPs were fitted in a multimarker model. Indeed, only position 4734 still exerts a significant effect, with the minor allele (base C) additively increasing NDFD with 0.3 units and decreasing TL/NDF with 0.08 units. The SNP at position 4734 is therefore more likely to be functional than the one at 3455, or at least closer linked to the true causative variant.

Conclusion

SNPs spread over the entire *LpCCR1* gene are significantly associated with NDFD and TL/NDF of stem, making this gene likely to be a major gene involved in monolignol biosynthesis in the stem of perennial ryegrass. Even though blade shows more variation in NDFD and TL/NDF than stem, no significant associations were found for this organ. Breeding for an improved cell wall digestibility could now be complemented by use of a marker in the last intron of *LpCCR1*. Fixing its minor allele could improve NDFD up to 0.6 units. We hope to identify more markers in other genes that play a role in cell wall digestibility, to fix more beneficial alleles and combine their effects.

References

Frame J (1991) Herbage production and quality of a range of secondary grass species at five rates of fertilizer nitrogen application. Grass Forage Sci 46:139–151.

Goering HK, Van Soest PJ (1970) Forage fiber analyses (apparatus, reagants, procedures, and some applications). Agric Handb 379:1–24.

Humphreys MO, Feuerstein U, Vandewalle M, Baert J (2010) Ryegrasses. In: Boller B, Posselt UK, Veronesi F (eds) Fodd. Crop. Amenity Grasses. Springer New York, New York, NY, pp 211–260

Lipka AE, Tian F, Wang Q, Peiffer J, Li M, Bradbury PJ, Gore M a, Buckler ES, Zhang Z (2012) GAPIT: genome association and prediction integrated tool. Bioinformatics 28:2397–9. doi: 10.1093/bioinformatics/bts444

Oba M, Allen MS (1999) Evaluation of the importance of the digestibility of neutral detergent fiber from forage: effects on dry matter intake and milk yield of dairy cows. J Dairy Sci 82:589–96. doi: 10.3168/jds.S0022-0302(99)75271-9

Raj A, Stephens M, Pritchard JK (2014) fastSTRUCTURE: Variational Inference of Population Structure in Large SNP Data Sets. Genetics 197:573–89. doi: 10.1534/genetics.114.164350

Ruttink T, Veeckman E, Haegeman A, van Parijs F, Van Glabeke S, Muylle H, Vandepoele K, Byrne S, Asp T, Roldán-Ruiz I (2015) Probing genome diversity in natural and breeding populations of Lolium perenne. Plant Anim. Genome Conf. PAG XXIII

Tu Y, Rochfort S, Liu Z, Ran Y, Griffith M, Badenhorst P, Louie G V, Bowman ME, Smith KF, Noel JP, Mouradov A, Spangenberg G (2010) Functional analyses of caffeic acid O-methyltransferase and cinnamoyl-CoA-reductase genes from perennial ryegrass (Lolium perenne). Plant Cell 22:3357–73. doi: 10.1105/tpc.109.072827

Van Parijs FRD, Ruttink T, Boerjan W, Haesaert G, Byrne SL, Asp T, Roldán-Ruiz I, Muylle H (2015) Clade classification of monolignol biosynthesis gene family members reveals target genes to decrease lignin in Lolium perenne. Plant Biol 1–32. doi: 10.1111/plb.12316

VanRaden PM (2008) Efficient methods to compute genomic predictions. J Dairy Sci 91:4414–23. doi: 10.3168/jds.2007-0980

Marker-Trait Associations for Flavonoids and Biomass in White Clover (*Trifolium repens* L.)

35

W.L. Ballizany, A.G. Griffiths, B.K. Franzmayr, M.Z.Z. Jahufer, R.W. Hofmann, and B.A. Barrett

Abstract

White clover, a primary forage legume in temperate permanent pasture, is limited by poor adaptation to abiotic stress factors such as water scarcity. Flavonoids contribute to abiotic stress tolerance in plants. Genetic analysis of flavonoid accumulation may help in understanding its relationship to plant growth and morphology in white clover. The objective of this research was to discover marker-trait associations for biochemical and morphological traits previously identified as associated with drought tolerance, using clonally replicated white clover plants. Parents and 131 progeny of a bi-parental cross between the cultivar 'Grasslands Kopu II' (K2) and the ecotype 'Tienshan' (T) were genotyped with 104 microsatellite (SSR) markers and a Diversity Array Technology (DArT) assay, revealing 320 polymorphisms segregating from parent K2 and 208 from parent T. Markers on linkage group (LG) 1–2 were significantly ($p<0.005$) associated with concentrations of the flavonols quercetin (Q) and kaempferol (K) and the Q:K ratio (QKR). A cluster of linked markers including prs406 accounted for 21 %, 167 % and 53 % change in Q, K, and QKR trait values, respectively. Polymorphic loci on LGs 6–1 and 7–1 in parent K2 influenced shoot and root dry matter. Loci on LGs 7–2 and 8–1 influenced root dry matter but not shoot dry matter. Root to shoot ratio was influenced by loci associated with markers from parent T on LGs 4–1 and 8–1. These results support a hypothesis that flavonoid metabolism is under close genetic control and largely independent of genetic factors influencing growth, suggesting that it might be possible to improve both abiotic stress tolerance and growth potential in white clover. These marker-trait

W.L. Ballizany • R.W. Hofmann
Lincoln University, PO Box 84
Lincoln, 7647, New Zealand

A.G. Griffiths • M.Z.Z. Jahufer • B.A. Barrett (✉)
AgResearch Grasslands, Private Bag 11008,
Palmerston North 4442, New Zealand
e-mail: brent.barrett@agresearch.co.nz

B.K. Franzmayr
AgResearch Grasslands,
Private Bag 11008,
Palmerston North 4442, New Zealand

Slipstream Automation, PO Box 1979
Palmerston North, 4410, New Zealand

I. Roldán-Ruiz et al. (eds.), *Breeding in a World of Scarcity: Proceedings of the 2015 Meeting of the Section "Forage Crops and Amenity Grasses" of Eucarpia*,
DOI 10.1007/978-3-319-28932-8_35

associations revealed deleterious alleles in an elite cultivar, and indicate the potential value of diversity from wild germplasm for white clover improvement.

Keywords
Flavonoids • Molecular marker • Stress tolerance • White clover

Introduction

White clover (*Trifolium repens* L.) is a stoloniferous perennial legume commonly sown or naturalized in mixed-species pastures and grasslands of temperate latitudes. It is valued for its compatibility with grass species, provision of high quality feed to ruminants, fixation of atmospheric nitrogen, and enhancement of soil quality; however, it is susceptible to seasonal drought (Hofmann and Campbell 2011).

For nearly a century, white clover breeding has focused on expanding adaptation and improving yield and plant survival under a range of biotic and abiotic stresses. A number of phenotypic characteristics contributing to drought adaptation and yield across environments have been identified (Jahufer et al. 2012; Jahufer et al. 2013). Recent advances in inter-specific hybridisation may extend these efforts via introgression of root characteristics and enhanced physiological response to water deficit (Nichols et al. 2014; Williams 2014).

White clover is allotetraploid ($2n=4x=32$) and exhibits disomic inheritance (Williams et al. 1998), with microsatellite (SSR) markers and genetic linkage maps (Griffiths et al. 2013) enabling genetic insight and analysis.

Flavonols including quercetin and kaempferol are implicated in tolerance of both ultra-violet stress and moisture limitation in white clover (Ballizany et al. 2012a, b; Hofmann and Jahufer 2011). However, no specific genetic analysis of factors influencing plant growth and the stress tolerance-associated flavonols quercetin and kaempferol has been conducted. The aim of this study was to identify molecular marker-trait associations for biochemical and morphological traits in white clover associated with plant response to abiotic stress.

Materials and Methods

Plant Material

The population and specific methods are fully described in Ballizany et al. (2014), and are briefly summarized here. Two heterozygous parent plants, one from the stress-resistant ecotype 'Tienshan' (parent T) with high constitutive accumulation of quercetin (Q) and kaempferol (K) and one from a high yielding, large leaved cultivar 'Grasslands Kopu II' (parent K2) with low levels of Q and K were crossed to generate a full-sib F_1 population. Randomly sampled seedlings ($n=130$) were clonally propagated and grown in three replicates in 7.5 L containers with non-limiting irrigation as part of a wider series of experiments including moisture stress in containers and field plots. Clones of both parent plants were included in all phenotypic and molecular marker assays.

Trait Measurements

As reported in Ballizany et al. (2014), for plants grown in the irrigated container experiment, total concentrations of glycosides of the flavonols Q (mg g^{-1} Dry Matter) and K (mg g^{-1} DM) were measured using high performance liquid chromatography according to the methods of Ballizany et al. (2012b). The Quercetin:Kaempferol Ratio (QKR) was calculated for each sample. Morphological traits measured were Shoot Dry Matter (g) (SDM); Root Dry Matter (g) (RDM); and the Root:Shoot Ratio (RSR). Pearson product-moment coefficient (r) was used to calculate correlations and test their significance.

DNA Markers

Following the methods of Griffiths et al. (2013), DNA of parents and progeny was purified and genotyped with five genomic microsatellite markers (SSRs), 26 EST-SSRs (Barrett et al. 2004), and 73 GeneThresher®-derived SSRs distributed evenly throughout the genome (Griffiths et al. 2013). Following capillary electrophoresis, each allele was scored for presence or absence in parents and progeny. A DArT array from Diversity Arrays Technology Pty Ltd. (Canberra. Australia) was developed and used to genotype parents and progeny using standard protocols (Jaccoud et al. 2001). DArT marker location was estimated by linkage analysis with SSRs, using JoinMap 3.0 (Van Ooijen and Voorrips 2001).

Marker-Trait Analysis

Each marker allele was tested for trait association using the trait Best Linear Unbiased Predictors (BLUPs) estimated by Restricted Maximum Likelihood in GenStat. MapQTL 4.0 (Van Ooijen et al. 2002) was used to classify BLUPs into present and absent treatment means for each marker allele, and implement a Kruskal–Wallis non-parametric test for treatment mean differences. The risk of false associations was reduced by setting the significance threshold at $p<0.005$. The effects reported are trait means and calculated percentage difference in trait value for progeny with and without the marker allele.

Results and Discussion

Phenotypic assessment indicated genetic variation for flavonols Q and K and for biomass production as estimated by SDM and RDM (for summary see Ballizany et al. 2014). For SDM and RDM, most progeny exhibited trait values greater than the parents. Parent T exhibited higher RSR and lower SDM and RDM values than parent K2 and their F_1 progeny. Moreover, parent T exhibited significantly ($p<0.05$) higher Q, K, and QKR values than parent K2. There were weak negative correlations among progeny between Q and RDM, and Q and SDM ($r=-0.24$, p=.006 and −0.30, p<0.001; respectively) and a weak positive correlation between Q and RSR ($r=0.20$, p=0.021). Despite the moderate correlation between Q and K ($r=0.56$, p<0.001), K was not correlated with SDM or RDM (p=0.427 and p=0.384, respectively).

The DArT array contained 2561 *Trifolium* features. However, only 729 features were identified in this study, of which 211 were polymorphic between the parents and segregated in the F_1 progeny. Sixty SSRs were polymorphic in both parents, 33 uniquely polymorphic in parent K2, and 11 in parent T. Parent K2 exhibited 320 SSR and DArT polymorphic alleles compared to 208 for parent T. Among K2 and T-derived alleles, 23 % and 12 % exhibited significant ($p<0.05$) segregation distortion, respectively; this implies that the loci that influence fitness are polymorphic within each parent. These data suggest germplasm from a region such as the Tienshan alpine zone of China has lower allelic diversity than the elite cultivar used in this experiment, perhaps due to extreme and unique environmental conditions and/or geographical isolation.

Twenty genome regions with significant ($p<0.005$) marker-trait associations were detected (Table 35.1), of which seven originated from parent T. The concentration of Q was influenced by alleles from loci on LGs 1–1, 1–2, 2–2, 4–2 and 5–1, each accounting for 13–21 % of the phenotypic variation. The concentration of K was strongly associated with a cluster of markers located on LG 1–2 including markers prs406 and 843887, with a 167 % change associated with the prs406 allele segregating from parent K2. The region 66–72 cM on LG 1–2 was significantly ($p<0.005$) associated with Q, K and QKR accounting for 21 %, 167 % and 53 % change in BLUPs, respectively. For each trait, prs406 was associated with a significant effect.

Marker-trait associations for SDM, RDM and RSR were identified in linkage groups 4–1, 6–2, 7–1, 7–2 and 8–1, with each accounting for up to 21 % of the phenotypic variation. The SDM and

Table 35.1 Summary of significant ($p<0.005$) marker-trait associations in white clover related to accumulation of Quercetin glycosides (Q) (mg/g DM), Kaempferol glycosides (K) (mg/g DM), Q:K ratio (QKR), Shoot Dry Matter (SDM) (g), Root Dry Matter (RDM) (g), and Root: Shoot Ratio (RSR) as measured in container plants

Trait	Marker allele[a]	Linkage group[b]	cM	Parent	Allele absent	Allele present	% Effect[c]	Supporting markers[d]
Q	gtrs366d	1–1	49	T	2.54	2.19	−14	–
	846087	1–2	72	K2	2.18	2.63	21	4
	845458	2–2	0	K2	2.17	2.52	17	–
	gtrs355b	2–2	40	K2	2.20	2.55	16	–
	prs592g	4–2	63	T	2.21	2.53	14	1
	gtrs273b	4–2	68	T	2.23	2.52	13	–
	gtrs245a	5–1	65	K2	2.14	2.45	14	–
K	prs406e	1–2	66	K2	0.46	1.23	167	11
	gtrs967g	4–2	45	T	0.91	0.67	−26	–
	gtrs250d	6–1	57	T	0.69	0.93	35	–
QKR	prs406e	1–2	66	K2	4.89	2.29	−53	11
	845742	6–1	0	K2	3.46	4.31	25	–
SDM	842847	6–2	36	K2	3.90	3.14	−20	1
	gtrs873b	7–1	39	K2	3.94	3.13	−21	–
	872195	7–2	51	K2	3.24	3.78	17	–
RDM	842847	6–2	36	K2	1.35	1.15	−15	1
	gtrs873b	7–1	39	K2	1.38	1.13	−18	–
	843925	8–1	14	K2	1.16	1.41	21	1
RSR	gtrs168c	4–1	27	T	0.36	0.41	13	1
	820291	8–1	54	T	0.41	0.36	−10	–

[a]Six numeral identifiers are DArTs. SSR locus nomenclature as per Griffiths et al. (2013)
[b]Genome nomenclature as per Griffiths et al. (2013)
[c]Percentage change in mean phenotypic value for presence versus absence of the marker allele
[d]Number of significantly associated loci linked to the reported marker allele

RDM associated polymorphisms were in parent K2. The drought-adaptive RSR effects originated from parent T and were in genome locations independent of SDM and RDM.

These findings indicate LG 1–2 includes a genome region that substantially influences flavonol accumulation independent of plant growth, and which may contribute to an adaptive drought response. This region may be a focus point to map at higher resolution to identify candidate genes for flavonol accumulation.

The low phenotypic correlations and distinct genome locations that influence the stress tolerance and biomass traits studied here, suggest that parallel improvement is possible. While this experiment has identified marker-trait associations related to flavonoids and morphological traits, it was limited by the use of clonally propagated plants in containers, different from farm swards where a mix of clover and grasses are normally used. Therefore, further research should evaluate plants in grazed field conditions. Another limitation of this experiment was the evaluation of a bi-parental population which restricts allelic diversity for trait detection and subsequent breeding from these plants. Association studies or genomic selection using complex and well-structured populations could alleviate this issue.

This is the first study examining molecular markers associated with levels of Q and K in temperate pasture plants. The finding that wild germplasm harbors adaptive biochemical and morphological features may contribute to future improvement of white clover. Further research is needed to confirm and accurately identify specific genomic regions associated with agronomic traits, elucidate the genetic regulation and consequences of Q glycosides, and examine the inheritance of Q glycosides and agronomic features in other genetic backgrounds and under influence of environmental factors such as high and low temperature. These data are useful for breeding programs aiming to develop populations with Q-mediated abiotic stress tolerance and high pasture yield. Collectively, these findings have positive implications for improving stress tolerance and yield potential in white clover.

Acknowledgments WLB was supported by fellowship PGGS0801 "Stress Tolerant White Clover" from the Ministry of Science and Innovation via Grasslands Innovation Ltd

References

Ballizany WL, Hofmann RW, Jahufer MZZ, Barrett BA (2012a) Genotype x environment analysis of flavonoid accumulation and morphology in white clover under contrasting field conditions. Field Crops Res 128 (0): 156–166.

Ballizany WL, Hofmann RW, Jahufer MZZ, Barrett BA (2012b) Multivariate associations of flavonoid and biomass accumulation in white clover (Trifolium repens) under drought. Funct Plant Biol 39 (2): 167–177.

Ballizany WL, Hofmann RW, Jahufer MZZ, Barrett BA (2014) Variation for constitutive flavonols and morphological traits in a new white clover population. Environ Exp Bot 105 (0): 65–69.

Barrett B, Griffiths A, Schreiber M, Ellison N, Mercer C, Bouton J, Ong B, Forster J, Sawbridge T, Spangenberg G, Bryan G, Woodfield D (2004) A microsatellite map of white clover. Theor Appl Genet 109 (3): 596–608.

Griffiths AG, Barrett BA, Simon D, Khan AK, Bickerstaff P, Anderson CB, Franzmayr BK, Hancock KR, Jones CS (2013) An integrated genetic linkage map for white clover (Trifolium repens L.) with alignment to Medicago. BMC Genomics 14 (1): 388.

Hofmann RW, Campbell BD (2011) Response of Trifolium repens to UV-B radiation: morphological links to plant productivity and water availability. Plant biology (Stuttgart, Germany) 13 (6): 896–901.

Hofmann RW, Jahufer MZZ (2011) Tradeoff between biomass and flavonoid accumulation in white clover reflects contrasting plant strategies. PLoS ONE 6 (4).

Jaccoud D, Peng K, Feinstein D, Kilian A (2001) Diversity arrays: a solid state technology for sequence information independent genotyping. Nucleic Acids Res 29 (4): E25.

Jahufer MZZ, Dunn A, Baird I, Ford JL, Griffiths AG, Jones CS, Woodfield DR, Barrett BA (2013) Genotypic Variation for Morphological Traits in a White Clover Mapping Population Evaluated Across Two Environments and Three Years. Crop Sci 53 (2): 460–472.

Jahufer MZZ, Ford JL, Widdup KH, Harris C, Cousins G, Ayres JF, Lane LA, Hofmann RW, Ballizany WL, Mercer CF, Crush JR, Williams WM, Woodfield DR, Barrett BA (2012) Improving white clover for Australasia. Crop Pasture Science 63 (9): 739–745.

Nichols SN, Hofmann RW, Williams WM (2014) Drought resistance of Trifolium repens × Trifolium uniflorum interspecific hybrids. Crop and Pasture Science 65 (9):911–921.

Van Ooijen JW, Voorrips RE (2001) JoinMap 3.0 software for the calculation of genetic linkage maps. Plant Research International Wageningen, the Netherlands.

Van Ooijen JW, Boer MP, Jansen RC, Maliepaard C (2002) MapQTL ® 4.0, Software for the calculation of QTL positions on genetic maps. 4.0 edn. Plant Research International, Wageningen, The Netherlands.

Williams WM (2014) Trifolium interspecific hybridisation: widening the white clover gene pool. Crop Pasture Science 65 (11): 1091–1106.

Williams WM, Mason KM, Williamson ML (1998) Genetic analysis of shikimate dehydrogenase allozymes in Trifolium repens L. Theor Appl Genet 96 (6-7): 859–868.

Breeding for Improved Nonstructural Carbohydrates in Alfalfa

36

A. Claessens, Y. Castonguay, A. Bertrand, G. Bélanger, and G.F. Tremblay

Abstract

Extensive degradation of proteins combined with insufficient readily fermentable energy in alfalfa (*Medicago sativa* L.) results in poor forage N utilization by ruminants and substantial N losses into the environment. Selection for increased nonstructural carbohydrate (NSC) concentration in forage crops is considered to be an effective approach to improve the feeding value of forage. Our objective was to assess the extent of genetic variability for NSC concentration and to identify DNA polymorphisms associated with that trait in alfalfa. A divergent selection was applied for forage NSC concentration among 560 alfalfa genotypes. Ten plants with high (NSC+), 10 plants with low (NSC-) and 10 plants randomly selected within the base population (NSC0) were selected to make three populations. Genetic variability for NSC concentration in alfalfa was observed (ranging from 73 to 230 mg/g DM under field conditions). Genotypes that consistently maintained high or low NSC concentrations were singled out under repeated assessments and were subsequently confirmed with clonal propagules. Although it is effective, phenotypic selection for NSC is resource intensive. Our identification of DNA polymorphism associated with this trait suggests the possibility of developing marker-assisted approaches to facilitate the identification of superior genotypes.

Keywords

Nonstructural carbohydrates • Genetic variation • Alfalfa • Breeding

Introduction

Alfalfa (*Medicago sativa* L.) is the most important forage legume crop cultivated in North America. Rapid and extensive ruminal degradation of proteins in alfalfa generally leads to

A. Claessens (✉) • Y. Castonguay • A. Bertrand
G. Bélanger • G.F. Tremblay
Agriculture and Agri-Food Canada, Québec Research and Development Centre,
Québec, QC, Canada
e-mail: annie.claessens@agr.gc.ca

I. Roldán-Ruiz et al. (eds.), *Breeding in a World of Scarcity: Proceedings of the 2015 Meeting of the Section "Forage Crops and Amenity Grasses" of Eucarpia*,
DOI 10.1007/978-3-319-28932-8_36

decreased protein efficiency (Broderick et al. 1995). With a greater supply of readily fermentable carbohydrates in the rumen, more non–protein N and amino acids can be incorporated into microbial protein, resulting in improved animal productivity (Parsons et al. 2011; Bryant 2012). Late-lactation dairy cows fed with high NSC alfalfa baleage had higher milk yield (Brito et al. 2008) and N use efficiency (Brito et al. 2009).

Forage water soluble carbohydrate (WSC) concentration of perennial ryegrass (*Lolium perenne* L.) has been increased by as much as 40 g kg^{-1} DM through genetic selection with a demonstrated positive impact on animal performance (Humphreys 1989). Even though this trait is strongly influenced by the environment, Humphreys (1989) reported that ryegrass genotypes with contrasted WSC accumulation potential can be reliably identified under low genotype × environment interactions.

Our objective was to assess the extent of genetic variability for forage NSC concentration in an alfalfa cultivar and to identify DNA polymorphisms associated with that trait.

Material and Methods

We seeded 560 alfalfa genotypes from the AC Caribou cultivar in the fall of 2003 in 13 cm pots and maintained them (fertilized twice per week and cut three times) under environmentally-controlled conditions set to 22 °C/17 °C (day/night) temperatures throughout the winter. Plants were sampled for NSC analysis under controlled conditions in April 2004 and subsequently transplanted to a field nursery where they were sampled in September 2004 and July 2005. All samples were taken at the late bud stage of development, dried at 55 °C in a forced-draft oven for 3 days and ground to pass through a 1-mm screen of a Wiley mill (Model 3. Arthur H. Thomas Co., Philadelphia, PA, USA). Ground samples were scanned using a NIR Systems 6500 monochromator (Foss, Silver Spring, MD, USA) as described by Nie et al. (2009). Seventy-five samples were selected based on their spectrum to form a calibration set (n = 60) and a validation set (n = 15). All samples were chemically analyzed for NSC fractions (glucose, fructose, sucrose, pinitol, and starch) as described in Bertrand et al. (2008). Based on NIRS predictions, 10 genotypes were selected for high (NSC+) or low (NSC−) NSC concentrations, while 10 genotypes were randomly selected within the base population (NSC0).

The selected genotypes were intercrossed in 2006 and progenies from each of the three populations (NSC+, NSC−, and NSC0) were vegetatively propagated. Four cuttings of each genotype were transplanted in 13 cm pots and maintained under the environmentally–controlled conditions previously described. The experiment was conducted as a randomized complete block design with four replicates. Plants were cut twice at the late bud stage of development. The harvested material was dried, ground, scanned, and NSC concentrations were predicted using NIRS as described above.

These progenies were also sampled for genomic DNA extraction. The DNA samples from the NSC+, NSC− and NSC0 genotypes were pooled within each group and used to uncover polymorphisms associated with NSC using the sequence-related amplified polymorphisms (SRAP) PCR-based marker system. The PCR amplifications of bulked samples were performed according to a protocol previously described (Castonguay et al. 2010).

Statistical Analysis

Data were analyzed using SAS software (SAS Institute 2002). Three procedures were used to determine the effect of the three populations: PROC MIXED for the analysis of variance, PROC PLOT and PROC UNIVARIATE for the graphic residue analysis and verification of the homogeneity of variance assumptions and normality of the data distribution. The likelihood ratio test was used for the analysis of variance between the NSC+ and NSC0 distribution curves.

Results and Discussion

Identification of Contrasting NSC Genotypes

The NSC concentration among the 560 genotypes varied from 5 to 170 mg/g DM at the first sampling date under controlled conditions and from 73 to 230 mg/g DM at the second sampling date under field conditions (Fig. 36.1). Thirty percent of the genotypes that were allocated to the low or to the high NSC groups at the first sampling date were not confirmed at the second sampling date. Genotype miscalling was even higher (50 %) between the second and third sampling date and resulted in a low correlation coefficient (r = 0.13). Variable ranking of genotypes at different cuts underscores the significant impact of plant development and environment on NSC concentration. Our findings are consistent with studies in tall fescue that demonstrated the effect of seasonal variations on NSC concentration (Shewmaker et al. 2006; Mayland et al. 2000). Large differences in NSC concentration could be attributable to the complex interaction between endogenous (e.g. stage of development) and exogenous (e.g. irradiance level, temperature, water status) factors affecting photosynthesis and plant metabolism. Notwithstanding these factors, 10 genotypes that consistently combined high DM yield with high (blue dots) or low (red dots) NSC concentration under repeated assessments were singled out (Fig. 36.1).

Clonal Evaluation of Stability of NSC Trait

The validity of the groupings based on NSC concentration was confirmed using the mean of four clonal replicates of progenies of high and low NSC crosses exposed to low genotype×environment interactions in environmentally-controlled chambers. Distribution of NSC concentrations showed significant differences between the three NSC groups (*P* <0.001; Fig. 36.2). On average, NSC+ clones had significantly higher NSC concentration (102.0 mg/g DM) compared to the NSC− (91.4 mg/g DM) and NSC0 (85.7 mg/g DM) clones. Furthermore, the variance of the NSC+ distribution was significantly narrower compared to

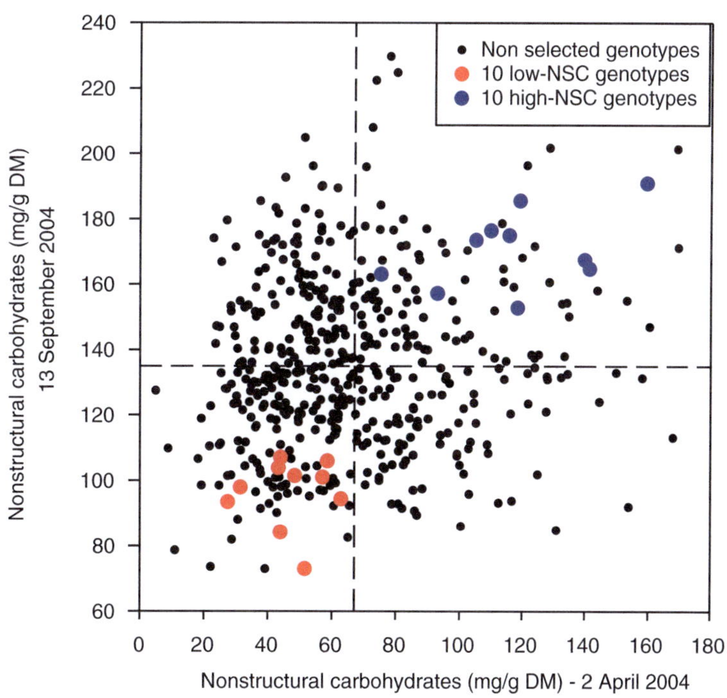

Fig. 36.1 Concentration of nonstructural carbohydrates in 560 alfalfa genotypes (cv. AC Caribou) at two sampling dates

that of the NSC0 distribution, indicating an enrichment in genotypes with high NSC concentration.

How Could Selection for NSC be Improved?

Genetic gains are tributary to the availability of screening techniques for the precise identification of superior genotypes. Although our results show that genotypes with high NSC concentration can be identified using multiple assessments of growth cycles, this approach is too resource intensive for breeding applications. Therefore, alternatives or complementary approaches need to be considered.

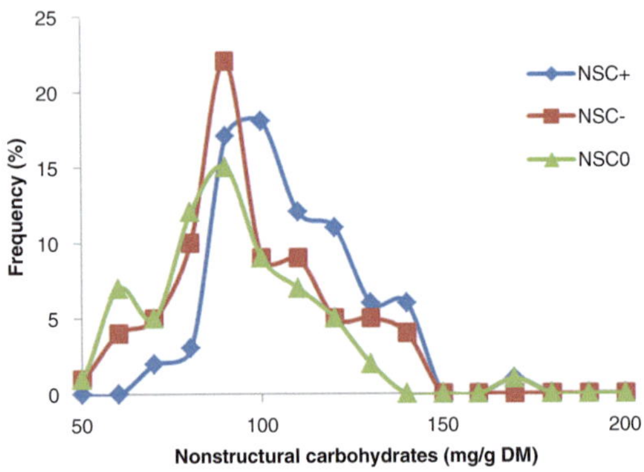

Fig. 36.2 Frequency distribution of NSC concentration in clonal replicates of progenies from populations divergently selected for NSC concentrations (NSC+ and NSC−) and a randomly selected population (*NSC0*). Plants were grown in an environmentally-controlled chamber and harvested twice at the late bud stage of development

Breeding alfalfa cultivars with superior NSC concentration could be accelerated by the identification of molecular markers associated to this trait. Sequence-related amplified polymorphism (SRAP) is a highly reproducible PCR amplification technique that has been successfully applied for the identification of polymorphisms associated to complex traits in alfalfa (Castonguay et al. 2010; Dubé et al. 2013). Using this approach, we identified several polymorphic fragments that vary in intensity in response to divergent selection for NSC (Fig. 36.3). These preliminary observations suggest the presence of genomic regions affecting NSC accumulation. Markers linked to these regions could be used to assist phenotypic selection in alfalfa.

In our study, selection was performed using NSC extracted from total aboveground material. However, it has been shown that NSC composition and accumulation vary between stems and leaves (Terry and Tilley 1964; Lechtenberg et al. 1971; Davies 1976; Wilman et al. 1996; Repetto et al. 2003). This is mostly due to an imbalance

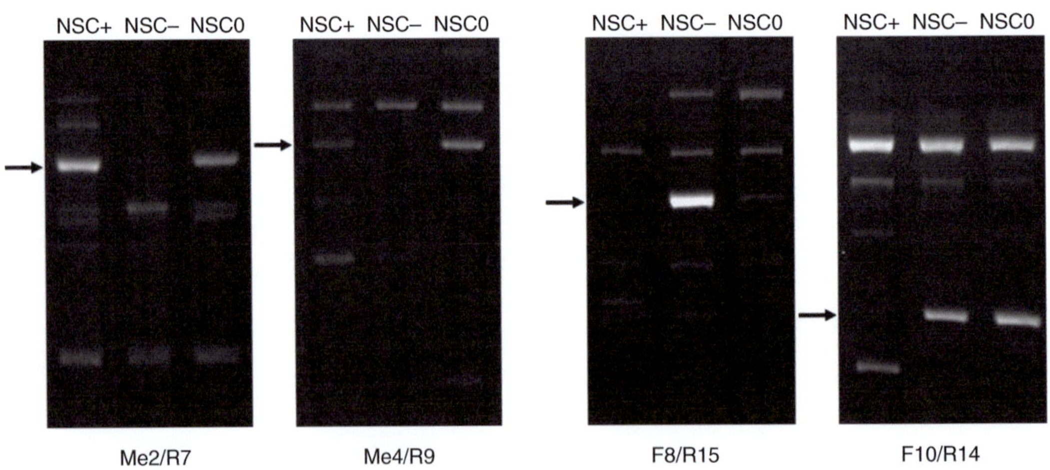

Fig. 36.3 Amplification profiles of four SRAP primer pairs on bulk segregant analysis (BSA) of DNA samples (96 genotypes) from three alfalfa populations (NSC+, NSC−, and *NSC0*). Arrows indicate polymorphic bands (Source: Dubé et al. (2010))

between photosynthesis and transport of carbohydrates, which leads to accumulation of labile pools of starch in leaves during the day (Bowden et al. 1968). Although selection based on stem NSC would still be labor intensive, it could be less prone to environmental fluctuations and should be investigated as an alternative selection criterion for NSC improvement in alfalfa.

Conclusion

Our results demonstrate that although phenotypic selection for NSC in alfalfa forage is feasible, current approaches are resource intensive due to the strong genotype × environment interaction. Methods for accurate identification of superior genotypes are necessary for the development of breeding programs aimed at increasing NSC concentration in alfalfa. In that perspective, our preliminary results suggest that marker assisted selection could be investigated as a complementary approach to phenotypic selection.

References

Bertrand A, Tremblay GF, Pelletier S, Castonguay Y, Belanger G (2008) Yield and nutritive value of timothy as affected by temperature, photoperiod, and time of harvest. Grass Forage Sci 63:421–432

Bowden DD, Taylor K, Davis WEP (1968) Water soluble carbohydrates in orchardgrass and mixed forages. Can J Plant Sci 48:9–15

Brito AF, Tremblay GF, Bertrand A, Castonguay Y, Bélanger G, Michaud R, Lapierre H, Benchaar C, Petit HV, Ouellet DR, Berthiaume R. (2008) Alfalfa cut at sundown and harvested as baleage improves milk yield of late-lactation dairy cows. J Dairy Sci 91:3968–3982

Brito AF, Tremblay GF, Lapierre H, Bertrand A, Castonguay Y, Bélanger G, Michaud R, Benchaar C, Ouellet DR, Berthiaume R. (2009) Alfalfa cut at sundown and harvested as baleage increases bacterial protein synthesis in late-lactation dairy cows. J Dairy Sci 92:1092–1107

Broderick GA, Gregorini P, Edwards GR (1995) Desirable characteristics of forage legumes for improving protein utilization in ruminants. J Anim Sci 73:2760–2773

Bryant (2012) Effects of N fertilisation, leaf appearance and time of day on N fractionation and chemical composition of Lolium perenne cultivars in spring. Anim Feed Sci Technol 173:210–219

Castonguay Y, Cloutier J, Bertrand A, Michaud R, Laberge S (2010) SRAP polymorphisms associated with superior freezing tolerance in alfalfa (Medicago sativa spp. sativa). Theor Appl Genet 120:1611–1619

Davies I (1976) Developmental characteristics of grass varieties in relation to their herbage production. 1. An analysis of high-digestibility varieties of Dactylis glomerata at three stages of development. J Agr Sci 87:25–32

Dubé MP, Castonguay Y, Duceppe MO, Bertrand A, Michaud R (2013) SRAP polymorphisms associated to cell wall degradability in lignified stems of alfalfa. Bioenerg Res 6:644–650

Dubé MP, Chouinard-Michaud C, Castonguay Y, Michaud R, Bertrand A (2010) SRAP polymorphisms associated to increased nonstructural carbohydrates in alfalfa. In: Abstracts of the 2nd Annual General Meeting of the Cellulosic Biofuel Network (CBN), Ottawa Marriott Hotel, Ottawa, Canada, 11-12 March 2010

Humphreys MO (1989) Water-soluble carbohydrates in perennial ryegrass breeding. II. Cultivar and hybrid progeny performance in cut plots. Grass Forage Sci 44:237–244

Lechtenberg VL, Holt DA, Youngberg HW (1971) Diurnal Variation in Nonstructural Carbohydrates in Vitro Digestibility and Leaf to Stem Ratio of Alfalfa. Agron J 63:719–724

Mayland HF, Shewmaker GE, Harrison PA, Chatterton J (2000) Nonstructural carbohydrates in tall fescue cultivars relationship to animal preference. Agron J 92:1203–1206

Nie Z, Tremblay GF, Bélanger G, Berthiaume R, Castonguay Y, Bertrand A, Michaud R, Allard G, Han J (2009) Near-infrared reflectance spectroscopy prediction of neutral detergent-soluble carbohydrates in timothy and alfalfa. J Dairy Sci 92:1702–1711

Parsons AJ, Edwards GR, Newton PCD, Chapman DF, Caradus JR, Rasmussen S, Rowarth JS (2011) Past lessons and future prospects: Plant breeding for yield and persistence in cool–temperate pastures. Grass Forage Sci 66:153–172.

Repetto, JL, Gonzalez J, Cajarville C, Avir MR, Rodriguez CA (2003) Relationship between ruminal degradability and chemical composition of dehydrated lucerne. Anim Res 52:27–36

SAS Institute (2002) Software Version 9.1 of the SAS System for Windows. SAS Institute Inc., Cary, NC., USA

Shewmaker GE, Mayland HF, Roberts CA, Harrison PA, Chatterton NJ, Sleper DA (2006) Daily carbohydrate accumulation in eight tall fescue cultivars. Grass Forage Sci 61:413–421

Terry RA, Tilley JMA (1964) The digestibility of the leaves and stems of perennial ryegrass, cocksfoot, timothy, tall fescue, lucerne and sainfoin, as measured by an in vitro procedure. J Brit Grassland Soc 19:363–372

Wilman D, Gao Y, Altimimi MAK (1996) Differences between related grasses, times of year and plant parts in digestibility and chemical composition. J Agr Sci 127:311–318

37 Breeding of Alfalfa 'Lines' with Divergent Earliness

C. Depedro, D. Kertikova, G. Gnocchi, M. Carelli, A. Tava, and C. Scotti

Abstract

Alfalfa stem development, particularly the switching to the reproductive phase, is related to the content of fiber fractions (concentration of cell wall material and lignification). We used a breeding method based on two generations of selfing to obtain S_2 alfalfa families and individual plants within family with high dry matter production and divergent earliness ('early' and 'late'). By crossing four of these families in a diallelic scheme, we obtained two subgroups of six $S_2 \times S_2$ simple hybrids (SHs) with 'early' (E) and 'late' (L) profile. We advanced the SHs for two generations (syn2 and syn3) to have 2-constituent synthetics genetically in equilibrium. The comparison of the two subgroups in the three generations for stem phenological stage showed that SHs maintained the earliness profile of the parents. Besides, E and L profile were consistently expressed by the syn2 and syn3 generations. A lower NDF content was evidenced in stems of the L SH subgroup compared to the E subgroup. These 'lines' ($2S_2$ synthetics) having a common genetic background and contrasting earliness profile can be used to identify markers whose polymorphisms are linked to the divergent phenological stage and as donors of specific earliness profile in variety constitution.

Keywords

Alfalfa • Earliness • Phenological stage • Selfing • Stem

C. Depedro • G. Gnocchi • M. Carelli • A. Tava
C. Scotti (✉)
Council for Agricultural Research and Economics, Fodder and Dairy Production Research Centre (CRA-FLC), Lodi, Italy
e-mail: carla.scotti@entecra.it

D. Kertikova
Breeding and Seed Production of Forage Crops, Institute of Forage Crops, Pleven, Bulgaria

Introduction

Selfing (in this case, two generations, with selection for dry matter (DM) production among and within selfed families in dense conditions) has been proposed in alfalfa by Rotili et al. (1999) as a powerful tool to improve parent genetic value

I. Roldán-Ruiz et al. (eds.), *Breeding in a World of Scarcity: Proceedings of the 2015 Meeting of the Section "Forage Crops and Amenity Grasses" of Eucarpia*,
DOI 10.1007/978-3-319-28932-8_37

(General Combining Ability or GCA), to homogenize plant material for important physiological traits and to decrease negative genetic load.

Earliness is an important bio-agronomic trait in alfalfa because variation for fiber content of stems occurs in relation with the switching to the reproductive phase and its dynamics (Kalu and Fick 1983). The uncoupling of DM production and earliness is then expected to improve stem quality and to provide plant materials with higher management flexibility, i.e. whose cutting schedule could be delayed without significant losses in forage quality.

A breeding program was started in two alfalfa populations of different origin – Mediterranean (southern Europe. northern Africa) and continental (continental Europe. North America) – and fall dormancy (FD) degree (FD 7–8 and 5–6, respectively). A positive selection for vigor (DM yield in a defined time interval) and divergent selection for earliness were applied for two generations of selfing. We obtained S_2 mother plants of comparable vigor and divergent earliness (early/late) that we manually crossed following a diallelic scheme with four parents (two from the Mediterranean population and two from the continental one).

We report here on part of the results concerning the performances of $S_2 \times S_2$ simple hybrids (SHs) and of their development in 2-constituent synthetics (Scotti and Brummer 2010). Our final aim is to capitalize the genetic gain obtained in the selfing phase by producing 'lines' (2-constituent synthetics) genetically in equilibrium, indefinitely reproducible and consistently expressing the specific earliness profile (early/late).

Material and Methods

Plant Material

S_2 selected plants from a Mediterranean population (parents 1 and 2) and continental population (parents 5 and 7) were manually crossed according to a diallelic scheme within each earliness subgroup (early/late) to produce six $S_2 \times S_2$ SHs for subgroup. The 2-constituent synthetics derived from SHs – syn2 and syn3 generations – were produced by the manual polycross of plants (6–8 per family) chosen at each generation (conservative selection).

Plant Growth Conditions

SH: plants were grown outside at Lodi Institute (45° 19' N. 9° 30' E; 81 m asl) in tube-plots 80 cm high × 20 cm diameter (20 plants/plot, density equivalent to 600 plants m^{-2}). Each SH was represented by 8–20 plots arranged in a randomized block design with 4 blocks; 3120 total plants were examined in a 2 years trial 2008–2009 (5 harvests in the sowing year and 2 in the subsequent year).

Syn2 and syn3 generations: plants (60/family) were grown individually under a rain-out shelter in tube-plots 80 cm high × 5 cm diameter (density equivalent to 500 plants m^{-2}) arranged in a randomized block design with 4 blocks; 720 total plants were examined in a 1 year trial (5 harvests for syn2 in 2011 and 6 for syn3 generation in 2013).

For all the trials, irrigation was not limiting; the average interval between cuts was 30 days.

Traits Recorded on Individual Plant Basis at Harvests 2–5

SH: DM yield and phenological stage (estimated by earliness index. the sum of the reproductive nodes each weighted by the phenological stage, from flower to green bud, see Fig. 37.1) on the two highest stems/plant were recorded together with total stem height and stem height at the first reproductive node.

Syn2 and syn3 generations: the same traits indicated for SH were recorded on the three highest stems/plant.

Near-Infrared Reflectance Spectroscopy (NIRS) Analysis

SH: for each SH. earliness subgroup and block combination, a sub-sample of plants homogenous for DM yield ($\geq$ mean + 0.75 SD) in the 3rd (July) and 4th (August) productive cycle was used for leaf and stem separation (96 total samples). Stem material was analyzed for total nitrogen (TN) content by an NA1500 elemental analyzer according to the Dumas method (Kirsten 1983) and crude protein (CP) calculated as TN × 6.25; the same material was analyzed by NIRS (FOSS 6500) and fiber fractions NDF, ADF and ADL determined according to Goering and Van Soest (1970).

Statistical Analysis

Diallelic analysis (according to Griffing method 4 model I) and analysis of variance (ANOVA) were performed using the general linear model (GLM) procedure of the SAS software version 8e.

Results and Discussion

Earliness index and the evaluation of the reproductive part of the stem, expressed as percentage of the total stem height, analytically depict stem and plant phenological stage (Fig. 37.1).

Diallel analysis on the six SHs of 'early' (E) and 'late' (L) subgroups indicated GCA was the most significant source of variation driving SH performances for the two traits, DM yield and earliness index, object of selection in the selfing phase (Table 37.1).

The genetic gain obtained in the selfing phase was therefore transmitted to SH progenies and, conversely, SHs are effective in capitalizing the parental genetic values. Besides, the prevailing GCA allows to orientate the crossing plan for SH seed production on the basis of the parental values as the best SHs are expected to originate from the best S_2 parents.

The SHs of the two subgroups E and L did not significantly differ for DM production. On the contrary, they displayed significant differences for all the traits estimating stem phenological stage (Table 37.2). Moreover, these differences were consistently maintained in syn2 and syn3 generations obtained by multiplication of SHs (Table 37.2). This confirms that selfing with selection in density condition is effective in capturing genetic structures (genes and groups of genes or linkats) underlying complex traits as DM yield and earliness (Rotili 1976, 1977). A significant decrease in stem height of the L subgroup in the 5th harvest (October) for SHs and syn3 suggested a common genetic basis for lateness in flowering and reduced internode elongation rate in response to fall conditions.

Stem chemical analysis was performed on a subsample of plants (as homogenous as possible for DM yield) in the two subgroups and in the summer productive cycles (July and August) where all the traits representing stem phenological stage had the highest values. A reduced but significant difference between the subgroups E

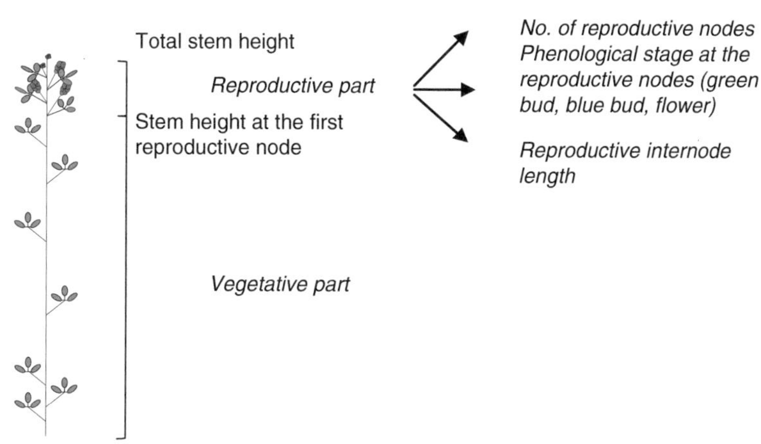

Fig. 37.1 Estimation of the phenological stage of an alfalfa stem

Table 37.1 Diallelic analysis of $S_2 \times S_2$ simple hybrids (SHs) for DM yield and earliness index (means of harvests 2–5): variances and significance

	Early SHs			Late SHs		
Sources	Crosses	GCA	SCA	Crosses	GCA	SCA
DM yield	1.20*	1.97***	0.05 *ns*	3.12*	4.44*	0.56 *ns*
Earliness index	1.52****	2.42****	0.17 *ns*	0.70****	0.98****	0.29 *ns*

*P<0.05; **P<0.01; ***P<0.005; ****P<0.001; *ns* not significant

Table 37.2 Comparison of the subgroups early (E) and late (L) in SH generation and in the syn2 and syn3 generation of multiplication (means of harvests 2–5 in different years)

	SH		Syn2		Syn3	
Generation	E	L	E	L	E	L
DMY (g)	1.31 ns	1.21	2.27 ns	2.36	1.64*	1.86
Stem height (cm)	48.26*	45.04	56.10 ns	52.63	51.11 ns	50.59
Stem reproductive part (%)	2.49**	0.92	3.33***	0.68	4.13****	1.23
Earliness index	2.89****	1.50	10.05***	2.89	12.03****	4.40
No. of reproductive nodes	2.44****	1.83	2.35***	1.14	2.39****	1.59
Plants with reproductive stage (%)	40.51**	23.77	67.10***	41.35	55.86***	29.89

*P<0.05; **P<0.01; ***P<0.005; ****P<0.001; *ns* not significant

Table 37.3 Stem chemical analysis of 'early' (E) and 'late' (L) SHs in two productive cycles. Data are expressed in percentage of DM

Subgroup	Third cycle				Fourth cycle			
	NDF	ADF	ADL	CP	NDF	ADF	ADL	CP
E	65.50 a	51.44 a	14.39 a	8.64 a	65.34 a	50.97 a	13.79 a	9.39 a
L	64.41 b	50.65 a	14.15 a	8.69 a	64.03 b	50.00 a	13.58 a	9.43 a

Means sharing the same letter are not significantly different (P<0.05)

and L was present only for NDF content in both cycles (Table 37.3).

As both crosses, averaged on the two subgroups, and the interaction subgroup × cross except for ADF did not significantly differ for the content of any of the fiber fractions, the difference in NDF was mainly to ascribe to the stem phenological stage. Currently, the syn4 generation (240 plants for each $2S_2$syn4; 2880 total plants) is being grown in the same conditions described for SH trial; the evaluation of DM yield, stem phenological stage and fiber fraction content will be performed.

Conclusions

The selection method applied (two generations of selfing with positive selection for DM yield and divergent selection for stem phenological stage) and the subsequent development of 2-component synthetics resulted in narrow genetic-based families with a common genetic background and consistently expressing divergent stem phenological stages. These 2-component synthetics can be used to identify markers (e.g. Single Nucleotide Polymorphisms. SNPs) differentiating the E and L subgroups, possibly linked or insisting in genetic structures involved in earliness determination. In breeding program, the 2-component synthetics can function as parents able to contribute specific traits (DM production associated to divergent earliness) to the final varietal product (either 4-constituent semi-hybrids or synthetics).

References

Goering HK, Van Soest PJ (1970) Forage fiber analyses (apparatus, reagents, procedures, and some applications). U.S. Agricultural Research Service.

Kalu BA, Fick GW (1983) Morphological stage of development as a predictor of alfalfa herbage quality. Crop Science 23: 1167–1172.

Kirsten WJ (1983) Rapid, automatic, high capacity Dumas determination of nitrogen. Microchem. J. 28: 529–547.

Rotili P (1976) Performance of diallel crosses and second generation synthetics of alfalfa derived from partly inbred parents. I. Forage yield. Crop Science 16: 247–251.

Rotili P (1977) Performance of diallel crosses and second generation synthetics of alfalfa derived from partly inbred parents. II. Earliness and mortality. Crop Science 17: 245–248.

Rotili P, Gnocchi G, Scotti C, Zannone L (1999) Some aspects of breeding methodology in alfalfa. ET Bingham (ed.), The Alfalfa Genome Conference.www.naaic.org/TAG/TAGpapers/rotili/rotili.html (on line).

Scotti C, Brummer EC (2010) Creation of heterotic groups and hybrid varieties. C. Huyghe (ed.), Sustainable Use of Genetic Diversity in Forage and Turf Breeding. Springer Science+Business Media, Dordrecht, pp 509–518.

Generative Development in Red Clover in Response to Temperature and Photoperiod

38

Å. Ergon, S. Solem, A.K. Uhlen, and A.K. Bakken

Abstract

Stem elongation and flower bud emergence in response to photoperiod (16 and 20 h) and temperature (10, 14 and 18 °C) were characterized in three Norwegian (Lea, Nordi, Reipo) and one Swiss (Fregata) red clover variety. The Norwegian varieties were slower in generative development than Fregata, and were, unlike Fregata, responsive to an increase in photoperiod from 16 to 20 h. Lea was later than Reipo, while Nordi was intermediate. Across all varieties, the earliness of stem elongation in response to temperature was saturated at 14 °C, whereas earliness of flower bud emergence also responded to an increase from 14 to 18 °C. Photoperiod and temperature had additive effects on timing of development prior to stem elongation and complementary effects on timing of development prior to flower bud emergence. Developmental rates calculated on a thermal time basis varied between temperature treatments.

Keywords

Developmental rate • Bud emergence • Day length • Flowering • Phenology • Stem elongation • Thermal time • *Trifolium pratense* L

Introduction

Generative development is important for radiation capture and yield formation in red clover, but reduces its digestibility and energy concentration as forage. Red clover requires long days to flower. The day length requirement varies with latitudinal origin of the material from 12–14 h at low latitudes to 16–18 h at high latitudes (Lunnan 1989 and references therein). Vernalization is not required, but may stimulate flowering in material

Å. Ergon (✉) • S. Solem • A.K. Uhlen
Department of Plant Sciences, Norwegian University of Life Sciences, P.O. Box 5003, Ås NO-1432, Norway
e-mail: ashild.ergon@nmbu.no

A.K. Bakken
Norwegian Institute of Bioeconomy Research, Kvithamar, Stjørdal NO-7512, Norway

I. Roldán-Ruiz et al. (eds.), *Breeding in a World of Scarcity: Proceedings of the 2015 Meeting of the Section "Forage Crops and Amenity Grasses" of Eucarpia*,
DOI 10.1007/978-3-319-28932-8_38

of high latitudinal origin (Lunnan 1989 and references therein). With climate change, temperatures will increase but photoperiods will remain stable, leading to shifts in seasonal patterns of temperature-photoperiod combinations. This will be particularly pronounced at northern latitudes where photoperiod changes rapidly in spring and autumn. A better understanding of the relative importance of temperature and photoperiod in regulating developmental processes can aid our breeding efforts to optimize phenology for high performance in future environments. The aim of this study was to characterize the developmental response to photoperiod and growth temperature in three Norwegian red clover varieties.

Material and Methods

Three Norwegian varieties – Lea (Graminor, 2x). Nordi (NMBU, 2x) and Reipo (Graminor, 4x) were studied. As a contrast, a more southern. Swiss variety, Fregata (Agroscope, 4x), was also included. The plants were sown in pots and grown in a greenhouse at 18 °C/15 °C (day/night) and 12 h photoperiod (natural light supplied with 130 μmol m^{-2} s^{-1} PAR) for one month, followed by two months in a growth chamber at 6 °C and 8 h photoperiod with 150 μmol m^{-2} s^{-1} PAR. At this stage the plants had on average 5 vegetative/non-elongated shoots. The plants were then distributed among six growth chambers with 16 h or 20 h photoperiod and 10 °C, 14 °C or 18 °C, in all combinations. The plants in chambers with 16 h photoperiod were given a photon flux of 200 μmol m^{-2} s^{-1} PAR. The photon flux in the 20 h photoperiod chambers were adjusted to 160 μmol m^{-2} s^{-1} PAR in order to have the same amount of light energy given per day in all chambers.

The timing of the initiation of stem elongation was recorded as the number of days from the start of the differential treatments until the length of the longest internode was minimum 1 cm (days to stem elongation). Similarly, the timing of bud emergence was recorded as the number of days until the first flower bud was palpable (days to flower bud emergence). For all chambers, observations were conducted until flower buds had emerged on all plants or to a maximum time of 20 weeks. To study whether any response to temperature was purely a response to thermal units or not, developmental rate on a thermal time basis was calculated as 1/the number of degree-days to stem elongation or flower bud emergence.

Results

All plants formed one or more elongated branches during the course of the experiment. There was a significant effect of photoperiod, temperature and variety on days to stem elongation (Table 38.1). On average. Fregata started elongating 3.2 days earlier than Reipo, which again started 1.5–2

Table 38.1 Analysis of variance of days to stem elongation and flower bud emergence using the model Y = Photoperiod (P) + Temperature (T) + Variety (V) + P × T + P × V + T × V + residual and average values of 22 plants of each variety and treatment

		MS	
Source	D.f.	Days to stem elongation[a]	Days to flower bud emergence
Photoperiod	1	8.2***	2986***
Temperature	2	21.5***	6619***
Variety	3	32.5***	1965***
Photoperiod × temperature	2	0.93*	70**
Photoperiod × variety	3	0.21ns	203***
Temperature × variety	6	1.23ns	99**
Residual	6	0.16	5.7

Only plants that reached flower bud emergence were used to calculate days to flower bud emergence. *D.f.* degrees of freedom, *MS* mean sum of squares; *, $P<0.05$; **, $0.05<P<0.01$; ***, $0.01<P<0.001$; *ns* not significant

[a]One outlier plant was removed before analysis: it elongated after 98 days as compared to an average of 16 days for the other plants of the same variety in the same growth treatment

Table 38.2 The effect of photoperiod, temperature and variety on days to stem elongation and flower bud emergence

Factor/level	Average number of days to stem elongation	Average number of days to flower bud emergence[1]
Photoperiod (N=12)		
16 h	14[a]	78[a]
20 h	13[b]	56[b]
Temperature (N=8)		
10 °C	16[a]	98[a]
14 °C	13[b]	61[b]
18 °C	13[b]	42[c]
Variety (N=6)		
Fregata	10[c]	40[c]
Lea	16[a]	79[a]
Nordi	15[b]	75[a]
Reipo	14[b]	74[b]

In each case the average value across levels of the other factors are given. [a, b]Values not followed by the same letter are significantly different

[1]At 16 h photoperiod combined with 10 °C or 14 °C, flower buds did not emerge on all plants of the Norwegian varieties during the course of the experiment. The percentage of plants not reaching bud emergence at 16 h were 50–36 at 10 °C, and 5–18 at 14 °C. These plants were not included in the analysis, most probably leading to some underestimation of days to flower bud emergence in these treatments

days earlier than Nordi and Lea (Table 38.2), and plants at 20 h photoperiod started elongating 1.2 days earlier than plants at 16 h. There was no significant difference between plants at 14 and 18 °C, but they started elongating 3.2 days earlier than plants at 10 °C. There was a weak but significant interaction between photoperiod and temperature on stem elongation - the temperature response from 10 to 14 °C was stronger at 20 h photoperiod than at 16 h and the photoperiod response was stronger at 14 and 18 °C than at 10 °C (Table 38.1; Fig. 38.1a), indicating additive effects of temperature and photoperiod. On a thermal time basis the rate of development to stem elongation decreased with increasing temperature (Table 38.3). The decrease was close to linear from 10 to 14 to 18 °C at 16 h photoperiod, but at 20 h there was no significant decrease from 10 to 14 °C, only from 14 to 18 °C.

There was also a significant effect of photoperiod, temperature and variety on days to flower bud emergence (Table 38.1). On average, plants at 18 °C had flower buds 19.3 days earlier than plants at 14 °C, which again were 37.3 days earlier than plants at 10 °C (Table 38.2). In plants at 20 h photoperiod, flower buds emerged 22 days earlier than in plants at 16 h. Fregata was 34–35 days earlier than Reipo and Nordi, which again were 4–5 days earlier than Lea. The three two-way interactions were also significant (Table 38.1). Flower buds emerged 26–31 days earlier at 20 h photoperiod than at 16 h ($P=0.01$–0.03) in the three Norwegian varieties, while there was no significant difference between photoperiods in Fregata (Fig. 38.1d). The number of days to flower bud emergence differed significantly between all temperatures for all varieties ($P=0.006$–0.02), but the reduction caused by increasing the temperature from 10 to 18 °C was stronger in the Norwegian varieties (61–67 days) than in Fregata (36 days) (Fig. 38.1c). The temperature response between 14 and 18 °C was weaker at 20 h photoperiod than at 16 h and the photoperiod response was weaker at 18 than at 10 and 14 °C (Fig. 38.1b), indicating complementary effects of photoperiod and temperature. On a thermal time basis the rate of development to flower bud emergence at 20 h photoperiod increased with increasing temperature (Table 38.3).

Discussion

Both increasing photoperiod and increasing temperature reduced the number of days to stem elongation and flower bud emergence. However, the varieties responded differently to photoperiod – flower bud emergence was close to being saturated in Fregata at 16 h. while it was not in the Norwegian varieties. The day length requirement in red clover has been found to increase with latitude (Lunnan 1989 and references therein), and can be interpreted as a result of evolutionary adaptation to long days in the growing season in the north (King and Heide 2009). Within a certain range, temperature increase affects development passively through a proportional increase in metabolic rates, cell

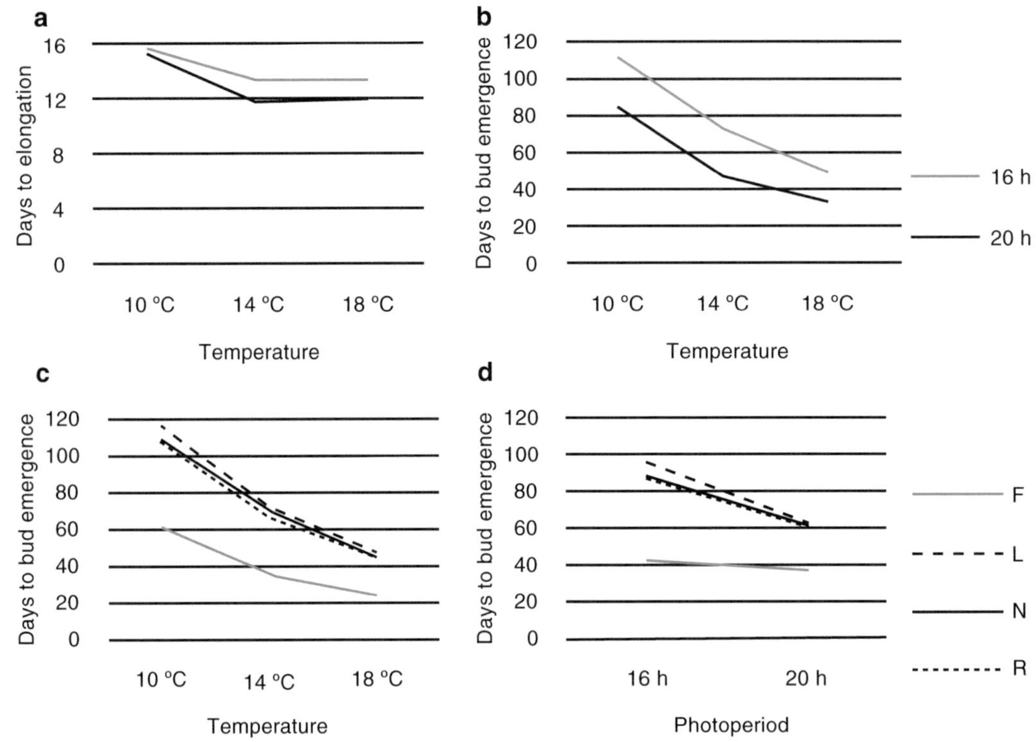

Fig. 38.1 Significant interactions between factors (see Table 38.1) on days to stem elongation (**a**) and days to flower bud emergence (**b**–**d**). Photoperiod and variety is indicated in the legends. *F* Fregata, *L* Lea, *N* Nordi, *R* Reipo

Table 38.3 Developmental rate based on thermal time (base temperature 0 °C) from the start of the treatment until stem elongation and flower bud emergence

	Developmental rate up to stem elongation (degree-day^{-1})		Developmental rate up to flower bud emergence (degree-day^{-1})[1]
Temperature (°C)	16 h	20 h	20 h
10	0.0067[a]	0.0068[a]	0.0012[c]
14	0.0054[b]	0.0062[a]	0.0016[b]
18	0.0042[c]	0.0047[b]	0.0017[a]

In each case the average value across levels of the other factors are given. [a–c]Values in the same column not followed by the same letter are significantly different. N=4

[1]Not calculated for the 16 h photoperiod treatments, as some plants did not reach this stage during the course of the experiment in the 16 h treatments

division and cell elongation, leading to a constant rate of development expressed per thermal time unit. On top of this, temperature can affect developmental rates through other mechanisms. e.g., mechanisms that involve temperature sensing and signaling pathways (Franklin 2009; Wigge 2013). We found that the stimulating effect of increasing temperature on days to flower bud emergence also involved an increase in developmental rate on a thermal time basis, at least in the 20 h photoperiod treatments. This may be explained by less competition for radiation at higher temperatures because of faster development and fewer tillers and leaves (Lunnan 1989), although molecular regulatory mechanisms may also play a role. On the contrary, the stimulating effect of increasing temperature on days to stem elongation was partly counteracted by a decrease in developmental rate on a thermal time basis, which is in line with the finding that the temperature response of stem elongation was saturated at 14 °C.

Conclusion

The Norwegian varieties were slower in generative development than Fregata, and were, unlike Fregata responsive to an increase in photoperiod from 16 to 20 h. Lea was later than Reipo, while Nordi was intermediate. Across all varieties, the earliness of stem elongation in response to temperature was saturated at 14 °C, whereas earliness of flower bud emergence also responded to an increase from 14 to 18 °C.

Photoperiod and temperature had additive effects on timing of development prior to stem elongation and complementary effects on timing of development prior to flower bud emergence.

References

Franklin KA. 2009. Light and temperature signal cross-talk in plant development. Curr Opin Plant Biol 12: 63–68.

King RW, Heide OM. 2009. Seasonal flowering and evolution: the heritage from Charles Darwin. Functional Plant Biology 36: 1027–1036.

Lunnan T. 1989. Effects of photoperiod, temperature and vernalization on flowering and growth in high-latitude populations of red clover. Norwegian Journal of Agricultural Sciences 3: 201-210.

Wigge PA. 2013. Ambient temperature signalling in plants. Curr Opin Plant Biol 23: 661–666.

Part IV

Scarcity of Focus

Improving the Focus of Forage Breeding Research

39

P. Annicchiarico, B. Boller, E.C. Brummer, and D. Reheul

Abstract

Forage crop breeding is particularly exposed to the risks of insufficient focus, owing to challenges and dilemmas arising from the wide diversity of possible bred species, target traits, target environments, crop utilizations, breeding approaches and seed markets, as well as from contrasting demands by farmers, the seed industry and the society. Despite the large number of satisfactory results from forage crop breeding, pitfalls associated with insufficient focus have not been rare. They have included different types of mismatch between selection and target environments, overlooking important traits (e.g., seed yield) or fairly inefficient multi-trait selection, and suboptimal exploitation of genetic diversity, among others. Dilemmas have arisen also from poor research efforts on various crucial issues and occasionally from shortcomings of variety registration procedures. Selecting simultaneously for several traits is hindered by high costs that could be alleviated by marker-based approaches. A focused choice of methods, plant material and phenotyping conditions is indispensable for the successful implementation of genomic selection and other marker-based selection procedures for higher forage yield. As a trend, we expect increasing breeding attention for legumes and. insofar as the application of breakthrough technology depends on seed market opportunities, less breeding effort for minor forage crops.

P. Annicchiarico (✉)
Centre for Fodder Crops and Dairy Productions, Council for Agricultural Research and Economics, Lodi, Italy
e-mail: paolo.annicchiarico@crea.gov.it

B. Boller
Fodder Crop Breeding, Agroscope, Zürich, Switzerland

E.C. Brummer
Department of Plant Sciences, University of California, Davis, CA, USA

D. Reheul
Department of Plant Production, Ghent University, Ghent, Belgium

I. Roldán-Ruiz et al. (eds.), *Breeding in a World of Scarcity: Proceedings of the 2015 Meeting of the Section "Forage Crops and Amenity Grasses" of Eucarpia*,
DOI 10.1007/978-3-319-28932-8_39

Keywords

Breeding methods • Crop utilization • Forage grass • Forage legume • Genetic diversity • Genotype-environment interaction • Molecular breeding • Seed market • Stress tolerance

Introduction

Forage crops are characterized by much lower rates of genetically-based yield improvement than major grain or oilseed crops (Chaves et al. 2009; Laidig et al. 2014; Brummer and Casler 2014; Annicchiarico et al. 2015a), owing to lower breeding investment, longer selection cycles, lack of possibility to capitalize on harvest index, an outbreeding mating system associated with severe inbreeding depression, low narrow-sense heritability, and large interaction of genotypes with cropping conditions and crop utilizations. Scarcity of breeding is undoubtedly a major factor: in the USA, for example, several hundred breeders are committed to maize alone, in comparison with just a handful of breeders working on forage crops (often on more than one species). Psychologists state that scarcity tends to change how we think and act, emphasizing the need for increased focus and productivity of work (Mullainathan and Shafir 2013). If we translate this knowledge to breeding activities, increased focus can lead to more efficient breeding strategies and procedures, or to the production of varieties better matched to criteria set by the seed industry, the farmers, or social demand.

As a matter of fact, forage breeders are challenged by the difficulty of addressing contrasting demands by farmers (higher forage yield), the seed industry (higher seed yield) and society (e.g., higher species diversity and other multifunctional attributes for landscape preservation and improvement). Other remarkable challenges, uncertainties and dilemmas come from the high number of possible target species, target traits, cropping conditions, crop utilizations and breeding schemes, which all have to be addressed using modest budgets. Forage production remains the main target trait, both for livestock and dairy production, also to safeguard the economic competitiveness of these crops in areas where intensively-managed cereals are widely grown for silage. However, the relative value for forage yield of parental material or breeding populations may be elusive or difficult to assess, as it may vary largely depending on the test environment and conditions, the time scale of the assessment, and the generation of seed multiplication (Parsons et al. 2011). In addition, possible genetically-based trade-offs of higher forage yield with other important or emerging traits, such as seed yield, drought tolerance, forage quality, or nutrient-use efficiency, all require careful consideration. Additional challenges and opportunities arise from the possible introduction of novel genomic tools into current breeding programs (Annicchiarico et al. 2015a). In this paper, we briefly discuss some crucial decisions, major shortcomings, and emerging issues that may result in insufficient focus of forage improvement research and selection procedures, using examples based largely on our own experience.

Species Prioritization

Even for a forage crop breeder, consulting the CPVO Variety Finder (http://www.cpvo.europa.eu/main/en/home/databases/cpvo-variety-finder) is simply overwhelming. The diversity of forage grass and legume species and varieties bred and used in the EU is impressive. There is also a substantial diversity in the way forage crops are bred. Few forage breeders breed just a single species; some are working on more than five. Experience teaches that both focusing on a single species and breeding several species have led to successful varieties, which may sound quite surprising at a first glance. One would expect that a greater abundance of species, which implies a reduced dimension of breeding programs for single species, would jeopardize the breeding success. Reality proves that the *métier* of the breeder partly overrules scientific doctrine.

In general, the breeding effort on a given species tends to be proportional to the size of the potential seed market for its varieties. Investment in expensive biotechnologies would justify a breeding focus on just a few major crops with a large market size (Bouton 2007). In Europe, where particular consideration is given to the multifunctional role of forage crops (including non-forage uses) and new consumer demands, the species portfolio may be broadened to respond to a more expansive seed market. An interest for diversification of bred species may emerge also in some regions of the USA, particularly for developing cultivars of native species with better tolerance than introduced species to rainfed cropping in dry, hot climates (Muir et al. 2005). Continued attention to several bred species may be justified by the ability of multi-species mixtures to out-yield pure stands or binary mixtures in various environments (e.g., Nyfeler et al. 2009), although their ability to also show better tolerance to climatic stresses is controversial (Van Peer et al. 2004). If multi-species mixtures confirm the reported greater production stability across climatically-variable environments (Hector et al. 2010), the progressive increase of climate variability due to climate change would widen the interest of multiple species and species mixtures. New species such as chicory are of particular interest for multi-species grassland because they add functional diversity. Breeding various species targeted to cultivation in mixture may allow for a reduction in the number of traits to be selected in the individual species; for example, associating a rapidly developing, high-yielding species with one that develops more slowly reduces the need for selecting the rapidly-developing species for better persistence and the slowly-developing one for higher initial growth rate.

The expected move towards more sustainable crop-livestock systems implies wider cultivation of forage legumes, either as legume-grass mixtures subjected to modest or zero nitrogen fertilization (in place of intensively-managed grasses), or as pure stand, high-protein crops. This has obvious implications for perspective species prioritization by breeders. For example, crop improvement research and funded projects at Lodi (Italy) in the last decade have progressively targeted legumes (perennials and annuals) rather than forage grasses. Alfalfa (lucerne), in comparison with other legumes, can provide higher feed protein yield per unit area (Huyghe 2003) along with comparable or somewhat greater agronomic and environmental advantages in terms of soil fertility, rate of nitrogen fixation per year, and reduction of energy and greenhouse gas emissions (COPA-COGECA 2007). Indeed, alfalfa breeding programs are spreading in countries of northern Europe (e.g., the Netherlands), also because of increased species adaptation to these regions that would arise from climate change. Annual legumes such as vetches or pea (grown for silage or hay in pure stand or in mixture with cereals), subterranean clover or medics may increase their importance in regions where extreme climatic stresses, particularly summer drought, could jeopardize the persistence of perennials.

The current evolution of grassland farming can offer substantial hints for forage species prioritization. In Europe, dairy farming tends to concentrate around the coastal regions of northwest Europe with a trend toward larger herds. High land prices are pushing farmers towards intensive systems, where high animal performances are supported by a very efficient use of high-quality forages. These intensive systems will need species and varieties with high levels of short- and long-term forage yield, disease tolerance, feeding value, nutrient use efficiency, and eco-efficiency (Reheul et al. 2012). While ryegrass will continue to play a central role in these regions, we expect an emerging role for high-yielding species with better use efficiencies and better tolerance to prevalent stresses emphasized by climate change, such as drought. Tall fescue is undoubtedly one such species (Annicchiarico et al. 2013a). This species offers additional advantages associated with a high proportion of rumen escape protein and high potential for carbon sequestration (Baert et al. 2012; Cougnon 2012; Cougnon et al. 2014a, b, c).

Forage production in relatively favorable regions for plant production is expected to keep relying on a restricted number of well-known species. As restrictions regarding residual soil nitrogen become tighter, intensively-managed grass monocultures could be pushed towards mowing management (owing to its higher nutrient export and the opportunity for a more equal distribution of nutrients). In grazed grassland, these restrictions would imply a declining content of nitrogen in the grass, reinforcing the interest of legume-grass mixtures as a means of attaining sufficient forage protein content. Grazing is expected to persist in many areas for reasons of animal welfare, consumer demand, or wherever mowing costs are economically unsustainable (Van den Pol-van Dasselaar et al. 2015; Pinxterhuis et al. 2015).

Extensively-used grassland is expected to dominate in environmentally vulnerable regions (Pflimlin and Faverdin 2014), with an emphasis on grazing management for meat production. Legume-grass mixtures of persistent species with high local adaptation will continue to be necessary under those conditions.

A fairly innovative means of getting an indication of which species, plant type within species and target management (e.g.,. pure stand vs. mixed stand) would be preferred is through farmers' participatory evaluations of forage crops embedded into regional experiments (Thami-Alami et al. 2016). This approach is advantageous because it takes the farmers' socio-economic factors into account alongside production and forage quality.

Successful, reliable, efficient and cost-effective genomic selection may partly change the described context. Displacing some of the selection work in the lab may offer additional field work resources and opportunities for breeding more species. However, economic reasons may rather suggest using these resources to increase the selection intensity or the ability to cope with genotype × environment (GE) interactions for the same number of target species, as predicted by Cooper et al. (2014). The successful completion of on-going genome sequencing projects, such as that on alfalfa, will probably contribute to increased breeding focus on a few major species.

Target Trait Prioritization

Breeders who are able to effectively define target traits, target markets and potential economic returns at the outset of a breeding program are more likely to develop successful cultivars. Market production of fodder (except for the relatively minor production of marketed hay) is not the primary reason for farmers to grow these crops, however. Market production relates more to animal products, but the correlation of selected plant traits with animal production is largely unknown. In most cases, forage breeding focuses on conceptually important traits (e.g., fiber concentration) that will hopefully result in optimal animal production per field unit area.

While the number of species affects the breeding effort rather linearly, the number of selected traits tends to exponentially increase the breeding effort. According to selection theory, a higher number of selected traits implies smaller probabilities to find the ideal individual (Simmonds. 1979). Selecting for a single trait with a selection rate of s among N plants results in a selected subpopulation of $N \times s$. This subpopulation shrinks with every additional selected trait. Assuming a similar selection rate for each of m traits, the final selected population equals $N \times s^m$. It takes a population of $N = (1/s)^m$ individuals to find the ideal plant. For example, even assuming quite reasonable values such as a selection rate of 0.10 and $m = 5$, the working population equals to $(1/0.10)^5 = 100.000$ individuals, which is hardly a workable number (particularly when involving at least some time-consuming evaluated traits, such as forage yield). Other elements could be taken into account, including different s values and the extent of genetic correlations between traits. In any case, a higher number of selected traits clearly requires larger evaluated plant populations to make significant progress – while implying smaller genetic gain for each trait for a fixed budget. A clear advantage of reliable, low-cost marker-assisted or genome-enabled selection procedures would be greater opportunities for multi-trait selection.

The number of possibly useful traits in forage crops tends to increase, given the trends to consider (i) forage crops increasingly important as multifunctional crops (with *inter alia* non-forage destinations), (ii) new consumer demands, and (iii) issues related to environmental protection/improvement or climate change. A survey of research activities presented in the European Grassland Federation meetings between 2010 and 2015 highlights a series of new targets (or targets with a clearly renewed interest) for forage crops, such as contents of fatty acids in grasses, specific chemical compositions (e.g.,. phytoestrogen and tannin content in legumes), rooting density and depth, eco-efficiency, phosphate use efficiency, water use efficiency, drought tolerance, persistence under conditions of keeping chickens outdoors, and cell wall digestibility. However, it is not sure whether new varieties selected for some of these traits would be granted a higher seed market price (particularly in relation to farmers' or market appreciation of higher forage quality).

Perennial forages provide not only animal feed but also valuable ecosystem services, such as improved levels of water infiltration, carbon sequestration and soil protection against erosion, as compared with annual cropping systems (Olmstead and Brummer 2008; Isselstein and Kayser 2014). They also make positive contributions to the productivity of following crops in a rotation. e.g., by releasing nitrogen fixed by legumes or by disrupting pest cycles. Although some attempts have been made to value ecosystem services provided by agricultural systems (e.g., Swinton et al. 2015) and pasture-based systems (Rodríguez-Ortega et al. 2014), the inclusion of these valuations in trait prioritization by forage breeding programs appears to be remote.

Broadly speaking, the key traits for all forage crops are forage yield, seed yield, forage nutritive value, and persistence (both intrinsic and as determined by tolerance to major stresses). Each of these traits includes numerous components. Grass nutritive value, for example, includes water soluble sugar, crude protein, cell wall concentration and forage digestibility, as well as other constituents (Kingston-Smith and Thomas 2003; Casler 2001). A clearer and more univocal definition of the ideal ideotype by animal nutritionists would help breeders to select for improved nutritive value. Various diseases and pests require resistance breeding attention in individual species (Stewart and Hayes 2011; Annicchiarico et al. 2015a). Several diseases and insects are problematic for alfalfa production in the USA (Monteros et al. 2014), and even if a particular disease is of limited importance in certain regions, cultivars without resistance suffer a marketing disadvantage. Consequently, most cultivars developed today maintain high resistance to as many diseases as possible. Actually, alfalfa breeding for increased forage yield has been far less successful than breeding for tolerance to biotic stresses (Annicchiarico et al. 2015a), despite the greater practical importance of yield. Drought tolerance is probably the single trait of greater prospective importance for forage legumes, because of decreased and more erratic rainfall and reduced irrigation water that are expected from climate changes (Annicchiarico et al. 2015a).

Developing a clear focus to improve certain traits while maintaining minimum acceptable levels of the other traits is the breeder's challenge. Multi-trait selection is probably most successfully accomplished using an index that would include economic values for each trait and, ideally, the heritabilities of and the genetic correlation among traits (see Bernardo [2010. p. 330]). In practice, however, obtaining precise estimates of trait heritabilities and (especially) genetic covariances is difficult, and generating economic values for each trait may be even more problematic.

Some recent studies have suggested economic values for particular traits (e.g., Smith et al. 2014). The relative value of biomass production, nutritive value and persistence traits were estimated for perennial ryegrass in Ireland (McEvoy et al. 2011). Based on the phenotypic performance and economic values, the overall genetic merit of 20 cultivars ranged from −197.95 to +128.54 €/ha/

year. Such weightings could easily be incorporated into a selection index. Whole farm planning models could help identify the impact that changes in trait values would have on profitability; for example, higher metabolizable energy from forage has a positive effect on farm profitability (Ludemann et al. 2015). On the other hand, one should consider that cropping conditions such as grass pure stand vs. grass-legume mixed stand (Stewart and Hayes 2011) or crop exploitation choices such as pasture vs. silage (McEvoy et al. 2011) can affect and complicate the estimation of trait economic values in the index (Chapman et al. 2014). Defining trait economic weights thus calls for clear identification of the target growing conditions for the future cultivar. Collectively, both modeling and empirical research could help derive values for various traits in a selection index, or at a minimum, to set goals for a breeding program. In any case, we need more information on the impact of different breeding approaches on animal performances.

One last issue regarding traits is the ability to collect sufficiently precise phenotypic data. Most forage crops are harvested multiple times per year (as many as 10–12 times in some regions), requiring repeated measures of yield and nutritive value both within and across years. Even if the equipment used is sufficiently sensitive, the time needed for data collection is daunting. Improvements in phenotyping technologies, including proximal and/or remote sensing, could make measuring these traits easier and faster, enabling breeders to either increase the size of selection nurseries, devote effort to other forage species, or do a better job throughout the breeding program. Sensor technologies have been used to estimate biomass yield and plant height in several forage species with promising results (Pittman et al. 2015), although complete evidence on their reliability is still lacking.

Matching Selection and Target Environments

Genetic gain in the target agricultural environments (ΔG. for one selection cycle) depends on the genetic variation available (s_A. standard deviation of breeding values), the selection intensity (*i.* standardized selection differential), the square root of narrow-sense heritability of the trait (h), and the genetic correlation (r_g) for the population (e.g., half-sib family) response between selection and target environments (Falconer. 1989):

$$\Delta G = i\, h\, r_g\, s_A.$$

Reaching high r_g values, especially for traits such as yield, is particularly challenging for forage breeders for two reasons. First, high experiment costs arising from long crop cycles and multiple harvests hinder the adoption of selection environments (e.g., swards) that may more faithfully reflect the target environments. Second, the breeding value of forage crop ecotypes and derived material – which represent the core genetic base of many breeding programs – is strongly affected by GE interaction, as highlighted by Turesson's (1922) pioneering work on cocksfoot. Poor breeding progress may derive from a mismatch between selection and target conditions (whose negative impact is proportional to departure from unity of r_g), or from failure to use a genetic base that is highly adapted to the target environments.

For example, alfalfa breeding performed in Lodi in the 1970s and 1980s was based on phenotypic selection for high biomass yield under optimal water supply (Rotili 1988). This breeding approach produced valuable varieties for cropping under irrigation that were, at the same time, ill-adapted to areas with rainfed cropping (prevalent in part of northern Italy and in central Italy). This was shown particularly clearly by an Additive Main effects and Multiplicative Interaction (AMMI) analysis of GE interactions (Annicchiarico 1992). Even soil type exhibited some influence on variety adaptation. Four artificial environments created by the factorial combination of two drought stress levels (almost nil vs. high) and two soil types (sandy-loam vs. silt-clay) allowed us to both successfully reproduce the adaptive responses of a set of reference varieties (Annicchiarico and Piano 2005) and identify optimal genetic resources for two regions characterized, respectively, by irrigated and rainfed cropping.

Breeding specifically-adapted material for these subregions has produced top-yielding registered varieties in their regions of specific adaptation, allowing for higher genetic gains for yield than when forages are bred for wide adaptation (Annicchiarico 2007a). Both Italian (Annicchiarico and Piano 2005) and foreign (Annicchiarico et al. 2011) alfalfa ecotypes and varieties displayed specific adaptation to the drought stress level under which they evolved or were selected, highlighting the importance of exploiting distinct genetic bases for moisture-favorable and stress environments. For both germplasm types, specific adaptation was consistently associated with adaptive traits of environment-specific value (Annicchiarico 2007b; Annicchiarico et al. 2013b). Other results summarized in Annicchiarico et al. (2015a) for this germplasm or Mediterranean material highlighted the low or zero genetic correlation for genotype responses between drought-stressed and moisture-favorable environments.

Forage crops are usually selected in pure stands while in many cases they are meant for cultivation in legume-grass mixtures. This can hinder breeding progress particularly for poorly-competing species, such as white clover in association with vigorous grass cultivars under a mowing regime. Selection work performed in Lodi on a genetic base of Ladino ecotypes and natural populations reached the target legume proportion of about 30 % on total dry matter only when performed entirely in competition with grass companions (Table 39.1). GE interactions for white clover yield increased from clover pure stand to clover-grass mixed stand as a function of the grass competitive ability, regardless of grass species (Annicchiarico and Piano 1994). This finding justified a low-cost phenotypic selection procedure to breed for higher clover's general competitive ability. By placing two clonal cuttings between densely-planted rows of two species of vigorous grass companions, a variety ('Giga') was produced that showed much greater competitive ability against associated grasses of several different species (Annicchiarico and Proietti 2010).

Table 39.1 Three-year clover and total dry matter (DM) yield, and clover proportion on total DM. for three white clover populations selected in Lodi in pure stand (PS) and/or mixed stand (MS) with vigorous grass companions and one top-performing control variety, averaging results across four binary mixtures with vigorous grass varieties of different species

Type of selection	Clover DM yield (t/ha)[a]	Total DM yield (t/ha)[a]	Clover % on total DM[a]
Phenotypic in MS (cv. 'Giga')[b]	13.7 a	44.7 a	30.9 a
Phenotypic in PS + Genotypic in MS[c]	10.1 b	39.6 a	25.6 b
Phenotypic in PS[d]	7.8 bc	42.4 a	18.4 c
Best control variety (cv. 'Espanso')	6.3 c	43.2 a	14.7 c

Details of the experiment are reported in Annicchiarico and Proietti (2010) (which provides results only for 'Giga' and 'Espanso')

[a]Column means with different letter differ at $P<0.05$ according to Duncan's test

[b]Seven selected parents out of 165 cloned individuals

[c]Sixteen individuals selected phenotypically out of 285 cloned ones; 6 parents selected out of 16 based on half-sib progeny responses

[d]Seven selected parents out of 285 cloned individuals

Most breeders select from space-planted nurseries of plants. Studies on alfalfa (Annicchiarico 2006a) and white clover (Annicchiarico and Piano 2000) indicated just fair genetic correlations (r_g=0.60–0.72) between spaced and dense planting conditions for both forage yield and seed yield, along with much lower selection efficiency of spaced planting, particularly for forage yield (only about 60 % efficiency relative to selection under dense planting). This finding also applies to forage grasses such as perennial ryegrass. based on poor phenotypic correlations observed for forage yield (Hayward and Vivero 1984) or seed yield (Elgersma et al. 1994). Breeding schemes that facilitate the adoption of dense planting (e.g., genotypic selection based on sown half-sib progeny responses) thus increase the match between selection and target environments.

Various studies for perennial legumes summarized in Annicchiarico et al. (2015a) provide evidence for negligible genetic correlation or high GE interaction of cross-over type for genotype responses under mowing and grazing conditions (particularly for sheep grazing). Indeed, direct selection under grazing conditions has been crucial in the USA (Bouton 2012) and in Lodi (Pecetti et al. 2008) to select alfalfa cultivars that are highly tolerant to grazing. Recent work by Cashman (2014) showed indirect evidence for large GE interaction between these exploitation conditions also for grasses. He found a directional selection within perennial ryegrass cultivars under mechanical defoliation in Ireland, concluding that "the plants that survived had a smaller phenotype in comparison with the original seed lines, suggesting that plants with a lower yield potential survived".

The need for specific forage breeding when targeting organic systems is disputed. The consistent yield responses displayed by alfalfa cultivars grown with and without weed competition (Annicchiarico and Pecetti 2010) suggest that in favorable organic environments, where competitive success is determined by high relative growth rate, germplasm selected for high biomass yield in the absence of competition should suffice. On the other hand. Boller et al. (2008) found significant GE interaction for biomass yield in Italian ryegrass cultivars grown under conventional and organic conditions, thus suggesting the scope for specific selection under organic conditions. Comparative selection programs under both conventional and organic systems could demonstrate whether selection directly in organic systems can lead to superior cultivars for those systems.

Information on the consistency of genotype responses between selection and target environments is relevant not only to develop selection procedures that optimize selection gains but also to devise efficient phenotyping conditions for genomic selection modeling or QTL (Quantitative Trait Loci) discovery research (Annicchiarico et al. 2015a).

Diversity-Related Breeding Issues and Breeding Schemes

The main perennial forage species are strongly allogamous and prone to severe inbreeding depression. Various diploid species (ryegrasses. meadow fescue. red clover) can be artificially polyploidized. These and other factors lead to a wide range of possible cultivar types and breeding schemes, in contrast to most other crops. The outbreeding nature of forage plants and their widespread occurrence in natural or semi-natural grassland makes it very easy for the breeder to intentionally introduce foreign germplasm into the breeding pool. However, a balance must be found in the diversity level that is accepted in the genetic base and in the selected populations. This involves considerable dilemmas with respect to cultivar type and breeding schemes (see detailed discussion below). Also, requirements for adequate distinctness, uniformity and stability (DUS) limit the level of acceptable intra-cultivar variation.

Strategies to maintain diversity of the breeding pool. Recurrent selection entails the danger of narrowing down the genetic diversity of the breeding pool, thereby reducing the potential for continued genetic gain and possibly increasing inbreeding depression. Therefore, breeders aim to refresh their breeding pools by introducing new genetic material, such as cultivars issued from other breeding programs, or ecotypes. When doing so, they must find a balance between minimizing the loss of performance in characters under strong selection in the breeding pool, and maximizing the added diversity from wild material or accessions from distant environments without jeopardizing homogeneity. Complete passport data and thorough characterization of genetic resources maintained *ex situ* can help to make a good choice. However, taking advantage of natural selection and further development of ecotype populations during *in situ* maintenance can be just as efficient.

About 25 % of the components of all synthetics produced in the breeding program on tall fescue at Ghent University have their roots in very intensively selected ecotypes. The potential value as genetic resources of landraces or other ecotype material for environments and conditions similar to those in which they evolved has probably been overlooked by many breeding programs. Forage yield responses comparable with those of best-yielding modern varieties were reported for landraces of alfalfa (Annicchiarico 2006b), white clover (Annicchiarico 1993) and red clover (Boller et al. 2010b), as well as for wild populations of ryegrass (Boller et al. 2009). However, a thorough exploitation of the richness in adaptive genes of landraces or wild populations is only possible through studies that allow for an ecological understanding of their specific-adaptation pattern (Annicchiarico and Piano 2005). Another key issue for a thorough exploitation of genetic resources is the proper focus of screening procedures on intra-population diversity, whose variance for forage and seed yield traits exceeds largely the between-population variance (e.g., Julier et al. 2000). On the other hand, the concern about losing genetic variation in breeding programs might be overestimated; for example, the IBERS perennial ryegrass selection program has progressed for about 40 years based on seven initial parent genotypes and the subsequent incorporation of two more that contributed specific traits. Despite the fact that only 18 haplotypes have entered the program, continual progress for yield, water soluble carbohydrate concentration, and stand persistence has been documented [Wilkins and Humphreys (2003) and R Hayes 2015, personal communication]. On the whole these findings highlight the advantage of introducing into the core genetic base only elite new germplasm carefully evaluated for specific traits.

Impact of official variety testing requirements. Getting a new cultivar listed and recommended officially is an inevitable requirement for its market success in Europe. The DUS and VCU (Value for Cultivation and Use) testing schemes, however do not always match the actual breeding targets. For example when aiming for wide adaptability to diverse environmental conditions one may end up with too much intra-cultivar diversity to successfully pass the DUS test for uniformity. DUS results are more difficult to predict than VCU because the candidate cultivar has to compete with many little-known commercial cultivars. Indeed, the threshold for being "sufficiently uniform" is impossible to derive from published results. Besides, this threshold tends to move toward greater uniformity with time, as it is fixed according to the average of all pre-existing cultivars. Therefore, breeders hesitate to combine distantly related and morphologically diverse parents in a synthetic, thus narrowing down its number of parents and its richness in adaptive genes. Even the distinctness requirement can be difficult to achieve for candidate varieties of crops featuring a high number of registered varieties (e.g., alfalfa and perennial ryegrass). Some revisions of DUS testing criteria leading to tolerate somewhat wider intra-population diversity and allowing, if technically feasible, for molecular marker-based distinctness, could eliminate any possible negative impact of DUS requirements on breeding progress.

Concerning VCU requirements, a dilemma exists between breeding for wide adaptation, which is usually necessary for a cultivar to pass the VCU test (based on average performances over contrasting environments), and breeding for maximum performance in a more specific target environment. For alfalfa, this situation has led Italy to grant sufficient VCU to a candidate variety, even if it is specifically adapted to one major cropping region. Another possible limitation of VCU testing is assessing variety performance only under a mowing regime, which hinders the acceptance of grazing-type cultivars. For example, a special grazing test under the guidance of French testing authorities had to be set up for the grazing-type red clover 'Pastor' to overcome its forage yield handicap under mowing, so that it could be listed for sale in France (GEVES 2011). Results of grazing trials are taken into account as a standard procedure in VCU testing in Belgium and the Netherlands.

Breeding scheme. Several selection schemes have been proposed for synthetic varieties of forage crops, particularly alfalfa. They may imply the selection of parent material based on mass selection, clonal evaluation, half-sib or full-sib progenies, or selfed progenies with varying numbers of selfing generations. In addition, selection may be performed among cloned parents or among plants of their evaluated progenies. Selecting and recombining partially inbred genotypes could increase the accumulation of favorable alleles with additive genetic variation, reduce the susceptibility to inbreeding, reveal new genetic variation by uncovering recessive alleles, and promote greater morphological homogeneity (Rotili et al. 1994). However, empirical comparisons of selection schemes have been rare and limited to just a few schemes, whereas theoretical comparisons (e.g., Casler and Brummer 2008) indicate that different values for relevant genetic parameters would likely lead to different methods being most effective. An empirical comparison of nine selection schemes for alfalfa forage yield is ongoing in Lodi. Breeders should also consider that optimal selection procedures may differ depending on the narrow-sense heritability (h_N^2) and genetic variation of the target trait. In alfalfa, for example, mass selection on large plant numbers is preferable for high leaf/stem ratio, because of high h_N^2 and only moderate genetic variation of this trait; whereas a final stage of genotypic selection (as allowed for by half-sib progeny-based parent selection) is preferable for high forage yield, because of its low h_N^2 arising from high non-additive genetic variance (Annicchiarico 2015).

Breeding real hybrids is still precluded for forage crops, despite progress in this direction. Systems to produce hybrids using self-incompatibility are possible in clovers (Riday and Krohn 2010). whereas a stable source of *cms* was identified and appears promising for *Lolium perenne* (Islam et al. 2014). "Hybrid" alfalfa varieties that exploit a patented male sterility system to attain about 75 % hybridity have entered the USA seed market (Veronesi et al. 2010), but their superiority over conventional synthetic cultivars is not clear. A simple and feasible alternative for exploiting heterosis is represented by semi-hybrids (or chance-hybrids), where hybridity between genetically-contrasting, narrowly-based populations reaches 50 % (Brummer 1999; Scotti and Brummer 2010). In alfalfa, semi-hybrids between *sativa* and *falcata* subspecies displayed substantial yield heterosis but negative traits including slower regrowth due to the *falcata* background, and could fail to satisfy DUS requirements (particularly distinctness and uniformity). The value of semi-hybrids within *sativa* germplasm awaits thorough assessment (Annicchiarico et al. 2015a).

Size of synthetics. Synthetic varieties built by cloning and polycrossing a fixed number of elite parent genotypes remain the most common forage variety type. When composing a synthetic, breeders must consider the trade-off between selection intensity and inbreeding depression, which depends on the number of parental genotypes and the extent of the depressive effect of selfing on the progeny performance. No clear rules exist that would point to the optimal size of a synthetic. Posselt (2010) modeled the performance of synthetics of diploid perennial ryegrass based on experimental yield data, finding an optimal parent number between 7 and 11. A similar study on tall fescue indicated even lower optimal parent number (4–6 parents) (Piano et al. 2007). However, the results from such studies may depend on the size, genetic diversity and genetic value of the test genetic base. At least 7–8 parents should be used in alfalfa to minimize inbreeding effects arising from selfing rates up to or over 20 % according to Busbice (1969), but breeders often use much larger (and probably excessive) parent numbers.

Both narrowly-based and widely-based synthetics have become successful cultivars. Only 31 % of the 89 cultivars bred by Agroscope that reached recommendation between 1991 and 2014 featured 7–11 parents, their parent number varying from 5 to 99 (Fig. 39.1). An analysis of the relationship between synthetic size and forage yield in the Agroscope breeding program of *Lolium perenne* (Boller et al. 2016) revealed no optimum number of parents, but greater tolerance of tetraploid than diploid synthetics to reducing the number of parents below 8. The vast majority of the registered synthetic varieties of forage grasses (Fig. 39.2) and clovers (Fig. 39.3) bred by ILVO (formerly RvP) in Belgium are composed by less than ten components. For clarity of the graph, three grass varieties are not presented in Fig. 39.2; they are composed by 39, 55 and 57 components, respectively. Although there is not necessarily a causal relationship between the composition and the commercial success of a variety, it is striking that the five best-sold varieties of the three forage grass spe-

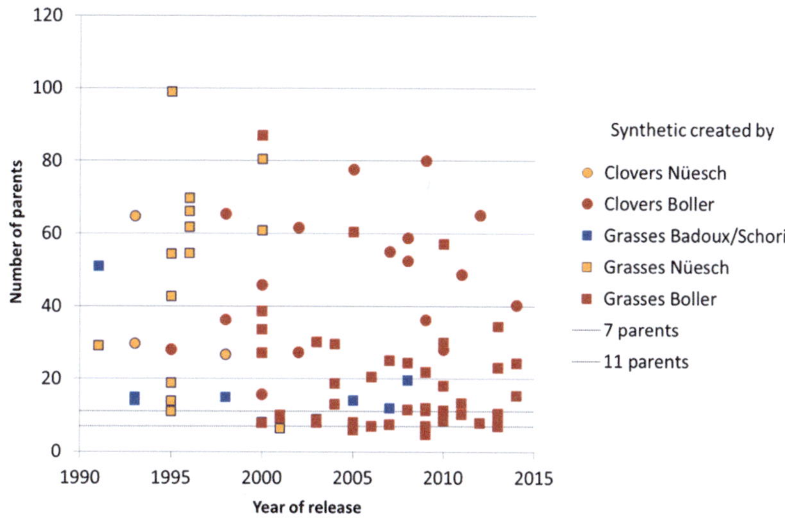

Fig. 39.1 Parent number of 89 cultivars of the breeding program of Agroscope released between 1991 and 2014, based on synthetics created by Bruno Nüesch. Beat Boller. Samuel Badoux and Arnold Schori

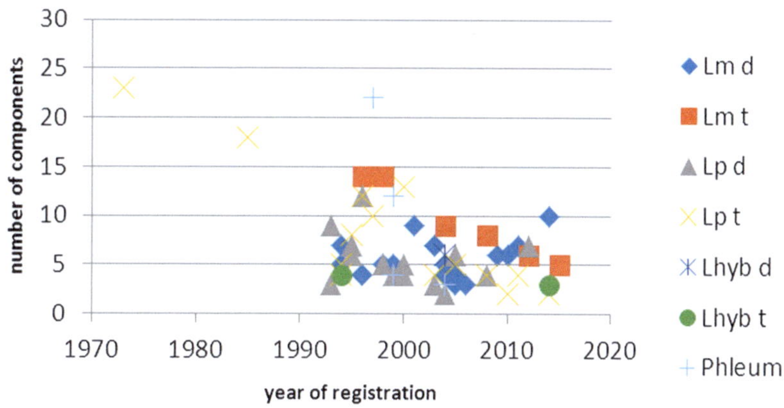

Fig. 39.2 Number of components in 66 registered forage grass synthetic varieties bred in Belgium between 1973 and 2015. *L Lolium*, *m multiflorum*, *p perenne*, *hyb hybridum*, *d* diploid, *t* tetraploid, *Phleum Phleum pratense*

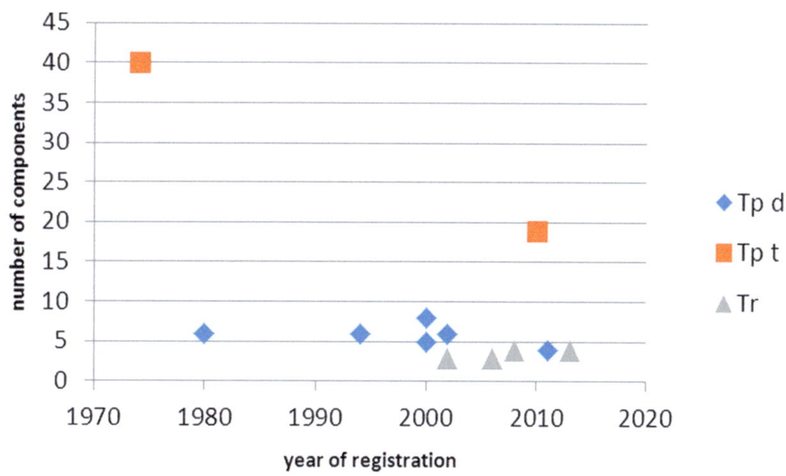

Fig. 39.3 Number of components in 12 registered clover synthetic varieties bred in Belgium between 1974 and 2013. *Tr Trifolium repens*, *Tp Trifolium pratense*, *d* diploid, *t* tetraploid

cies in 2004–2014 feature very different parent numbers. i.e., 12, 14, 22, 55 and 57 components. Apart from synthetics, ILVO has bred a number of registered varieties that are not based on clones but rather on a very restricted number of F_1. F_2 and further developed families.

Diploids vs. induced tetraploids. In all species able to be artificially polyploidized (e.g., by colchicine treatment), some desirable traits are inherently more pronounced at either the diploid or the tetraploid level. In ryegrass, diploid plants are characterized by higher tillering capacity and sward density, and tetraploid plants by the higher cell content to cell wall proportion (which leads to higher digestibility and soluble carbohydrate content). Typically, diploid cultivars of red clover have higher seed yield, and tetraploids higher forage yield. The lower sward density of tetraploids makes them less susceptible to diseases spreading from leaf to leaf through mycelium, such as snow mold in ryegrass and *Sclerotinia* in red clover. The higher dry matter content of diploids is of advantage for hay and silage making, and increases dry matter uptake per bite during grazing.

The contrasting properties of diploids and tetraploids lead most breeders to conduct two breeding programs per species in parallel, which either doubles the workload or halves the possible selection intensity. Market demand often dictates a predominance of one ploidy level. When expanding the breeding pool of the tetraploids, breeders must decide whether introducing new genetic diversity by exploiting unreduced gametes, by making crosses at the tetraploid level, or by using colchicine to incorporate pre-selected and hitherto not polyploidized diploid breeding material. The third approach was advocated by Mansat et al. (1966), who found a close correlation between agronomic characters of diploid and isogenic tetraploid lines of hybrid ryegrass and red clover. However, artificial polyploidization is more time-consuming: at least two selection cycles (usually 4 years) are needed to create stable tetraploids ready to be introduced in the usual breeding scheme, while F_1 progenies obtained by crossing tetraploids are readily usable even as parents of a synthetic variety the next year. Among the 38 tetraploid cultivars of the Agroscope breeding program that reached listing and recommendation between 1991 and 2014, no trend is manifest in favor of either of these crossing strategies (Table 39.2). Nineteen cultivars originated directly from colchicine treatment. 18 from 4x × 4x crosses, and one from both types of parent plants in the synthetic. Within the cultivars from colchicine treatment, clearly the C_2 generation was preferred, whereas within those from crosses. F_1 and F_2 progenies were used to build synthetics at similar frequencies.

Seed Yielding Ability and Market Exploitation

The need for a compromise between high yield of optimal quality forage and sufficient seed yield to make a cultivar commercially acceptable has been a prominent issue in forage breeding since its beginning. Hertzsch (1959) stated that the negative correlation between production of leaves and production of seed-bearing stems forces the breeder to take a "very unpleasant" decision, because improving seed yield is possible only at the expense of forage quality. Estimates of genetic correlation between forage and seed yield have usually confirmed this view. e.g., for white clover (Annicchiarico et al. 1999). An exception

Table 39.2 Number of tetraploid (4x) cultivars of the Agroscope breeding program listed officially between 1991 and 2014, grouped by origin of elite parent plants used to build the respective synthetics

Origin	Colchicine treatment				4x × 4x crosses				Mix	
Generation of synthetic parents	C_1	C_2	C_3	Sum	F_1	F_2	F_3	Sum		Total
Lolium multiflorum	1	3	1	5		3		3		8
Lolium perenne	2	5		7	3	3		6		13
Lolium x boucheanum	1	2	1	4	3	2	2	7	1	12
Trifolium pratense		2	1	3	1		1	2		5
Total	4	12	3	19	7	8	3	18	1	38

is represented by alfalfa, where an inverse correlation between these traits exists in the *Medicago sativa* complex (pooling subsp. *sativa* and the much less cultivated subsp. *falcata*) (Pecetti et al. 2008) but not within subsp. *sativa* (Bolaños-Aguilar et al. 2002).

Boelt and Studer (2010) concluded that the best way to improve seed yield without compromising forage quality is to target efficient use of the seed yield potential, rather than increasing the size of the reproductive system. Achieving this goal is complicated by the mentioned modest correlation for genotype seed yield between spaced planting and dense planting. Besides, correct estimates of genotype seed yields are hindered by differences in ripening time and narrow windows of optimal harvest date among the tested varieties. The true seed yield potential of a new variety often becomes manifest only after several years of large-scale seed production in different environments.

Seed yield components may offer different opportunities with regard to trade-offs between seed yield and forage yield. For example, breeding for higher forage and seed yield in white clover could be pursued by selecting for high numbers of florets per inflorescence and seeds per floret, neglecting high number of inflorescences because of its physiologically-based negative association with vegetative growth (Thomas 1987).

VCU trials of forage crops do not consider seed yielding ability. Various registered varieties with excellent forage yield failed to become a market success due to lack of competitive seed production. Farmers are usually less willing to pay a higher seed price for superior varieties of forage crops than they are for those of cash crops, because the economic benefit of a superior forage variety is less obvious. A further difficulty arises from the high degree of specialization of seed producers. They are often used to a particular variety and have difficulties to adapt to specific properties of a new variety. The very early-maturing perennial ryegrass varieties of Agroscope are a good example of how this could be overcome. Seed producers first had to learn that such perennial ryegrass types can be ready to harvest before the farmers even think of sharpening the mowing bar of their combines. Seed yields were low in the first years after release, and the end user price for seed of the very early varieties was 15–30 % higher than those of other varieties. Nowadays, seed producers are familiar with them, seed yields are reasonable, and there is no longer a price difference.

Strategies and Extent of Molecular Breeding

Over the past two decades, considerable investment worldwide has been made in technologies for molecular breeding of various forage crops, beginning with rudimentary genetic linkage maps in the early 1990s for alfalfa and perennial ryegrass and moving to extensive maps for many forages today. Nevertheless, markers appear to have hardly contributed materially to the development of any forage cultivars to date. The question may be raised as to why this lag is occurring and what to do about it.

Undoubtedly, part of the reason markers have been slow to be applied directly in breeding programs is the result of peculiarities of forage crops relative to the major grain and oilseed crops. For most forage crops, cultivars are synthetic populations and consequently, fixing QTL is more difficult than for inbred lines. Until recently, marker systems used in forage programs were limited to relatively few markers that were not trivial to assay and reasonably expensive in the context of a forage breeding program. Because forage yield and other important traits are genetically complex, they usually require a complete genome scan by high marker numbers to identify many relevant QTL (mostly with relatively small effects). In addition, most important traits are expensive and/or time-consuming to phenotype, and require proper coping with GE interactions. Finally, the choice of the mapping population is not trivial, since narrow-based material may provide results of limited applicability to the actual genetic base – which is a contributing reason for the lack of exploitation in breeding programs of published useful markers. Broad-based material, however, complicates QTL detection because of shorter linkage disequilibrium (Annicchiarico et al. 2015a).

The marker landscape is changing rapidly, however. High-density marker arrays based on single nucleotide polymorphisms (SNP) and sequencing-based marker methods have been developed and applied to forage grasses (Byrne et al. 2013; Fé et al. 2014) and alfalfa (Li et al. 2013, 2014). Genotyping-by-sequencing procedures (Elshire et al. 2011) are making thousands of SNP markers available (e.g., over 7,000–11,000 in different alfalfa mapping populations: Annicchiarico et al. 2015b), at a very low cost per data point. If marker assessment becomes cost-effective, then application to forage programs will proceed quickly. Numerous QTL have been mapped in various forage crops, and discussion of that literature is beyond the scope of this paper [for a summary of forage legume results. see Annicchiarico et al. (2015a)]. However, most of those experiments were not done in manner that makes their application to synthetic cultivar development straightforward. In most cases, mapping populations were not derived from or directly related to breeding populations, so translation of QTL results for selection is fraught with the usual concerns of QTL-marker phase and the extent of QTL effect in a different genetic background. One exception is represented by the successful application of one major QTL-based selection project for improving white clover seed yield (Barrett et al. 2009). This result suggests that other direct applications of markers for selection should be possible, when good marker-trait relationships are identified and a facile marker assay is developed. Further progress in this direction can be expected in the long term from developments of genome sequencing projects (such as the on-going one on alfalfa) followed by extensive gene annotation.

In alfalfa, for instance, a suite of disease and insect pest resistances have been incorporated into cultivars, and in many cases, a greenhouse screening procedure is well developed. Finding markers for these various resistances may not be required for continued selection, because the phenotypic assays are reasonably fast and inexpensive. However, assaying resistance loci tagged with markers for which dosage information can be obtained would be very useful to confirm that parental genotypes identified through phenotypic or genomic selection programs have multiple copies of desirable alleles, thereby leading toward fixation of these alleles in populations, or at least, expression of resistance in virtually all individuals in the synthetic population (Brummer and Monteros 2014).

Perhaps more compelling for forage crops is the use of genome-wide markers to develop prediction equations for yield (or other complex traits) that could facilitate multiple cycles of marker-only selection, theoretically enhancing genetic gain, especially for traits that require multi-year evaluations. Recently, a first analysis of genome prediction models in an alfalfa clonal selection program suggested that a model could be developed that explains sufficient variation to warrant application to a breeding program (Li et al. 2015). Encouraging results have been obtained also for predicting breeding values of alfalfa candidate parents (Annicchiarico et al. 2015b), which is of special interest for substituting the time-consuming stage of parent selection based on half-sib progeny yield responses. Another method that relates phenotypes to SNP allele frequencies of full-sib populations (rather than to genotypes of individual plants) is being studied in perennial ryegrass (Byrne et al. 2013). These experiences raise the expectation that marker-facilitated selection of new cultivars will soon be on the horizon.

The need for genome-wide marker coverage required by genomic selection approaches is likely to be satisfied by the development of genotyping-by-sequencing procedures (ideally optimized for the specific species). The other major requirement is extensive phenotyping to generate high quality data on a reasonably sized population (to build a high-confidence model). If models can be developed that span plant material from different genetic pools, then a coordinated international program focused on collecting data across several contrasting cropping environments and pooling resources for marker development would be worth pursuing. How broadly prediction models can be applied is not clear, but the high costs imply that partnership among groups should be fostered. Another possible advantage of this perspective (currently pursued on alfalfa in the ERA-NET project REFORMA) is the development of genomic selection models that are fine-tuned to specific target environments characterized by different type and extent of prominent stress or alternatively suited to breeding for wide adaptation to contrasting environments (Annicchiarico et al. 2015a).

Conclusions

Forage crop breeding, in comparison to breeding of major arable crops, is far more likely to suffer from insufficient focus due to challenges and dilemmas deriving from the wide diversity of possible bred species, target traits, target environments, crop utilizations, breeding approaches and seed markets. In addition, genetic gains are limited by several biological factors and modest investment, whereas the intermediate position of forages in production chains of animal products implies modest visibility for its achievements (hence, restricted opportunities for selling premium seed of innovative varieties). Within these constraints, forage crop breeders have deployed creative and diverse, mostly very field-based breeding activities, often achieving satisfactory results. However, pitfalls associated with insufficient focus have not been rare, owing to mismatch between selection and target environments (e.g., due to parent selection under spaced planting or excessively favorable selection environments), overlooking of important traits (e.g., seed yield) or fairly inefficient multi-trait selection, suboptimal exploitation of genetic diversity, insufficient research on crucial issues (such as breeding schemes. or more univocal definitions of forage quality), and other reasons. Improvements may also be necessary for some aspects of DUS and VCU testing procedures that are susceptible to hinder forage breeding progress.

A step change in the breeding of forage crops could happen through novel opportunities for genomic selection or genome-wide association studies based on extensive, low-cost genome scanning. Successful implementation into breeding programs would still depend on a focused choice of methods, plant material and phenotyping conditions. The need for extensive, multi-environment phenotyping that is required to fully exploit the potential of genomic selection for yield and adaptation to diverse cropping environments could conveniently be met by greater international cooperation among institutions committed to breeding research. Potentially, marker-based selection procedures could greatly facilitate the simultaneous selection for several traits, which is currently impeded by excessively high numbers of required evaluated genotypes.

As a trend, we expect increasing breeding attention towards legumes for several environmental and economic reasons. The application of breakthrough technology, insofar as it depends on seed market opportunities, could ultimately lead to less actively bred forage species.

References

Annicchiarico P (1992) Cultivar adaptation and recommendation from alfalfa trials in northern Italy. Journal of Genetics and Breeding 46: 269–278.

Annicchiarico P (1993) Variation for dry matter yield, seed yield and other agronomic traits in Ladino white clover landraces and natural populations. Euphytica 71: 131–141.

Annicchiarico P (2006a) Prediction of indirect selection for lucerne seed and forage yield from space-planted evaluation. Plant Breeding 125: 641–643.

Annicchiarico P (2006b) Diversity, genetic structure, distinctness and agronomic value of Italian lucerne (*Medicago sativa* L.) landraces. Euphytica 148: 269–282.

Annicchiarico P (2007a) Wide- versus specific-adaptation strategy for lucerne breeding in northern Italy. Theoretical and Applied Genetics 114: 647–657.

Annicchiarico P (2007b) Lucerne shoot and root traits associated with adaptation to favourable or drought-stress environments and to contrasting soil types. Field Crops Reseach 102: 51–59.

Annicchiarico P (2015) Alfalfa forage yield and leaf/stem ratio: narrow-sense heritability, genetic correlation, and parent selection procedures. Euphytica 205: 409–420.

Annicchiarico P, Pecetti L (2010) Forage and seed yield response of lucerne cultivars to chemically weeded and non-weeded managements and implications for germplasm choice in organic farming. European Journal of Agronomy 33: 74–80.

Annicchiarico P, Piano E (1994) Interference effects in white clover genotypes grown as pure stands and binary mixtures with different grass species and varieties. Theoretical and Applied Genetics 88: 153–168.

Annicchiarico P, Piano E (2000) Response of white clover genotypes to evaluation environments of dense and spaced planting, and implications for selection. Euphytica 111: 111–120.

Annicchiarico P, Piano E (2005) Use of artificial environments to reproduce and exploit genotype × location interaction for lucerne in northern Italy. Theoretical and Applied Genetics 110: 219–227.

Annicchiarico P., Proietti S. (2010) White clover selected for enhanced competitive ability widens the compatibility with grasses and favours the optimization of legume content and forage yield in mown clover-grass mixtures. Grass and Forage Science 65: 318–324.

Annicchiarico P, Barrett B, Brummer EC, Julier B, Marshall AH (2015a) Achievements and challenges in improving temperate perennial forage legumes. Critical Reviews of Plant Science 34: 327–380.

Annicchiarico P, Nazzicari N, Li X, Wei Y, Pecetti L, Brummer EC (2015b) Accuracy of genomic selection for alfalfa biomass yield in different reference populations. BMC Genomics 16:1020.

Annicchiarico P, Pecetti L, Abdelguerfi A, Bouizgaren A, Carroni AM, Hayek T, Bouzina MM, Mezni M (2011) Adaptation of landrace and variety germplasm and selection strategies for lucerne in the Mediterranean basin. Field Crops Research 120: 283–291.

Annicchiarico P, Pecetti L, Abdelguerfi A, Bouzerzour H, Kallida R, Porqueddu C, Simões NM, Volaire F (2013a) Optimal forage grass germplasm for drought-prone Mediterranean environments. Field Crops Research 148: 9–14.

Annicchiarico P, Pecetti L, Tava A (2013b) Physiological and morphological traits associated with adaptation of lucerne (*Medicago sativa*) to severely drought-stressed and to irrigated environments. Annals of Applied Biology 162: 27–40.

Annicchiarico P, Piano E, Rhodes I (1999) Heritability of, and genetic correlations among, forage and seed yield traits in Ladino white clover. Plant Breeding 118: 341–346.

Baert J, Van de Walle M, De Riek J, De Boever J, Fievez V, Van Waes C (2012) Variation and heritability of alpha-linolenic acid content and rumen escape protein fraction in fodder grass and clover. In: Barth S, Milbourne D (eds), Breeding strategies for sustainable forage and turf grass improvement. Springer Science + Business Media, Dordrecht, the Netherlands, pp 181–188.

Barrett B, Baird I, Woodfield D (2009) White clover seed yield: A case study in marker-assisted selection. In: Yamada T, Spangenberg G (eds), Molecular breeding of forage and turf, Springer, New York, pp 241–250.

Bernardo R (2010) Breeding for quantitative traits in plants, 2nd Ed. Stemma Press, Woodbury, MN, USA.

Boelt B, Studer B (2010) Breeding for grass seed yield. In: Boller B, Posselt UK, Veronesi F (eds) Fodder crops and amenity grasses. Handbook of plant breeding, vol 5. Springer Science + Business Media, New York, pp 161–174.

Bolaños-Aguilar ED, Huyghe C, Ecalle C, Hacquet J, Julier B (2002) Effect of cultivar and environment on seed yield in alfalfa. Crop Science 42: 45–50.

Boller B, Grieder C, Schubiger FX (2016) Performance of diploid and tetraploid perennial ryegrass synthetics with variable numbers of parents. In: Baert, J, Reheul D, Roldan I (eds), Breeding in a world of scarcity. Springer Science + Business Media, Dordrecht, the Netherlands (in press).

Boller B, Peter-Schmid MKI, Tresch E, Tanner P, Schubiger FX (2009) Ecotypes of Italian ryegrass from Swiss permanent grassland outperform current recommended cultivars. Euphytica 170: 53–65.

Boller, B, Schubiger, FX, Kölliker, R. 2010b. Red clover. In: Boller B, Posselt UK, Veronesi F (eds) Fodder crops and amenity grasses. Handbook of plant breeding, vol 5. Springer Science + Business Media, New York, pp 439–456.

Boller B, Tanner P, Schubiger FX (2008) Breeding forage grasses for organic conditions. Euphytica 163: 459–467.

Bouton JH (2007) The economic benefits of forage improvement in the United States. Euphytica 154: 263–270.

Bouton JH (2012) Breeding lucerne for persistence. Crop and Pasture Science 63: 95–106.

Brummer EC (1999) Capturing heterosis in forage crop cultivar development. Crop Science 39: 943–954.

Brummer EC, Casler MD (2014) Cool-season forages. In: Smith S, Diers B, Specht J, Carver B (eds), Genetic gain in major U.S. field crops. CSSA Spec. Publ. 33. ASA, CSSA, and SSSA. Madison, WI.

Brummer EC, Monteros MJ (2014) Marker-assisted selection. Legume Perspectives 4: 17–18.

Busbice TH (1969) Inbreeding in synthetic varieties. Crop Science 9: 601–604.

Byrne S, Czaban A, Studer B, Panitz F, Bendixen C, Asp T (2013) Genome wide allele frequency fingerprints (GWAFFs) of populations via genotyping by sequencing. PLoS One 8: e57438.

Cashman PA (2014) Differential productivity and persistency responses to simulated and cattle grazing of perennial ryegrass genotypes. PhD thesis, Queens' University, Belfast.

Casler MD (2001) Breeding forage crops for increased nutritive value. Advances in Agronomy 71: 51–107.

Casler MD, Brummer EC (2008) Theoretical expected genetic gains for among- and within- family selection methods in perennial forage crops. Crop Science 48: 890–902.

Chapman DF, Edwards GR, Stewart AV, McEvoy M, O'Donovan M, Waghorn GC (2014) Valuing forages for genetic selection: what traits should we focus on? Proc 5th Australasian dairy science symposium. Available on-line http://www.adssymposium.com.au/inewsfiles/2014proceedings/36ChapmanADSS2014.pdf

Chaves B, De Vliegher A, Van Waes J, Carlier L, Marynissen B (2009) Change in agronomic performance of *Lolium perenne* and *Lolium multiflorum* varieties in the past 40 years based on data from the Belgian VCU trials. Plant Breeding 128: 680–690.

Cooper M, Messina CD, Podlich D, Totir LR, Baumgarten A, Hausmann NJ, Wright D, Graham G (2014) Predicting the future of plant breeding: Complementing empirical evaluation with genetic prediction. Crop and Pasture Science 65: 311–336.

COPA-COGECA (2007) Lucerne: an asset to the environment. COPA-COGECA, Paris.

Cougnon M (2012) Potential in mixed swards and breeding of tall fescue (*Festuca arundinacea* Schreb.). PhD thesis, University of Gent, Belgium.

Cougnon M, Baert J, Van Waes C, Poinsard L, Reheul D (2014b) Production de mélanges de fétuque élevée et de ray-gras d'Italie avec ou sans trèfle violet en Belgique. Fourrages 218: 185–190.

Cougnon M, Baert J, Van Waes C, Reheul D (2014c) Performance and quality of tall fescue (*Festuca arundinacea* Schreb.) and perennial ryegrass (*Lolium perenne* L.) and mixtures of both species grown with or without white clover (*Trifolium repens* L.) under cutting management. Grass and Forage Science 69: 666–677.

Cougnon M, De Koker J, Fievez V, Reheul D (2014a) Factors influencing animal preference of tall fescue genotypes. In Sokolovic D, Huyghe C, Radovic J. Quantitative traits breeding for multifunctional grassland and turf. Springer, Dordrecht, The Netherlands, 125–129.

Elgersma A, Winkelhorst GD, den Nijs APM (1994) The relationship between progeny seed yield in drilled plots and maternal spaced-plant traits in perennial ryegrass (*Lolium perenne* L.). Plant Breeding 112: 209–214.

Elshire RJ, Glaubitz JC, Sun Q, Poland JA, Kawamoto K, Buckler EC, Mitchell SE (2011) A robust, simple Genotyping-by-Sequencing (GBS) approach for high diversity species. PLoS One 6:e19379.

Falconer DS (1989) Introduction to quantitative genetics. Longman, Harlow, UK.

Fé D, Ashraf B, Byrne F, Szaban A, Roulund N, Lenk I, Asp T, Pedersen M, Janss L, Jensen J, Jensen CS (2014) Prospects for introducing genomic selection into forage grass breeding. Grassland Science in Europe 19: 830–832.

GEVES (2011) Plantes fourragères. Nouvelles variétés proposées à l'inscription sur la Liste A du Catalogue Officiel français. Résultats Agronomiques. Available on-line: http://cat.geves.info/CAT_WEB/Data/PLA_Fourrages_2011.pdf

Hayward MD, Vivero JL (1984) Selection for yield in *Lolium perenne*. II. Performance of spaced plant selections under competitive conditions. Euphytica 33: 787–800.

Hector A, Hautier Y, Saner P, Wacker L, Bagchi R, Joshi J, Scherer-Lorenzen M, Spehn EM, Bazeley-White E, Weilenmann M, Caldeira MC, Dimitrakopoulos PG, Finn JA, Huss-Danell K, Jumpponen A, Mulder CPH, Palmborg C, Pereira JS, Siamantziouras ASD, Terry AC, Troumbis AY, Schmid B, Loreau M (2010) General stabilizing effects of plant diversity on grassland productivity through population asynchrony and overyielding. Ecology 91:2213–2220.

Hertzsch W (1959) Die Gräser (Allgemeiner Teil). In: Kappert H, Rudorf W (eds) Breeding of forage plants, manual of plant breeding vol. 4, 2nd Ed. Paul Parey, Berlin/Hamburg, pp 346–376.

Huyghe C (2003) Les fourrages et la production de protéines. Fourrages 174: 145–162.

Islam MS, Studer B, Møller IM, Asp T (2014) Genetics and biology of cytoplasmic male sterility and its applications in forage and turf grass breeding. Plant Breeding 133: 299–312.

Kingston-Smith AH, Thomas HM (2003) Strategies of plant breeding for improved rumen function. Annals of Applied Biology 142: 13–24.

Isselstein J, Kayser M. (2014) Functions of grassland and their potential role in delivering ecosystem services. Grassland Science in Europe 19: 199–214.

Julier B, Huyghe C, Ecalle C (2000) Within- and among-cultivar genetic variation in alfalfa: Forage quality, morphology, and yield. Crop Science 40: 365–369.

Laidig F, Piepho HP, Drobek T, Meyer U (2014) Genetic and non-genetic term trends of 12 different crops in German official variety performance trials and on-farm yield trends. Theoretical and Applied Genetics 127: 2599–2617.

Li X, Han Y, Wei Y, Acharya A, Farmer AD, Ho J, Monteros MJ, Brummer EC (2013) Development of an alfalfa SNP array and its use to evaluate patterns of population structure and linkage disequilibrium. PLoS One 9: e84329.

Li X, Wei Y, Acharya A, Jiang Q, Kang J, Brummer EC (2014) A saturated genetic linkage map of autotetraploid alfalfa (*Medicago sativa* L.) developed using genotyping-by-sequencing is highly syntenous with the *M. truncatula* genome. Genes Genomes Genetics 4: 1971–1979.

Li X, Wei Y, Acharya A, Hansen JL, Crawford JL, Viands DR, Michaud R, Claessens A, Brummer EC (2015) Genomic prediction of biomass yield in two selection cycles of a tetraploid alfalfa breeding population. The Plant Genome (in press; doi: 10.3835/plantgenome2014.12.0090).

Ludemann CI, Eckard RJ, Cullen BR, Jacobs JL, Malcolm B, Smith KF (2015) Higher energy concentration traits in perennial ryegrass (*Lolium perenne* L.) may

increase profitability and improve energy conversion on dairy farms. Agricultural Systems 137: 89–100.
Mansat P, Picard J, Barthou, F (1966) Value of selection at diploid level before tetraploidization. Proc. X[th] International Grassland Congress, pp 671–676.
McEvoy M, O'Donovan M, Shalloo L (2011) Development and application of an economic ranking index for perennial ryegrass cultivars. Journal of Dairy Science 94: 1627–1639.
Monteros MJ, Han Y, Brummer EC (2014) Alfalfa (*Medicago sativa* L.): Biology, agronomy, genetics, genomics and molecular breeding. In: Cai H, Yamada T, Kole C (eds) Genetics, genomics, and breeding of forage crops. CRC Press, Boca Raton, FL, pp 187–219.
Muir JP, Taylor J, Interrante SM. (2005) Herbage and seed from native perennial herbaceous legumes of Texas. Rangeland Ecology and Management 58: 643–651.
Mullainathan S, Shafir E (2013) Scarcity: why having too little means so much. Henry Holt Books, New York.
Nyfeler D, Huguenin-Elie O, Suter M, Frossard E, Connolly J, Lüscher A (2009) Strong mixture effects among four species in fertilized agricultural grassland led to persistent and consistent transgressive overyielding. Journal of Applied Ecology 46: 683–691.
Olmstead J, Brummer EC (2008) Benefits and barriers to perennial forage crops in Iowa corn and soybean rotations. Renewable Agriculture and Food Systems 23: 97–107.
Parsons AJ, Edwards GR, Newton PCD, Chapman DF, Caradus JR, Rasmussen S, Rowarth JS (2011) Past lessons and future prospects: Plant breeding for yield and persistence in cool-temperate pastures. Grass and Forage Science 66: 153–172.
Pecetti L, Romani M, De Rosa L, Piano E (2008). Selection of grazing-tolerant lucerne cultivars. Grass Forage Science 63: 360–368.
Pflimlin A, Faverdin P (2014) Les nouveaux enjeux du couple vache-prairie à la lumière de l'agroécologie. Fourrages 217: 23–35.
Piano E, Annicchiarico P, Romani M, Pecetti L (2007) Effect of the number of parents and their combining ability on the performance of synthetic varieties in tall fescue. Australian Journal of Agricultural Research 58: 1100–1105.
Pinxterhuis JB, Beare MH, Edwards GR, Collins RP, Dillon P, Oenema J (2015) Eco-efficient pasture based dairy farm systems: a comparison of New Zealand, The Netherlands and Ireland. Grassland Science in Europe, 20, 349–346.
Pittman JJ, Arnall DB, Interrante SM, Moffet CA, Butler TJ (2015) Estimation of biomass and canopy height in bermudagrass, alfalfa, and wheat using ultrasonic, laser, and spectral sensors. Sensors 15: 2920–2943.
Posselt UK (2010) Breeding methods in cross-pollinated species. In: Boller B, Posselt UK, Veronesi F (eds) Fodder crops and amenity grasses. Handbook of plant breeding, vol 5. Springer Science+Business Media, New York, pp 39–87.
Reheul D, De Cauwer B, Cougnon M, Aper J (2012) What global and/or European agriculture will need from grasslands and grassland breeding over the next 10–15 years for sustainable agriculture. In: Barth S, Milbourne D (eds), Breeding strategies for sustainable forage and turf grass improvement. Springer Science+Business Media, Dordrecht, the Netherlands, pp 3–18.
Riday H, Krohn AL (2010) Genetic map-based location of the red clover (*Trifolium pratense* L.) gametophytic self-incompatibility locus. Theoretical and Applied Genetics 121: 761–767.
Rodríguez-Ortega T, Oteros-Rozas E, Ripoll-Bosch R, Tichit M, Martín-López B, Bernués A (2014) Applying the ecosystem services framework to pasture-based livestock farming systems in Europe. Animal 8: 1361–1372.
Rotili P (1988) The concept of vigour and its application in a breeding program of lucerne. In: Marji J (ed.) Breeding for vigour in forage crops. Research Institute for Irrigation, Szarvas, Hungary, pp 64–78.
Rotili P, Scotti C, Zannone L, Gnocchi G (1994) Lucerne stand system: Dynamics of forage production, quality and demography: Consequences on the variety constitution process. REUR Tech. Ser. 36: 54–62.
Scotti C, Brummer EC (2010) Creation of heterotic groups and hybrid varieties. In: Huyghe C (ed), Sustainable use of genetic diversity in forage and turf breeding. Springer, the Netherlands, pp. 509–518.
Simmonds NW (1979) Principles on crop improvement. Longman Scientific and Technical, New York.
Smith KF, Ludemann C, Lewis CD, Malcolm B, Banks RG, Jacobs JL, Fennessy PF, Spangenberg, GC (2014) Estimating the value of genetic gain in perennial pastures with emphasis on temperate species. Crop and Pasture Science 65: 1230–1237.
Stewart A, Hayes R (2011) Ryegrass breeding – balancing trait priorities. Irish Journal of Agriculture and Food Resources 50: 31–46.
Swinton SM, Jolejole-Foreman CB, Lupi F, Ma S, Zhang W, Chen H (2015) Economic value of ecosystem services from agriculture. In: Hamilton SK, Doll JE, Robertson GP (eds), The ecology of agricultural landscapes: long-term research on the path to sustainability. Oxford University Press, New York, pp 54–76.
Thami-Alami I, Pecetti L, Souihka A, Annicchiarico P (2016) Optimizing species and variety choice of annual legume-based forage crops in a dry Mediterranean region. In: Baert, J, Reheul D, Roldan I (eds), Breeding in a world of scarcity. Springer Science+Business Media, Dordrecht, the Netherlands (in press).
Thomas RG (1987) Reproductive development. In: Baker MJ, Williams WM (eds), White Clover. CAB International, Wallingford, UK, pp 63–123.
Turesson G (1922) The genotypical response of the plant species to the habitat. Hereditas 3: 211–350.
Van Peer L, Nijs I, Reheul D, De Cauwer B (2004) Species richness and susceptibility to heat and drought extremes in synthesized grassland ecosystems: com-

positional vs physiological effects. Functional Ecology 18: 769–778.

Van den Pol-van Dasselaar A, Aarts HFM, De Caesteker E, De Vliegher A, Elgersma A, Reheul D, Reijneveld JA, Vaes R, Verloop J (2015) Grassland and forages in high output dairy farming systems in Flanders and the Netherlands. Grassland Science in Europe 20: 3–25.

Veronesi F, Brummer EC, Huyghe C (2010) Alfalfa. In: Boller B, Posselt UK, Veronesi F (eds) Fodder crops and amenity grasses. Handbook of plant breeding, vol 5. Springer Science+Business Media, New York, pp 395–437.

Wilkins PW, Humphreys MO (2003) Progress in breeding perennial forage grasses for temperate agriculture. J. Agric. Sci. 140: 129–150.

Metabolomics for Methane Mitigation: A Case Study in Biserrula (*Biserrula pelecinus* L.)

40

K. Ghamkhar, S. Rochfort, B. Banik, and C. Revell

Abstract

Methane, a greenhouse gas, is an important driver of climate change. It also costs energy for the ruminant because it is produced in the rumen as a by-product of anaerobic microbial fermentation of feed. Biserrula (*Biserrula pelecinus* L.) is one among only a few pasture and forage species that have been identified as methane moderators when fermented by rumen microbes. At present, it is unclear whether metabolites in biserrula play a role in causing this effect. The objective of this study was to investigate this putative role. A preliminary relationship between metabolite profile and methanogenic potential was identified, and candidate metabolites were discovered for future reverse genetics approaches for gene discovery and for use in breeding.

Keywords

Climate change • Forage • Metabolomics • Methane • Rumen

K. Ghamkhar (✉)
Centre for Plant Genetics and Breeding, The University of Western Australia, 35 Stirling Hwy., Crawley, WA 6009, Australia

Gin Silico Pty Ltd, PO Box 1159, Blackburn North, VIC 3130, Australia

Grasslands Research Centre, AgResearch, Private Bag 11008, Palmerston North 4442, New Zealand
e-mail: kioumars.ghamkhar@agresearch.co.nz

S. Rochfort
Department of Economic Development, Jobs, Transport and Resources, 1 Park Drive, Bundoora, VIC 3083, Australia

B. Banik
School of Animal Biology, The University of Western Australia, 35 Stirling Hwy, Crawley, WA 6009, Australia

Centre for Plant Genetic and Breeding, The University of Western Australia, 35 Stirling Hwy., Crawley, WA 6009, Australia

C. Revell
Centre for Plant Genetics and Breeding, The University of Western Australia, 3 Baron-Hay Court, South Perth, WA 6151, Australia

Pasture Science, Department of Agriculture and Food Western Australia, 35 Stirling Hwy., Crawley, WA 6009, Australia

I. Roldán-Ruiz et al. (eds.), *Breeding in a World of Scarcity: Proceedings of the 2015 Meeting of the Section "Forage Crops and Amenity Grasses" of Eucarpia*,
DOI 10.1007/978-3-319-28932-8_40

Introduction

Climate change is partly caused by the emission of greenhouse gases such as carbon dioxide and methane (CH_4) (Pachauri and Meyer. 2014). About 70 % of CH_4 emissions arise from anthropogenic sources including methane production by livestock and about 30 % from natural sources (Mosier et al. 1998). Methane is a common by-product of anaerobic microbial fermentation of food in the rumen. To abate CH_4 emission and increase the efficiency of rumen processes in ruminants, the focus should be to divert some of the energy lost as CH_4 into animal products. Enteric fermentation can vary widely depending on factors such as type of the animal, the amount and type of feed, environment, feed additives and bodyweight of the animal (Moss et al. 2000). Improving pasture quality is viewed as a means of reducing CH_4 emissions because animal productivity may be enhanced with less dietary energy lost in the form of CH_4 due to reduced fiber content of the fodder. In New Zealand. Friesian and Jersey dairy cows grazing on sulla (*Hedysarum coronarium*), which contains condensed tannins, emitted less CH_4 per unit of dry matter than cows grazing on perennial ryegrass (Woodward et al. 2002).

Feed intake and feeding behavior is governed by factors including forage availability, palatability and feedback mechanisms (Rochfort et al. 2008). A recent study has reported on effects on methane production in ruminants that were fed a broad range of dietary pasture and forage species (Banik et al. 2013). Biserrula (*Biserrula pelecinus* L.) was identified as the species with the greatest effect on methane output reduction. Biserrula is an inbreeding annual legume native to the Mediterranean areas of Europe. North Africa and the Canary Islands. In the wheat belt of southern Australia, it is an important pasture species (Loi et al. 1997). Biserrula is well adapted to a wide range of soil types and pH levels and is able to survive on poor soils (Carr et al. 1999). It is also a potential candidate species for the pastures of the drier northeastern North Island and the Hill Country of New Zealand as it can grow within the range of 325–700 mm annual rainfall and it has a deep root system (over 2 m).

Recently, a core collection of 30 accessions of biserrula was developed from the genetic resource of 279 accessions in the Australian Trifolium Genetic Resource Centre (ATGRC. Perth), using molecular and non-molecular data (Ghamkhar et al. 2013). The aim of the current study was to characterize the metabolomic profiles of a subset of the accessions from the biserrula core collection, and evaluate the possible links between their metabolite profile and *in vitro* methanogenic potential.

Materials and Methods

Plant Material

Plant material was identical to that used by Banik et al. (2013). Briefly, 45 plants of all accessions of biserrula core collection (half of the accessions were replicated) were grown in individual pots and randomized in a glasshouse. Plant material was collected at 77 days post-sowing, the approximate timing for typical grazing, to analyze *in vitro* methanogenic potential. The edible parts of the plants (leaf and stem) were harvested from individual plants, placed on dry ice, freeze-dried in a bench-top freeze dryer (BenchTop. VirTis/SP Scientific. Warminster. PA. USA), and ground in a grinding mill (Glen Creston. Stanmore. UK) to pass through a 1.0-mm screen. A total of 12 plants of nine accessions were selected based on extremely high, extremely low and a medium level of methane output reported by Banik et al. (2013). The same ground herbage was used for metabolomics analysis (Table 40.1).

MS and Wet-Chemical Analyses

Freeze-dried samples were stored with desiccant at −80 °C until analysis. Ground freeze-dried sample (100 mg) was weighed into 2 mL Eppendorf tubes. These 100 mg samples were re-

Table 40.1 The subset selected from the *B. pelecinus* core collection for metabolomics analysis, their origin and *in vitro* methane output

Accession ID	Country of Origin	(Average) Methane output (mL/g DM)	Sample #
2004ERI56PEL[a]	Eritrea	7.71	4
2006MAR29PEL	Morocco	(6.31)	1, 35
143467	Italy	5.24	12
GEH71PEL[a]	Greek Islands	4.80	25
2006ISR20PEL	Israel	(4.00)	28, 42
143469[a]	Italy	2.57	13
143474	Italy	(2.16)	14, 32
93ITA45PEL[a]	Italy	2.06	24
2004ESP39PEL	Canary Islands	1.07	8

All samples had experimental replicates. All methane data were significant at $P \leq 0.05$
[a]Single biological replicate

suspended in a 1 mL volume of 80 % methanol:water. Samples were vortexed thoroughly for 60 s. The sample was sonicated for 10 min in a sonicating water bath. After briefly being vortexed, the sample solution was centrifuged for 10 min at 14.000 rpm and 20 °C. A total of 200 μL of supernatant from each sample was removed and placed into an individually labelled vial containing a 250 μL insert for LCMS analysis. LCMS data was acquired with positive/negative switching for greater metabolome coverage.

Data Analysis

The LCMS data for all samples was analyzed using GeneData Analyst™ software 4.1 (Genedata Solutions, In Silico, Basel, Switzerland). Principal components analysis was conducted in GeneData Analyst to plot the grouping of accessions based on their metabolomic profile. Data from the analysis of methanogenic potential (Banik et al. 2013) were used in Genedata to explore relationships between methanogenic potential and metabolomic profiles of the accessions.

Results and Discussion

Methane data have been provided in detail as per Banik et al. (2013) (Table 40.1). The first metabolomics test with 12 samples (1, 4, 8, 12, 13, 14, 24, 25, 32, 35, 38, and 42) clustered the nine accessions according to their methanogenic potential. The lower methane accessions (≤3 ml/gDM) have less metabolite diversity associated with their geographical origin, possibly longitudinal distribution. When compared to methanogenic data, accessions with medium (between 3 and 5 ml/gDM) and also accessions with high methane output (≥5 ml/gDM) grouped with their respective categories metabolically, except the accession from Eritrea (Fig. 40.1). Among low methane output accessions, one exception was noted. Accession 143467 (12 in Fig. 40.1) from Italy is among higher methane output accessions, but is grouped on the basis of metabolites with low methane output accessions. Hierarchical cluster analysis of metabolites (data not shown) also separated the higher methane output and lower methane output accessions, except accession 143467. The two analyses share the general results but cluster analysis revealed more details. In general, the hierarchical cluster analysis shows a pattern of LCMS metabolite profile being associated with methanogenic potential in this species. Eritrean accessions clustered with some accessions from Morocco and the Canary Islands; however, within this shared cluster hosting high methane output accessions, the Eritrean accessions make a distinct sub-cluster separate from the Morocco-Canary Islands sub-cluster.

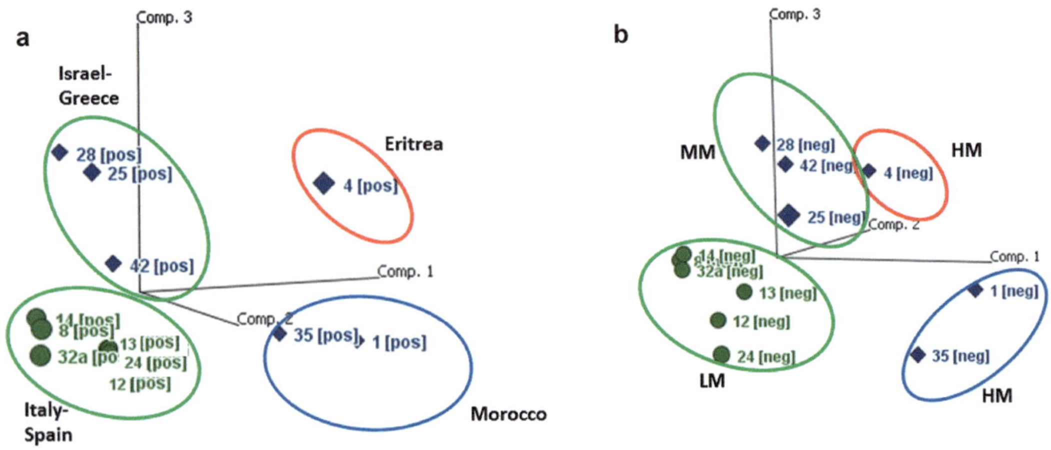

Fig. 40.1 Principal component analysis of LCMS data in a subset of high, medium and low methane output accessions of biserrula; (**a**) positive mode; (**b**) negative mode. *Green ellipse* and *green text* indicates accessions with low methane output (*LM*), *green ellipse* and *blue text* indicates medium methane output (*MM*), and *blue and red ellipses* indicate accessions with high methane output (*HM*), while *red ellipse* separates the Eritrean accession from the rest

Searching for Specific Metabolites for Methane Output

Of 417 metabolites showing pick diversity among the screened accessions, 52 candidate metabolic markers were identified for further data exploitation in biserrula to elucidate the metabolism of methane generation in the rumen. This subset was selected based on differences in metabolite concentration between high and low methane output accessions. This number was then further reduced to eight markers based on extreme metabolic differences (presence/absence) of these markers between high methane vs. low methan output accessions (Table 40.2). Other markers were eliminated from the list due to their partial incongruence with the methane data.

Table 40.2 Most statistically significant ions differentiating 'low' from 'high' methane output data set

m/z	RT (min)	P-Value	BH Q-Value	ST Q-Value
134.0	1.89	1.7E-5	0.004	0.002
945.3	7.97	2.1E-5	0.004	0.002
946.3	7.97	2.9E-5	0.004	0.002
572.3	18.74	2.8E-5	0.004	0.002
234.1^	1.72	8.0E-6	0.005	0.002
213.0^	2.18	4.3E-5	0.011	0.005
380.9^	2.01	5.2E-5	0.011	0.005
511.4^	14.31	1.4E-4	0.015	0.007

^ indicates ions are from positive LCMS analysis

Conclusion

Metabolomic and *in vitro* methanogenic profiles of the accessions of biserrula used in this experiment align well. This suggests that some metaboliltes discovered in this study might affect methanogenic potential in the rumen. In addition to validation *in vivo*, genetic analysis of high and low methane accessions will further improve biserrula for agricultural purposes and also increase knowledge of the metabolites and gene discovery for ruminal methane mitigation through transcriptomics and reverse genomics.

References

Banik BK, Durmic Z, Erskine W, Nichols P, Ghamkhar K, Vercoe P (2013). Variability of in vitro ruminal fermentation and methanogenic potential in the pasture legume biserrula (*Biserrula pelecinus* L.). Crop and Pasture Science 64: 409–416.

Carr SJ, Loi A, Howieson JH, Porqueddu C (1999). Attributes of Biserrula pelecinus L.(biserrula): A new pasture legume for sustainable farming on acidic sandy soils in Mediterranean environments. Cahiers Options Méditerranéennes, 39: 87–90.

Ghamkhar K, Revell C, Erskine W (2013). *Biserrula pelecinus* L.–genetic diversity in a promising pasture legume for the future. Crop and Pasture Science 63: 833–839.

Pachauri R K, Meyer LA (2014) Climate Change 2014: Synthesis Report. Contribution of Working Groups I, II and III to the Fifth Assessment Report of the Intergovernmental Panel on Climate Change. Geneva, Switzerland.

Loi A, Cocks P, Howieson JH, Carr SJ (1997). Morphological characterization of Mediterranean populations of *Biserrula pelecinus* L. Plant breeding 116: 171–176.

Mosier A, Duxbury J, Freney J, Heinemeyer O, Minami K., Johnson D (1998). Mitigating agricultural emissions of methane. Climatic Change 40: 39–80.

Moss AR, Jouany JP Newbold J (2000). Methane production by ruminants: its contribution to global warming. Annales de Zootechnie, 49: 231–253.

Rochfort S, Parker AJ, Dunshea FR (2008). Plant bioactives for ruminant health and productivity. Phytochemistry 69: 299–322.

Woodward S, Waghorn G, Lassey K, Laboyrie P (2002). Does feeding sulla (*Hedysarum coronarium*) reduce methane emissions from dairy cows? Proceedings-New Zealand society of animal production. 62: 227–230.

41 Silica Content, Leaf Softness and Digestibility in Tall Fescue (*Festuca arundinacea* Schreb.)

M. Cougnon, R. Shahidi, E. Struyf, C. Van Waes, and D. Reheul

Abstract

In tall fescue breeding, a great deal of effort is devoted to develop soft-leaved varieties as it is generally accepted that leaf softness is positively correlated with digestibility and animal preference. In advanced breeding programs, it becomes difficult to discriminate the leaf softness between genotypes. Moreover, there is evidence that the digestibility of the softest varieties is not necessarily higher compared to varieties with coarser leaves. We studied the presence of trichomes (dentation) on the leaf margins and the silica content of plants in relation to the leaf softness and digestibility on a selection of 19 tall fescue clones. On average, soft genotypes had fewer trichomes (2.74 mm^{-1}) on the leaf margins than coarse genotypes (9.03 mm^{-1}), but there was no relation between leaf softness or trichome number and digestibility ($R^2=0.05$), nor between silica content and softness or digestibility ($R^2=0.09$).

Keywords

Tall fescue • Silica • Coarse leaf • Leaf softness • Digestibility • Animal preference • Trichomes

Introduction

A coarse leaf texture is often presumed to be a main factor explaining animals' distaste for tall fescue (Fa). Fa leaves can be so coarse that they cause small wounds when touched with unprotected skin. It is believed that the coarse Fa leaves sting the muzzle of grazing animals, which would explain the animals' aversion to unwilted Fa forage. In early Fa breeding programs, clones were grazed (mostly by sheep) to select for palatability

M. Cougnon (✉) • R. Shahidi • D. Reheul
Department of Plant Production, Ghent University, Gent, Belgium
e-mail: mathias.cougnon@ugent.be

E. Struyf
Ecosystem Management Research Group, University of Antwerp, Antwerp, Belgium

C. Van Waes
Plant Sciences, Institute for Agriculture and Fisheries Research (ILVO), Merelbeke, Belgium

I. Roldán-Ruiz et al. (eds.), *Breeding in a World of Scarcity: Proceedings of the 2015 Meeting of the Section "Forage Crops and Amenity Grasses" of Eucarpia*,
DOI 10.1007/978-3-319-28932-8_41

(the amount of grass consumed under proper grazing conditions) (Craigmiles et al. 1964; Petersen et al. 1958). However, this pragmatic approach is labor intensive. Gillet and Jadas-Hecart (1965) showed that the flexibility of the leaves as determined by palpation with the hand was positively correlated with animal preference. In addition, the authors found that there was a high variation for this trait in wild Fa populations. Varieties bred from clones selected for leaf flexibility (e.g., Lubrette) proved to have an increased animal preference and digestibility in grazing trials with sheep (Jadas-Hecart 1982).

"Leaf flexibility" as used in the papers cited above refers to the perceived coarseness when the grass leaves are touched. We prefer to use "leaf softness" instead of leaf flexibility as the latter can also refer to the mechanical properties of grass leaves. Selection for soft leaf texture offers a limited progress in Fa quality. First, perceiving differences in leaf softness becomes more difficult as breeding progresses since the more the breeding material improves, the less variation for these traits remains. Second, the correlation between leaf softness and animal preference is lower in elite breeding material of tall fescue. Third, data of the official Swiss tall fescue variety trials indicated that the varieties with the highest digestibility were not necessarily varieties with the softest leaves (Suter et al. 2009). Previously we focused on leaf shear strength to explore a relation between leaf softness and digestibility but we found no convincing relationship (Cougnon et al. 2014). This study focused on a particular aspect of leaf morphology and on its mineral constitution to find out if these characteristics contribute to leaf softness and digestibility of tall fescue.

Silica in plant tissues can be either present as silica bodies in the lumen of epidermal cells or as constituents of the cell wall. Regardless of its location it negatively affects herbivorous vertebrates and invertebrate crop pests (Coocke and Leishman 2011). This was further evidenced for grass species by Massey et al. (2009): in sheep preference trials, sheep grazed longer, took more bites and had a higher bite rate on grass species with low silica content. Rafi et al. (1997) showed a positive relationship between the silica content of wheat plants and their coarseness. There are a number of indications that the presence of sharp silicified dentations (trichomes) on the leaf margins and midribs are responsible for the perceived coarseness of Fa leaves. Trichomes are hook-shaped epidermal cells, which can be almost completely filled with silica (Jones and Handreck 1967). Digestibility is also affected by silica content: in a study with several grass species including Fa, Van Soest and Jones (1968) found an average decline of 3.0 percentage points in digestibility per percent of silica in the dry matter.

In this study, we hypothesized that:

(a) Fa clones with softer leaves have fewer trichomes on the leaf margins and/or a lower silica content compared to clones with coarse leaves.
(b) There is a positive correlation between the number of trichomes on the leaf margins and the silica content and a negative relationship between the digestibility and the silica content.

Materials and Methods

A clonal nursery containing 458 genotypes was selected for good disease resistance and vigor from West European ecotypes and varieties obtained from breeders and gene-banks. All genotypes were planted in September 2012 in clonal rows of 1 m long consisting of 5 ramets. All clones were monitored for vigor, disease resistance and leaf softness as assessed via palpation at three occasions in 2013 (10/07, 15/08 and 08/10). Leaf softness was scored using a scale from 1 (very flexible and soft) to 5 (stiff leaves with sharp cutting edges). Based on the average coarseness score in 2013, the clones were classified into three classes: SOFT (1–1.5), MEDIUM (1.6–3.4) and COARSE (3.5–5). In May 2014 nineteen healthy and vigorous Fa clones from different origins were identified within the three softness classes (Table 41.1). Fertilization was 200 kg N ha^{-1}, 130 kg K ha^{-1} and 37 kg P ha^{-1} per year; the nurs-

Table 41.1 Origin, softness class based on scores of 2013, length, width, presence of trichomes on the edges of adult leaf blades, silica (Si) content of DM and digestibility of organic matter (DOM) averaged (standard deviation) over three cuts in the summer of 2014 for 19 tall fescue clones

Softness	Clone	Origin	L (cm)	W (mm)	Trichomes (mm^{-1})	Si-content (%)	DOM (%)
COARSE	3	Ecotype	36.6 (0.1)	10.1 (0.7)	10.7 (0.6)	0.76 (0.06)	74.6 (4.3)
COARSE	5	Ecotype	33.1 (4.2)	6.9 (0.8)	10.0 (0.3)	0.70 (0.09)	73.1 (6.3)
COARSE	7	Ecotype	28.6 (0.1)	8.2 (0.6)	7.4 (1.0)	0.93 (0.19)	73.0 (4.8)
COARSE	12	Cv. Hykor	27.9 (2.5)	7.5 (0.3)	11.8 (0.5)	1.00 (0.10)	69.8 (8.7)
COARSE	17	Ecotype	24.6 (3.7)	7.3 (0.1)	5.3 (0.7)	0.78 (0.07)	69.8 (2.0)
COARSE	19	Cv. Kora	31.2 (8.1)	7.5 (0.5)	9.0 (1.0)	0.63 (0.07)	78.2 (1.8)
MEDIUM	4	Ecotype	34.5 (1.4)	6.6 (0.1)	8.6 (1.5)	0.67 (0.10)	70.5 (2.0)
MEDIUM	6	Cv. Jordane	20.0 (2.9)	4.9 (0.8)	5.1 (1.8)	0.74 (0.06)	69.3 (4.8)
MEDIUM	8	Cv. Barolex	23.6 (1.1)	4.8 (0.6)	3.1 (1.8)	0.91 (0.16)	71.0 (4.7)
MEDIUM	10	Ecotype	28.0 (3.9)	5.3 (0.7)	6.4 (0.6)	0.80 (0.13)	65.9 (2.3)
MEDIUM	14	Ecotype	47.0 (18.6)	7.3 (0.2)	5.3 (0.5)	0.85 (0.07)	68.9 (5.3)
MEDIUM	15	Cv. Jordane	28.6 (4.1)	5.9 (0.3)	6.0 (1.3)	0.92 (0.06)	66.3 (5.1)
MEDIUM	16	Cv. Masterpiece	14.0 (2.1)	4.7 (0.8)	9.4 (1.0)	0.83 (0.01)	70.9 (2.0)
SOFT	1	Cv. Callina	28.2 (1.3)	5.6 (1.0)	4.5 (2.2)	0.69 (0.04)	67.9 (6.7)
SOFT	2	Cv. Aprillia	24.5 (2.1)	6.1 (0.5)	4.3 (0.8)	0.81 (0.12)	70.9 (5.8)
SOFT	9	Ecotype	27.0 (0.6)	6.5 (0.7)	0.9 (0.4)	0.50 (0.10)	71.0 (6.0)
SOFT	11	Cv. Dauphine	30.9 (1.3)	6.2 (0.8)	3.7 (2.2)	1.08 (0.29)	72.2 (4.6)
SOFT	13	Cv. Dauphine	23.0 (1.6)	4.9 (0.7)	0.4 (0.3)	1.23 (0.20)	69.0 (6.8)
SOFT	18	Cv. Dauphine	23.0 (0.7)	5.1 (0.6)	2.5 (1.1)	0.75 (0.09)	73.1 (2.5)

ery was cut five times per year. On three occasions in the summer of 2014 (11/06, 04/08 and 02/09) *ca* 0.5 m of the clonal rows of the selected clones was harvested. The grass was cut at 5 cm height using an electric long-reach hedge trimmer. The cut grass was collected and ten tillers with at least one complete adult leaf were selected from each clone. The remainder of the sample was rinsed with tap water (to remove soil particles), dried for 16 h at 75 °C and ground to pass through a 1 mm sieve. The ten adult leaf blades were used to measure leaf blade length, width, and a section of 3 mm was cut from the middle of each leaf blade and conserved in ethanol (70 %). Using a stereomicroscope, the number of trichomes on both sides of this section of 3 mm was counted. The dried and ground sample was analyzed for silica content using the wet chemistry method described in Smis et al. (2014). The NIRS spectra of the ground samples were collected and the digestibility of the organic matter (DOM) was calculated from an equation that contained 396 grass samples analyzed according to Tilley and Terry (1963).

ANOVA was performed to compare the effect of the softness classes and the cutting date on the measured parameters. Regression of DOM on softness and silica content included cutting date as a categorical variable. Statistics were performed using the free software package R (R core team 2013).

Results and Discussion

There was a significant effect of the softness class on the number of trichomes on the leaf margin ($p<0.001$): leaves from coarse clones had approximately three times more trichomes on the edges compared to soft clones (Fig. 41.1). The DOM and the silica content did not significantly differ between the softness classes. The soft-leaved clones did not necessarily have a high DOM, nor did they have a low Si content. Soft clone nr. 13, for example, had the highest silica content and a rather low digestibility (Table 41.1). Neither were all coarse clones high in silica content and low in digestibility: for example coarse clone nr. 19 combined the highest DOM with a rather low silica content (Table 41.1). Over all cuts, there was no effect of the number of trichomes on the leaf edges ($R^2=0.049$, $p=0.171$), nor of the silica content in the dry matter (DM) ($R^2=0.0905$, $p=0.0914$) on the DOM (Fig. 41.2a, b). Nor was there a relationship between the number of trichomes on the leaf edges and the silica content ($R^2=0.051$, $p=0.436$).

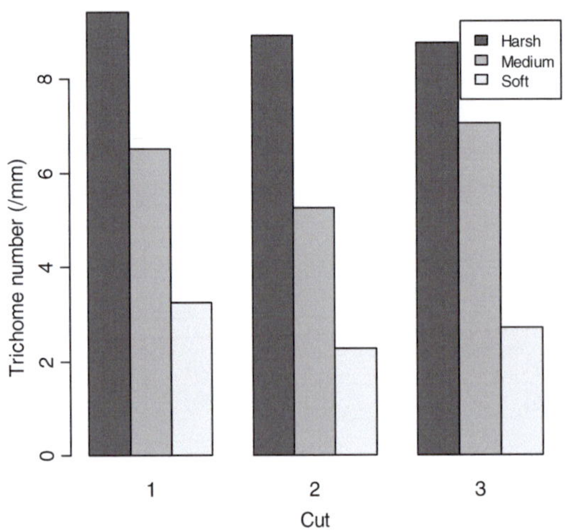

Fig. 41.1 Average number of trichomes on the margins of 19 tall fescue clones with contrasting coarseness, harvested on three occasions in 2014

Rafi et al. (1997) made similar observations: they raised wheat plants in hydroponics with and without silica. The latter plants contained ten times less silica and were softer than the former, but the number and the size of the trichomes on the spikes were not altered.

The number of trichomes on the leaf margins explained leaf softness at least partially. However, the presence of trichomes certainly cannot be the only component of leaf coarseness: in Italian ryegrass, generally perceived as very soft, genotypes can be found with a high number of trichomes on the edges. Neither leaf coarseness nor the presence of trichomes or silica content correlated well with DOM. Hence, selection against trichomes, leaf softness and silica content will not necessarily improve digestibility. Nevertheless, selection for a lower silica may improve animal preference of tall fescue, in line with the results reported by Massey et al. (2009).

Conclusion

We conclude that:

1. Fa clones with soft, fine leaf blades had a lower number of trichomes on the leaf margins, but their silica content was not lower and their DOM not higher compared to clones with coarser leaves.

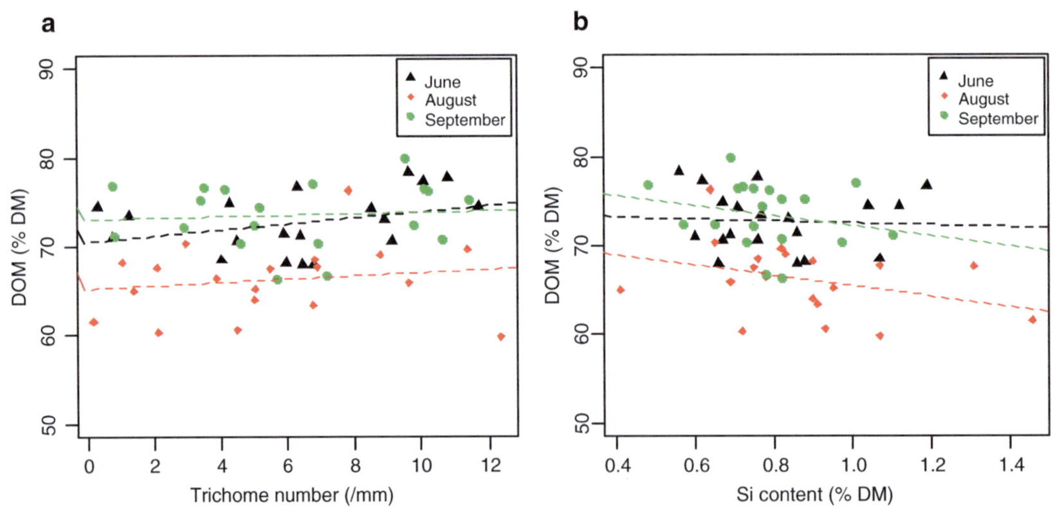

Fig. 41.2 Effect of the number of trichomes on the leaf edge (**a**) and the silica content of the dry matter (**b**) on the digestibility of the organic matter (*DOM*) of 19 tall fescue clones harvested in three cuts in the summer of 2014

2. No relationship was found between DOM, silica content and number of trichomes on the leaf blade edges of Fa clones.

References

Coocke J., Leishman M. (2011) Is plant ecology more siliceous than we realise? Trends in plant science 16: 61–68.

Cougnon M, De Koker J, Fievez V, Reheul D (2014) Factors influencing animal preference of tall fescue genotypes. D. Sokolovic et al. (eds.). Quantitative traits breeding for multifuncional grassland. Springer Science + Business Media, Dordrecht. pp 125–129.

Craigmiles J.P., Crowder L.V., Newton J.P. (1964) Palatability differences in tall fescue using leaf and plant type. Crop science 4:658–660

Gillet M, Jadas-Hecart J. (1965) La flexibilité des feuilles, critère de sélection de la fétuque élevée en tant que facteur d'appétibilité. Fourrages 22:6–11.

Jadas-Hecart J. (1982) Etude en pâture de l'appétibilité de fétuques élevées (Festuca arundinacea Schreb.) à l'aide de moutons. Agronomie 82:487–492.

Jones L., Handreck KA. (1967) Silica in soils, plants, and animals. Advances in Agronomy 19:107–49

Massey F., Massey K., Ennos A., Hartley S. (2009) Impacts of silica based defences in grasses on the feeding preferences of sheep. Basic and applied ecology 10:622–630.

Petersen R.G., Weswig P.H. and Cowan J.R., (1958) Measuring palatability differences in tall fescue by grazing sheep. Agronomy Journal, 50:117–119.

R Core Team (2013). R: A language and environment for statistical computing. R Foundation for Statistical Computing, Vienna, Austria. URL http://www.R-project.org/.

Rafi M, Epstein E, Falk R (1997) Silicon deprivation causes physical abnormalities in wheat (Triticum aestivum L.) Journal of Plant Pathology 151:497–501.

Smis A., Murguzur F., Struyf E., Soininen E., Jusdado J., Meire P., Brathen K (2014) Determination of plant silicon content with near infrared reflectance spectroscopy. Frontiers in plant science 5:1–9.

Suter D, Frick R, Hirschi H and Chapuis S (2009) Rohrschwingel- und Timotheesorten geprüft. Agrarforschung 16:250–255.

Tilley J.M.A and Terry R.A. (1963) A two stage technique for the in-vitro digestion of forage crops. Grass and Forage Science 18:104–111.

Van Soest P.J., Jones L.H.P. (1968) Effect of silica in forages upon digestibility. Journal of Dairy Science 51: 1644–1648.

Parameters Influencing Preference by Sheep in Soft Leaved Tall Fescue Genotypes

42

R. Shahidi, M. Cougnon, E. Struyf, C. Van Waes, M.C. Van Labeke, and D. Reheul

Abstract

An important disadvantage of tall fescue (*Festuca arundinacea* Schreb.) is its low voluntary intake, resulting in suboptimal performances of cattle grazing tall fescue. Ideally selection for this trait is done by animals themselves, but the use of grazing animals in large breeding programmes is laborious. Repeatable, stable and quantifiable parameters that can be linked to animal preference could ease tall fescue breeding. We established a trial to find relations between the grazing preference of sheep and other plant parameters. Twenty clones were selected from a breeding programme and swards of 2 m^2 were planted with three replications for each clone. On four different occasions in 2014, sheep were allowed to graze the clones and grazing preference was determined visually. Prior to the grazing, multiple morphological and chemical parameters were measured. Parameters that were correlated with sheep preference were digestibility (r=0.86), water soluble carbohydrate content (r=0.74) crude fiber content (r=−0.67), leaf blade width (r=0.57) and sward height (r=−0.53).

Keywords

Festuca arundinacea • Grazing preference • Digestibility • Water soluble carbohydrates (WSC) • Crude fiber content • Silica content

R. Shahidi (✉) • M. Cougnon • M.C. Van Labeke
D. Reheul
Department of Plant Production, Ghent University, Ghent, Belgium
e-mail: reihaneh.shahidi@ugent.be

E. Struyf
Ecosystem Management Research Group, University of Antwerp, Antwerp, Belgium

C. Van Waes
Plant Sciences Unit, Institute for Agricultural and Fisheries Research (ILVO), Merelbeke, Belgium

Introduction

Owing to its good drought resistance and high yield potential under cutting, tall fescue (*Festuca arundinacea* Schreb.; Fa) may fit into future demands for forage grasses in NW Europe. The current Fa varieties have a lower feeding value compared to perennial ryegrass (*Lolium perenne* L.; Lp). This problem is two-pronged.

I. Roldán-Ruiz et al. (eds.), *Breeding in a World of Scarcity: Proceedings of the 2015 Meeting of the Section "Forage Crops and Amenity Grasses" of Eucarpia*,
DOI 10.1007/978-3-319-28932-8_42

First the digestibility of the organic matter is lower: in a cutting trial with five cuts per year the digestibility of Fa was on average 8 % points lower compared to Lp (Cougnon et al. 2014a). Second, the voluntary intake of animals fed with Fa is lower than the intake of ryegrasses: averaged over three years of feeding trials, Luten and Remmelink (1984) found a 7 % lower dry matter intake (DMI) of Fa compared to Lp. When ensiled, the difference in intake between Fa and Lp decreased substantially (Luten and Remmelink 1984). Therefore, tall fescue varieties with an improved digestibility and intake are desirable when grazing is targeted or when freshly cut grass is provided. In early Fa breeding programs, clones were grazed (mostly by sheep) to select for palatability, the amount of grass consumed under proper grazing conditions (Petersen et al. 1958). This pragmatic approach is labor intensive, so reliable indirect selection would be desirable. Parameters that proved to affect preference and intake of Fa include leaf softness (Jadas-Hecart 1982), leaf mechanical properties (MacAdam and Mayland 2003), and chemical composition (Mayland et al. 2000). None of these parameters alone could explain the differences in preference or intake. In an earlier trial, comparing the preference of diverse Fa genotypes (both ecotypes and breeding material with diverging leaf softness) under grazing, we found that the most important parameters affecting sheep preference were rabbit preference and pre-grazing sward height (Cougnon et al. 2014b). The present trial is conceptually similar to the previous one, but it differs in two aspects. First it focused on breeding material, all with relatively soft leaves. Second, the chemical composition of the clones was studied in more detail.

Our research question was: which parameter(s) can be used to select indirectly for sheep preference?

Material and Methods

A grazing trial comparing 20 Fa clones was established in March 2013 on a sandy loam soil at the experimental farm of Ghent University in Melle, Belgium. Plant material was dug out from a clonal nursery to establish the trial. The clones were selected for vigorous growth and good resistance to crown rust (*Puccinia coronata*) and bacterial wilt (*Xanthomonas campestris*). Clones with particularly coarse or broad leaves were excluded. The plant material of each of the selected clones was divided into at least 200 ramets of 3–5 tillers. These ramets were used to plant "clonal plots" of 2 m^2; distances between ramets was 0.20 m between and within rows. The trial was planted as a randomized complete block design with three replicates, resulting in 60 plots of 2 m^2. Alongside and between blocks, 2 m wide alleys were sown with an amenity type of tall fescue. No measurements were performed in the year 2013 so the swards could establish. Weeds were removed manually during 2013 and 2014.

On four occasions in 2014 (15 April, 12 June, 29 July, 14 September), silica content, dry matter content, protein content, fiber content, water soluble carbohydrate content, digestibility of the organic matter (OM), grazing preference by sheep, softness, sward height, chlorophyll content, leaf length, leaf width and leaf shear strength were measured. Pre-grazing sward height was measured with a falling plate meter on four randomly selected spots per plot (Bransby et al. 1977). Leaf softness was scored by two plant breeders on an ordinal scale from 1 (very fine, soft and flexible leaf blades) to 5 (very coarse, rough and inflexible leaf blades). Per clone, 10 tillers were collected randomly in the plots of block 1. From each tiller, the youngest fully developed (auricles visible) leaf blade was used to measure leaf blade length, width, shear breaking load and chlorophyll content. Leaf blade width was measured with vernier callipers at 1/3 of the distance between the ligule and the tip, while leaf blade length from the ligule to the tip was measured using a metric ruler. Leaf shear breaking load, the maximum load needed to cause breakage at a 90° breaking angle to the length of the leaf, was measured using a Warner-Bratzler meat shear apparatus. Chlorophyll content was measured using a SPAD meter (Konica Minolta SPAD 502plus): it determines the relative amount of chlorophyll present by measuring the absorbance of the leaf in two wavelength regions. We used chlorophyll content in the present trial as a proxy for leaf color.

On the same dates, a sample of circa 125 g of fresh grass was clipped from each plot at a height of 5 cm. The samples were washed with tap water, dried during 16 h at 75 °C and ground to pass through a 1 mm sieve. Silica content and fiber content in these samples were determined using the protocols described by Smis et al. (2014) and Van Soest et al. (1991), respectively. Digestibility of the organic matter, water soluble carbohydrate content, and crude protein content were determined using NIRS.

After collecting the samples, the trial was stocked with sheep. These sheep had been grazing for at least two weeks on pastures sown with a mixture of tall fescue varieties to get the animals accustomed to the grass in the trial as recommended by Gillet et al. (1983). Stocking density was regulated to allow complete grazing of the trial in four days. The trial area of 4.5 a was stocked with 3 adult sheep with 4 lambs in the first grazing round up to 5 adult sheep in the last grazing round. In accordance with Shewmaker et al. (1997), grazing preference was scored every morning on a scale from 0 (no grazing at all), 1 (between 0 and 10 % of standing biomass eaten) to 9 (between 80 and 90 % of standing biomass eaten). As soon as one clone had reached a score of 9, measurements were stopped. The observation that discriminated most between the clones was withheld.

Data were averaged over the different grazing periods. Shapiro-Wilk normality tests were performed to test the normality of the measured parameters. As none of the parameters failed the normality test ($p>0.05$ for all parameters), correlation was calculated using Pearson's correlation coefficient. All statistics were done with the software package R (R core team 2013).

Results and Discussion

Correlation coefficients between the studied parameters for the four grazing periods together are shown in Table 42.1. Parameters with a significant positive correlation with preference by sheep were digestibility ($r=0.86$), water soluble carbohydrate content ($r=0.74$), and leaf blade width ($r=0.57$). Factors with a significant negative correlation with preference by sheep were crude fiber content ($r=-0.67$) and sward height ($r=-0.53$). The positive correlation between animal preference and water soluble carbohydrates is in line with the findings of Mayland et al. (2000) offering eight tall fescue varieties to cattle.

The large positive effect of leaf blade width on preference by sheep observed here is not well understood. In contrast to our results, Cougnon et al. (2014b) found no effect of leaf blade width on preference by sheep, even though the variation for leaf blade width was 2.5 times larger in their study compared to our study.

In contrast to our expectations, no correlations were found between preference and leaf shear strength (Macadam and Mayland 2003), leaf softness (Jadas-Hecart 1982), silica content (Massey et al. 2009) or leaf color (measured as chlorophyll content) (Jadas-Hecart 1982).

Because softness is related to the presence of trichomes on the leaves' surface (Coocke and Leishman 2011), we expected to observe a negative correlation between leaf softness and silica content. This correlation, however, was not significant. Nor did we find a correlation between leaf softness and digestibility, while Gillet and Jadas-Hecart (1965) found a negative correlation between these parameters. However, the elite breeding material studied in this trial may explain the non-effect of parameters reported to affect preference in earlier studies. Indeed, Rognli et al. (2010) found that relation between leaf softness and animal preference decreased in elite breeding material and Suter et al. (2009) showed that the varieties with the softest leaves were not necessarily those with the highest digestibility.

Although selection for soft-leaved genotypes allowed an important improvement of both preference and digestibility in the early tall fescue breeding programs (Gillet and Jadas-Hecart 1965), this trial suggests that selection for leaf softness does not further improve feeding quality in advanced breeding material. A similar

Table 42.1 Pearson correlation coefficients for parameters measured on 20 tall fescue clones grazed by sheep on four occasions in

		1	2	3	4	5	6	7	8	9	10	11	12
1	Silica content	1											
2	Dry matter content	−0.22	1										
3	Protein content	0.31	−0.38	1									
4	Fiber content	0.31	−0.03	0.12	1								
5	WSC content	−0.47*	0.37	−0.52*	−0.83**	1							
6	Digestibility of the OM	−0.37	0.08	0.10	−0.82**	0.82**	1						
7	Preference by sheep	−0.35	0.34	−0.11	−0.67**	0.74**	0.86**	1					
8	Softness	−0.27	0.59**	0.04	0.10	0.17	0.13	0.28	1				
9	Sward height	0. 01	−0.04	−0.24	0.36	−0.27	−0.37	−0.53*	0.09	1			
10	Chlorophyll content	−0.32	0.43	0.43	−0.10	0.11	0.17	0.30	0.73**	−0.27	1		
11	Leaf blade length	−0.09	−0.20	−0.25	0.41	−0.29	−0.28	−0.35	−0.05	0.68**	−0.36	1	
12	Leaf blade width	−0.44	0.26	0.05	−0.20	0.34	0.57*	0.57*	0.65**	0.16	0.41	0.30	1
13	Shear strength	−0.30	0.53*	−0.13	0.34	−0.06	−0.04	0.07	0.76**	0.34	0.45	0.33	0.65**

** and * indicate $p<0.01$ and $p<0.05$ respectively; *OM* organic matter; *WSC* water soluble carbohydrates conclusion can be drawn for silica content. While the literature suggests silica content to be a qualifying parameter for forage preference of herbivores (Massey et al. 2009; Coocke and Leishman 2011), our study found no such correlation, indicating that silica content does not affect preference by sheep in advanced breeding material of tall fescue.

Our results suggest that selection for low fiber content, high digestibility and high water soluble carbohydrate content will improve sheep's grazing preference. The determination of the chemical composition of breeding material is of course not as easy as an assessment of leaf softness, but if appropriate calibrations are available, NIRS can be used to estimate the chemical composition of a high number of samples at a relatively low cost.

Conclusion

The factors with the highest correlation with sheep preference were digestibility (r=0.86), water soluble carbohydrate content (r=0.74) crude fiber content (r=–0.67), leaf blade width (r=0.57) and sward height (r=–0.53). However, not all clones with high digestibility, high water soluble carbohydrate content, broad leaves, low crude fiber content and low sward heights were preferred by sheep. With the knowledge we have now, animal preference trials remain the best method to make progress in preference for grazed tall fescue.

References

Bransby DI, Matches AG and Krause GF (1977) Disk meter for rapid estimation of herbage yield in grazing trials. Agronomy Journal 69:393–396.

Coocke J, Leishman M (2011) Is plant ecology more siliceous than we realise? Trends in Plant Science 16: 61–68.

Cougnon M, Baert J, Reheul D (2014a) Dry matter yield and digestibility of five cool season grass species under contrasting N fertilizations. Grassland Science in Europe 19: 175–177.

Cougnon M, De Koker J, Fievez V, Reheul D (2014b) Factors influencing animal preference of tall fescue genotypes. D. Sokolovic et al. (eds.). Quantitative traits breeding for multifuncional grassland. Springer, Dordrecht. pp 125–129.

Gillet M, Jadas-Hecart J (1965) La flexibilité des feuilles, critère de sélection de la fétuque élevée en tant que facteur d'appétibilité. Fourrages 22:6–11.

Gillet M, Noël C, Jadas-Hecart J (1983) La cafétéria d'auges, méthode d'étude de l'appétibilité des fourrages. Agronomie 3 : 817–822.

Jadas-Hecart J (1982) Etude de l'appétibilité de fétuques élevées (*Festuca arundinacea* Schreb.) à l'aide de moutons. Agronomie 2:487–492.

Luten W, Remmelink GJ (1984) Intake of perennial ryegrass, tall fescue and Italian ryegrass by dairy cattle. Proefstation voor Rudveehouderij, Lelystad, Netherlands.

MacAdam JW and Mayland HF (2003) The relation of leaf shear strength to cattle preference in tall fescue cultivars. Agronomy Journal, 95:414–419.

Massey F, Massey K, Ennos A, Hartley S (2009) Impacts of silica based defenses in grasses on the feeding preferences of sheep. Basic and Applied Ecology 10:622–630.

Mayland HF, Shewmaker GE, Harrison PA and Chatterton NJ (2000) Non-structural carbohydrates in tall fescue cultivars: Relationship to animal preference. Agronomy Journal 92:1203–1206.

Petersen RG, Weswig PH and Cowan JR (1958) Measuring palatability differences in tall fescue by grazing sheep. Agronomy Journal 50:117–119.

R Core Team (2013) R: A language and environment for statistical computing. R foundation for statistical computing, Vienna, Austria. http://www.R-project.org/.

Rognli OA, Saha MC, Bhamidihari S, van der Heijden S. (2010) Fescues. In: Boller, B., Posselt, U.K. & Veronesi, F. (eds) Handbook of Plant Breeding Vol. 5. Springer, Dordrecht. 261–292.

Shewmaker GE, Mayland HF and Hansen SB (1997) Cattle grazing preference among eight endophyte-free tall fescue cultivars. Agronomy Journal 89:695–701.

Smis A, Murguzur F, Struyf E, Soininen E, Jusdado J, Meire P and Brathen K (2014) Determination of plant silicon content with near infrared reflectance spectroscopy. Frontiers in Plant Science 5:1–9.

Suter D, Frick R, Hirschi H and Chapuis S (2009) Rohrschwingel- und Timotheesorten geprüft. Agrarforschung 16:250–255.

Van Soest PJ, Robertson JB, Lewis BA (1991) Methods for dietary fibers, neutral detergent fiber and nonstarch polysaccharides in relation to animal nutrition. Journal of Dairy Science 74: 3583–3597.

No Trade-off Between Root Biomass and Aboveground Production in *Lolium perenne*

43

J. Deru, H. Schilder, J.R. Van der Schoot, and N. Van Eekeren

Abstract

Although grasses have dense rooting systems, further improvements to rooting may increase nutrient uptake and drought resistance and reduce N leaching. Improved rooting of grasses in agricultural systems should, however, not reduce aboveground biomass allocation and yield. In two field experiments on sandy soil in The Netherlands, we measured the variation in grass yield of 16 varieties of *Lolium perenne* (Lp) during three years, and the root dry matter (RDM) at the end of the experiments. The Lp- varieties differed in aboveground and genetic characteristics such as productivity (classified according to the measured yields in the actual experiments), grass cover and ploidy. Results of the experiments show that RDM of perennial ryegrass differed significantly between varieties, and that these differences were not linked to grass yield. Our results indicate that it is possible to select perennial ryegrass varieties that combine high aboveground productivity with high RDM. Considering challenges in the areas of climate change, pollution and soil degradation, high yielding grass varieties with improved root systems could contribute to an efficient use of nutrients and water, and to erosion control, soil improvement and carbon sequestration.

Keywords

Root mass • Grass yield • *Lolium perenne* varieties

J. Deru • N. Van Eekeren
Louis Bolk Institute,
Driebergen 3972 LA, The Netherlands

H. Schilder
Wageningen UR Livestock Research,
Wageningen 6708 WD, The Netherlands

J.R. Van der Schoot (✉)
Wageningen UR Applied Plant Research,
Lelystad 8219 PH, The Netherlands
e-mail: janrinze.vanderschoot@wur.nl

Introduction

Grasses have dense rooting systems as compared with other agricultural crops. Nevertheless, increased rooting may lead to increased uptake of nutrients, improved drought resistance and reduced N leaching (Van Loo et al. 2003).

I. Roldán-Ruiz et al. (eds.), *Breeding in a World of Scarcity: Proceedings of the 2015 Meeting of the Section "Forage Crops and Amenity Grasses" of Eucarpia*,
DOI 10.1007/978-3-319-28932-8_43

Improved rooting of grasses in agricultural systems should, however, not create a trade-off with aboveground biomass allocation and yield. Questions to explore are which management practices promote deeper rooting of grassland plants and the relation with yield. One possible management practice is sowing of varieties with a high root mass, making use of the high genetic variation available for this trait among grass varieties (Crush et al. 2007), without compromising aboveground yield. In this paper, we describe the relation between root biomass and grass yield of 16 *Lolium perenne* (*L. perenne*) varieties in two field experiments.

Materials and Methods

In Experiment I, eight *L. perenne* varieties were selected from a field experiment that is part of the value for cultivation and use (VCU) testing program for the Dutch variety list. This trial was sown in 2005 with 50 Lp varieties on a sandy soil in Raalte (52.50N, 6.40E) in three replicates with a plot size of 9 m^2. The *L. perenne* varieties differed in characteristics of productivity class, earliness class, and ploidy (diploid versus tetraploid) in an orthogonal way. Grass dry matter (GDM) yield was measured in 2006–2008. In October 2010, root samples were taken in three soil layers: 0–8, 8–16 and 16–24 cm. Per plot and per layer, three soil cores (8.5 cm diameter, 8 cm depth) were taken and pooled into one sample. The fresh samples were washed through a sieve (mesh size 2 mm) and non-root particles were removed. The root dry matter (RDM) was determined after drying in an air ventilated oven, first at 70 °C and then at 105 °C, during 24 h at both temperatures. During the VCU testing and subsequent years, the experiment was managed according to a cutting regime with five cuts per year.

In Experiment II, a different set of eight *L. perenne* varieties was selected. These varieties were evaluated in a field trial as part of the VCU testing program on sandy soil in Schalkhaar (52.31N, 6.22E). Here, 80*L. perenne* varieties had been sown in 2009 in four replicates with a plot size of 32 m^2. The varieties differed in their characteristics of productivity class, soil cover class and ploidy, also in an orthogonal way. GDM was measured in 2010–2012. Root sampling was done in September 2013 following the method of Experiment I. This trial was managed as a cutting/grazing pasture (two cuttings/four grazing events per year). More details on the experiments are given in Deru et al. (2014).

Variety effect on RDM was analyzed for each experiment separately using the ANOVA procedure in Genstat 13.3 (VSN International, 2012). RDM effect on GDM was calculated with linear regression analysis. Regression was carried out on the whole dataset with and without discerning different groups (experiments, ploidy, aboveground yield). RDM was expressed either as the total of the three soil layers combined or the layers separately, and GDM as the average of the three measured years.

Results and Discussion

The overall mean RDM in the 0–24 cm soil layer in Experiment II (2,663 kg ha^{-1}) was lower than in Experiment I (3,286 kg ha^{-1}). Total nitrogen applied, including slurry manure, artificial fertilizer and the estimated extra N input from dung and urine during grazing (Experiment II) was comparable in both experiments. Besides differences in abiotic conditions and sampling year between the experiments, the more intensive cutting/grazing regime in Experiment II may have contributed to reduced RDM. Similar results were reported by Ennik and Baan Hofman (1983), with lower RDM at higher harvesting frequencies.

RDM in the 3 soil layers for the 16 varieties are shown in Fig. 43.1. In Experiment I, there was a significant variety effect on $RDM_{0\text{-}24cm}$ ($P=0.004$). In Experiment II, the variety effect was close to significant on $RDM_{0\text{-}24cm}$ ($P=0.053$) and significant on $RDM_{8\text{-}16cm}$ ($P=0.008$).

Regression analysis revealed no significant relation ($P>0.05$) between GDM and RDM in any soil layer, either in the whole dataset (Fig. 43.2) or considering separate groups (Experiment I and II; high or low production category; diploid or tetraploid, results not shown). This lack of relation as measured in our experiments does not exclude the possibility

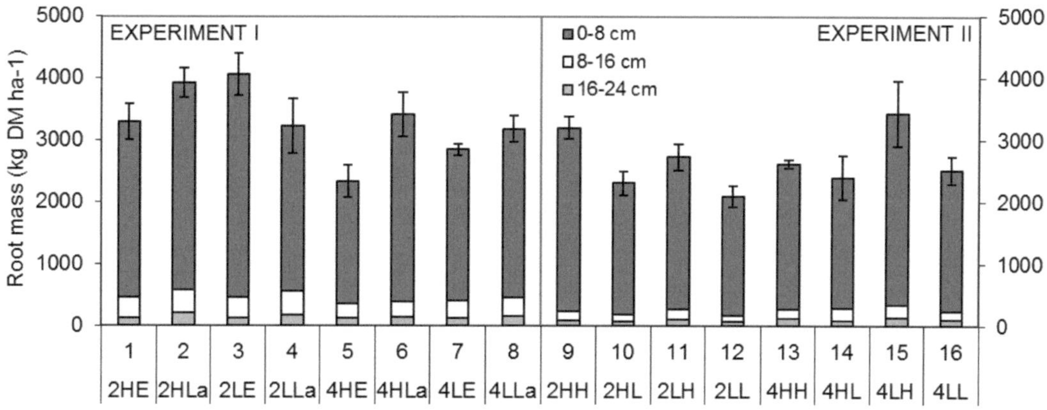

Fig. 43.1 Cumulated root mass (kg DM ha^{-1}) of sixteen *Lolium perenne* varieties with different aboveground characteristics. The results of two experiments are presented separately. Abbreviations for Experiment I: (i) 2 or 4: diploid or tetraploid; (ii) H or L: high or low productive; (iii) E or La: early or late. Abbreviations for Experiment II: (i) 2 or 4: diploid or tetraploid; H or L: high or low productive; (iii) H or L: high or low soil cover. Error bars represent + and – the standard error of RDM across the 0–24 cm soil layer

Fig. 43.2 Relation between total root dry matter (RDM_{0-24cm}) and grass yield (GDM) in the two experiments. Exp I: open circles, line: $R^2=0.10$, $P=0.079$; Exp II: crosses, line: $R^2=0.00$, $P=0.866$

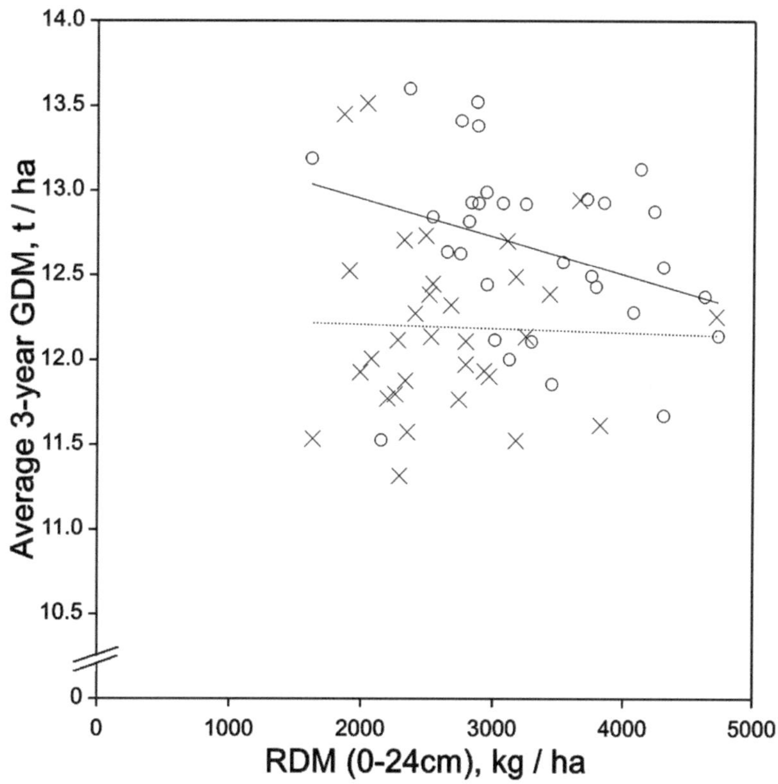

that varieties with high RDM would be less vulnerable to stress conditions (e.g., better drought resistance) and give higher aboveground yield compared with varieties with low RDM. Further, this indicates that the possibility of selecting perennial ryegrass varieties that combine high aboveground productivity with high root mass. The potential for improvement is certainly available, considering that Crush et al. (2006) reported a narrow-sense heritability value for root mass of 0.35 for perennial ryegrass, and Bonos et al. (2004) achieved higher root mass productions in the deeper soil layer of forage type perennial ryegrass in test tubes by 367 % after only two breeding cycles.

The results presented here should be interpreted with caution, however: differences in root morphology between genotypes (e.g., differing proportions of fine roots) could possibly have influenced the results, as the washing method applied is not suitable to separate all the roots from the soil. Another possible bias is that part of RDM can include dead roots, and that genotypes differ in their proportion of dead roots (or root decomposition rate). Therefore, for breeding purposes, techniques with a higher level of resolution and able to distinguish functional and dead roots should be used.

Conclusions

Based on the results of these two experiments, it can be concluded that root mass of *L. perenne* differs between varieties under field conditions. No trade-off between root mass and grass yield was detected.

Acknowledgments This research was part of the project Bufferboeren, financed by the following Dutch institutions: Agricultural Innovation Bureau (LIB), Dairy Board (PZ), Brabant Water, Water Board Aa en Maas, Rabobank Bernheeze, Foundation for Applied Water Research (STOWA), NCB-fund and Ministry of Infrastructure and Environment (I&M). We want to thank the Dutch breeders association Plantum for the opportunity to take root samples after ending of the VCU trials.

References

Bonos S.A., Rush D., Hignight K., Meyer W.A. (2004) Selection for deep root production in tall fescue and perennial ryegrass. *Crop Science* 44:1770–1775.

Crush J.R., Easton H.S. and Waller J.E. (2006) Genetic variation in perennial ryegrass for root profile characteristics. In: Mercer C.F. (ed.) *New Zealand Grassland Association Research and Practice* Series No. 12, pp. 63–65. Dunedin, New Zealand: New Zealand Grassland Association.

Crush J.R., Easton H.S., Waller J.E., Hume D.E., Faville M.J. (2007) Genotypic variation in patterns of root distribution, nitrate interception and response to moisture stress of a perennial ryegrass (*Lolium perenne* L.) mapping population. *Grass and Forage Science*, 62, 265–273.

Deru J., Schilder H., Van der Schoot J.R., Van Eekeren N. (2014) Genetic differences in root mass of *Lolium perenne* varieties under field conditions. *Euphytica*, 199: 223–232.

Ennik G.C., Baan Hofman T. (1983) Variation in the root mass of ryegrass types and its ecological consequences. *Netherlands Journal of Agricultural Science* 31:325–334.

Van Loo E.N., Dolstra O., Humphreys M.O., Wolters L., Luessink W., De Riek J., Bark N. (2003) Lower nitrogen losses through marker assisted selection for nitrogen use efficiency and feeding value in ryegrass (Nimgrass). *Vortrage fur Pflanzenzuchtung*, 59, 270–279.

Are Corolla Tube Dimensions the Reason for Low Seed Yield in Tetraploid Red Clover?

44

T. Vleugels, I. Roldán-Ruiz, B. Ceuppens, G. Smagghe, and G. Cnops

Abstract

Seed yield is considerably lower in tetraploid than in diploid red clover. The most commonly suggested reason is that tetraploids have longer corolla tubes that prevent short-tongued bumblebee species from reaching the nectar, thereby resulting in deficient pollination. However, recent studies contradict this generally accepted statement. Because there is no consensus on this topic, we investigated the relation between seed yield and corolla tube length on the same plants in two separate field trials with different plant material. One field was sown with the tetraploid cultivar 'Avanti', while a second field was planted with tetraploid elite families. No correlation was found between seed yield and corolla tube length in either trial ($p=0.480$ and $p=0.430$ respectively), which indicates that corolla tube length does not affect seed yield in tetraploids. To explore whether the corolla tube diameter can explain the lack of correlation observed here, we measured this trait in flowers of 25 diploid and 25 tetraploid plants grown in pots. The average corolla tube diameter was 19.0 % wider in tetraploids than in diploids. Possibly, greater corolla tube diameters compensate for longer corolla tubes by allowing bees to insert their heads deeper into the tubes. This hypothesis needs to be validated by direct observations, currently underway at ILVO.

T. Vleugels (✉) • I. Roldán-Ruiz • G. Cnops
Plant Sciences Unit, Institute for Agricultural and Fisheries Research (ILVO), Caritasstraat 39, Melle 9090, Belgium
e-mail: tim.vleugels@ilvo.vlaanderen.be

B. Ceuppens
Plant Sciences Unit, Institute for Agricultural and Fisheries Research (ILVO), Caritasstraat 39, Melle 9090, Belgium

Department of Crop Protection, Ghent University, Coupure Links 653, Ghent 9000, Belgium

G. Smagghe
Department of Crop Protection, Ghent University, Coupure Links 653, Ghent 9000, Belgium

I. Roldán-Ruiz et al. (eds.), *Breeding in a World of Scarcity: Proceedings of the 2015 Meeting of the Section "Forage Crops and Amenity Grasses" of Eucarpia*,
DOI 10.1007/978-3-319-28932-8_44

Keywords
Red clover • Corolla tube • Seed yield • tetraploids

Introduction

Numerous diploid and tetraploid red clover (*Trifolium pratense* L.) cultivars are available with excellent agronomic performance (Taylor 2008; Boller et al. 2010). Although tetraploid cultivars outperform diploid cultivars for most traits, their seed yield is often substantially lower, despite numerous breeding efforts to increase seed yield (Boller et al. 2010). In diploid cultivars, seed yields between 400 and 500 kg seed/ha are considered 'normal' (Taylor and Quesenberry 1996), while 'normal' seed yields in tetraploid cultivars range between 150 and 300 kg seed/ha (Boller et al. 2010). As a consequence, seed production of tetraploid cultivars is more expensive and seed prices are higher (Taylor 2008).

Red clover flowers have long corolla tubes with pistil and stamen at their top and nectaries located at their base. Multiple studies (Skirde 1963; Bond 1968) have reported longer corolla tubes in tetraploid red clover: 8.9 % (Bender 1999b) or 9.2 % (Vleugels et al. 2015) longer than in diploid red clover. Honeybees (*Apis mellifera* L.) and bumblebees (*Bombus* spp.) are both pollinators of red clover, but bumblebees are more effective pollinators (Jablonski 2001). Honeybees are short-tongued and cannot reach the nectar. Therefore they are less attracted to red clover and will only gather pollen from red clover when few other flowering crops are available (Julén 1959; Bender 1999b). In contrast, bumblebees can obtain nectar from red clover when they insert their tongues deeply into the corolla. Hence, bumblebees are more attracted to red clover than honeybees are (Bender 1999a; Brodsgaard and Hansen 2002). With this information in mind, the most commonly alleged reason for low seed yield in tetraploid red clover is that their longer corollae prevent short-tongued bumblebee species from reaching the nectar, thereby resulting in poor pollination (Julén 1950; Skirde 1963; Furuya 2001; Boller et al. 2010). However, studies measuring the corolla tube length and seed yield on the same plants strongly contradict this statement (Starling et al. 1950; Vleugels et al. 2015).

In previous research we have examined seed yield, corolla tube length and flowering traits in 244 plants from five diploid and five tetraploid cultivars that were grown in pots on a container field (Vleugels et al. 2015). Compared to diploids, tetraploid plants had equal numbers of flowers per flower head, but fewer seeds per flower head, fewer flower heads per plant and fewer seeds per plant. Tetraploid corolla tubes were on average 9.2 % longer than diploid corollae, but no correlation was detected between the corolla tube length and seed yield. Therefore we proposed that, rather than longer corollae, the lower number of flower heads per plant and the lower degree of determinacy (or degree of synchronous flowering) are responsible for low seed yield in tetraploids (Vleugels et al. 2015). However, results obtained in pot experiments are not always representative for field conditions. Moreover, little is known about the interaction between the corolla tube length and its width. Therefore, we conducted a new study that validates the relation between seed yield and corolla tube length in two field trials with different plant material, different sowing conditions and different soil types. We also explored the possible influence of the corolla tube width, in combination with its length, on seed yield.

Materials and Methods

The relation between corolla tube length and seed yield was studied in two trials with sandy loam and heavy loam. Field 1 was sown with the tetraploid cultivar 'Avanti' in September 2010 for multiplication purposes, while field 2 was planted with a tetraploid breeding population in April 2010.

In field 1, the cultivar 'Avanti' was sown in rows separated by 15 cm, at a sowing density of 12 kg/ha. The field size was 200 m × 50 m. A total of 50 plants were sampled in a grid pattern, with 10 plants along the long axis (200 m) and 5 plants along the short axis (50 m) of the field. Between 23 and 27 July 2012, two flower heads with well-opened flowers were sampled from each plant to measure the corolla tube length. This was done through image analysis using ImageJ software (Abramoff et al. 2004) on 80 flowers per plant, as described by Vleugels et al. (2015). All studied plants were marked, so that seeds could be harvested from the same plants. Seeds were harvested when each individual plant was mature for seed harvest, between 27 and 31 August 2012. Because it is difficult to identify individual plants in a sown field, seed yield was measured on five representative flower heads per plant. Seeds were cleaned, weighed and the thousand seed weight (TSW) was determined by weighing four times 100 seeds per plant.

In field 2, 1300 plants from 47 tetraploid families were planted for selection purposes. Families were established in rows of 25–30 plants, with 50 cm planting distance between rows and within rows. After 2 years of intense selection for vegetative traits, 45 elite plants remained in summer 2012. The corolla tube length from these 45 plants was measured on 16 July 2012 as described above. Seeds were harvested when each individual plant was mature: between 7 and 17 September 2012. To compare the results with the first trial, five ripe flower heads from each plant were harvested and cleaned separately, after which the remaining seeds were harvested per plant and the TSW was determined. Spearman's rank correlations coefficients (Spearman 1904) were calculated between seed yield and corolla tube length with Statistica v.12 (Statsoft Inc., Tulsa, OK, USA). Correlations were considered significant at $P < 0.05$.

The corolla tube diameter was measured in 25 diploid and 25 tetraploid plants, a subset of the plant used by Vleugels et al. (2015). Plants were grown in pots on a container field and watered frequently depending on the temperature and solar radiation. Fifty plants from five tetraploid cultivars 'Astur', 'Avanti', 'Larus', 'Rotra' and 'Taifun' and five diploid cultivars 'Arlington', 'Crossway', 'Global', 'Merviot' and 'Milvus' were studied (five plants per cultivar). During flowering, 30 flowers from three well-opened flower heads (10 flowers per head) were sampled from each plant, laid on their side and photographed. Pictures were analyzed using image analysis as described above, and the corolla tube diameter was measured at the opening of the corolla tube. The average corolla tube diameter in diploids and tetraploids was calculated.

Results and Discussion

In field 1, sown with the cultivar 'Avanti', plants yielded between 24 and 542 seeds on five flower heads, with an average of 296 seeds. The corolla tube lengths were between 8.41 and 11.55 mm, with an average of 10.01 mm. No correlation was observed between seed number on five flower heads and corolla tube length ($P = 0.480$). In field 2, planted with tetraploid families, seed yields were between 16 and 455 seeds on five flower heads, with an average of 214 seeds. Corolla tube lengths between 8.74 and 10.37 mm were recorded, with an average of 9.56 mm. Again, no correlation was found between seed number on five flower heads and corolla tube length ($P = 0.346$). Similar conclusions were drawn when seed number was expressed per plant: no correlation with corolla tube length was detected ($P = 0.430$). For both trials, scatter plots between corolla tube length and seed yield are given in Fig. 44.1.

Although both trials had different plant material, different establishment (sowing vs. planting), different planting densities, and different soil types, no correlation was found between corolla tube length and seed yield in either trial. This is consistent with Starling et al. (1950), and confirms our previous results obtained in pot experiments (Vleugels et al. 2015). As a consequence, it becomes increasingly evident that low seed yields in tetraploid red clover cannot be explained by longer corolla

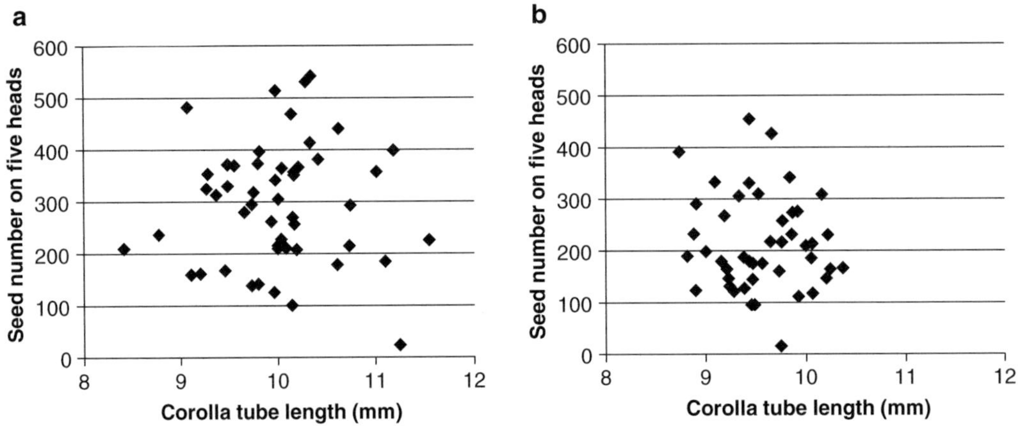

Fig. 44.1 Scatter plots of the seed number on five flower heads in function of the corolla tube length, measured in 50 plants from a field sown with the tetraploid cultivar 'Avanti' (**a**), and in 45 plants from a field planted with tetraploid elite families (**b**) (n = 80)

Fig. 44.2 Average corolla tube diameter for 25 diploid plants (*squares*) and 25 tetraploid plants (*triangles*) from 10 cultivars (n = 30)

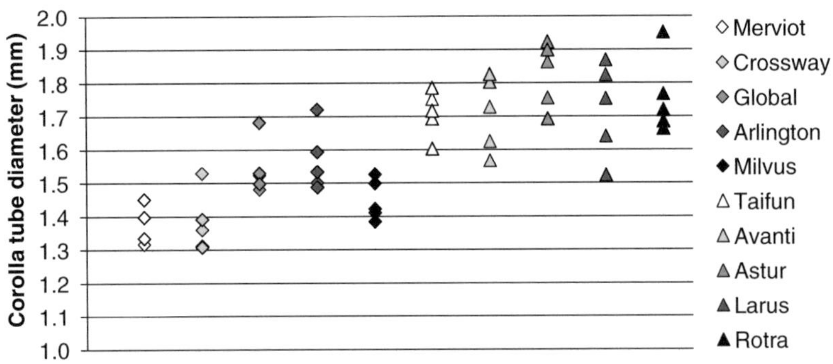

tubes. Other factors must be responsible. A lower number of flower heads per plant and a lower degree of determinacy have been suggested as partly responsible for low seed yields in tetraploids (Vleugels et al. 2015).

To determine whether the diameter of the corolla tube may explain the lack of correlation between seed yield and corolla tube length observed here, we measured the corolla tube diameter in a set of 25 diploid and 25 tetraploid plants from 5 diploid and 5 tetraploid cultivars. The average corolla tube diameter for each plant is given in Fig. 44.2.

Corolla tube diameters ranged between 1.31 and 1.72 mm in diploid plants, and between 1.52 and 1.95 mm in tetraploid plants. Diploids and tetraploids overlapped in corolla tube diameter between 1.52 and 1.72 mm. The average corolla tube diameters in diploids and tetraploids were 1.47 ± 0.01 mm and 1.75 ± 0.01 mm respectively, i.e., 19.0 % wider in tetraploids. This difference corresponds with that of Bender (1999b) who reported 33.7 % wider corollae in the tetraploid cultivar 'Ilte' compared to its diploid ancestor. In addition, Bender (1999b) observed 8.9 % longer corollae in the same set of plants. Possibly, the enlarged width of tetraploid corolla tubes allows bumblebees to insert their head deeper into the flower and reach the nectar anyway, in spite of the longer corolla tube. Hence, the enlarged width of tetraploid corolla tubes may compensate for their increased length. This may explain the lack of correlation between corolla tube length and seed yield observed here in tetraploid red clover. We aim to further investigate the role of the corolla tube diameter in a diverse set of diploid and tetraploid plants.

Conclusion

Our results indicated no correlation between corolla tube length and seed yield in tetraploid red clover, which confirms the results of previous studies by Starling et al. (1950) and Vleugels et al. (2015). Therefore, it is becoming increasingly evident that longer corolla tubes are not the determining factor (or inhibiting cause) for low seed yields in tetraploid red clover. Possibly, the increased width of tetraploid corollae, which are 19.0 % wider than diploid corollae, can compensate for increased length by allowing bumblebees to insert their heads deeper into the flowers and reach the nectar in spite of the increased corolla tube length. It remains unclear, however, whether low seed yields in tetraploids are due to deficient pollination, to ploidy-related factors such as embryo abortion, or to a combination of both. Therefore, future research should elucidate the role of pollination in seed yield by investigating the effects of flower morphology, such as the corolla tube diameter, the number of ovules per pod, pollen viability, nectar secretion, and flower color on pollinator behaviour in a diverse set of diploid and tetraploid red clover plants. Research is ongoing at ILVO to explore these questions.

Acknowledgments We thank Nancy Mergan, Frederik De Jonghe and Winnie Poncelet for technical assistance.

References

Abramoff MD, Magelhaes PJ, Ram SJ (2004) Image processing with ImageJ. Biophotonics International 11: 36–42.

Bender A (1999a) An impact of polyploidization on the dimensions of red clover flowers, species' composition and number of insect pollinators. Agraarteadus 4: 9–23.

Bender A (1999b) An impact of morphological and physiological transformations of red clover flowers accompanying polyploidization on the pollinators' working speed and value as a guarantee for cross pollination. Agraarteadus 4: 24–37.

Boller B, Schubiger FX, Kölliker R (2010) Red Clover. Boller B (ed.), Handbook of Plant Breeding. Springer, Dordrecht. pp 439–455.

Bond DA (1968) Variation between tetraploid red clover plants in corolla tube length and height of nectar. Journal of Agricultural Science 71:113–116.

Brodsgaard CJ, Hansen H (2002) Pollination of red clover in Denmark. DIAS Report 71: 1–50.

Furuya H (2001) Comparisons of seed weight and seedling characteristics of diploid and autotetraploid red clover. Dissertation. University of Florida, Florida, 96 p.

Jablonski B (2001) A preliminary appreciation of agricultural and beekeeping value of a red clover population with a shortened flower tube. Acta Horticultura 561: 215–217.

Julén U (1950) Fertility conditions of tetraploid red clover. Communications from the Swedish Seed Association 25: 151–160.

Julén G (1959) Rotklee, *Trifolium pratense* L. Parey P (ed.) Handbuch der Pflanzenzüchtung. Band IV, Berlin. pp 239–305.

Skirde W (1963) Studies of morphology and secretion in diploid and tetraploid clovers. Annales Agriculturae Fenniae 2: 90.

Spearman C (1904) The proof and measurement of association between two things. American Journal of Psychology 15: 72–101.

Starling TM, Wilsie CP, Gilbert NW (1950) Corolla tube length studies in red clover. Agronomy Journal 42: 1–8.

Taylor NL (2008) A century of red clover breeding developments in the United States. Crop Science 48: 1–13.

Taylor NL, Quesenberry KH (1996) Red clover science. Kluwer academic publishers, Dordrecht, 226p.

Vleugels T, Cnops G, Roldán-Ruiz I (2015) Influence of flower and flowering characteristics on seed yield in diploid and tetraploid red clover. Plant Breeding 134: 56–61.

Part V
Reports of Interactive Sessions

Workshop "Phenotyping" 45

P. Lootens, T. De Swaef, I. Roldán-Ruiz, and T. Altmann

Abstract

The aim of phenotyping is the evaluation of plant behavior in controlled environments and in the field. This is essential to keep pace with developments in plant genotyping. Thorough phenotyping of breeding germplasm enables understanding of the links between genotypes and environmental factors, as it helps the breeder to assess plant traits in a more objective way. Here, we report on the presentations and the subsequent discussions during the Phenotyping Workshop held at the 31st symposium of the Eucarpia section Forage crop and amenity grasses.

Keywords

Phenotyping • Controlled environment phenotyping • Field phenotyping Breeding • Pre-screening • Ecophysiology • Modelling • Airborne imagery

Even though the term 'phenotyping' was already discussed in 1909 (Johannson 1909), there is an urgent need for high-throughput plant phenotyping (HTPP) tools for the evaluation of plants both in controlled environments and in the field. This is essential to keep pace with developments in plant genotyping. Thorough phenotyping of breeding germplasm enables understanding of the links between genotypes and environmental factors, as it helps the breeder to assess/estimate plant traits in a more objective way. Because of the importance of advanced phenotyping tools, and the current limited use in forage crop breeding, five speakers were invited to share their view and experience on phenotyping in the frame of the Eucarpia Forage Crops and Amenity Grasses 2015 meeting. These five presentations reviewed current use of high-throughput above-ground and below-ground phenotyping in controlled, semi-controlled and field environments, combining perspectives of scientists and practical breeders.

In his introductory talk, Thomas Altmann (IPK Gatersleben, Germany) identified two

P. Lootens (✉) • T. De Swaef • I. Roldán-Ruiz
Plant Sciences Unit, Institute for Agricultural and Fisheries Research (ILVO), Melle, Belgium
e-mail: peter.lootens@ilvo.vlaanderen.be

T. Altmann
Department of Molecular Genetics, Leibniz Institute of Plant Genetics and Crop Plant Research (IPK) – Gatersleben, Stadt Seeland, Germany

I. Roldán-Ruiz et al. (eds.), *Breeding in a World of Scarcity: Proceedings of the 2015 Meeting of the Section "Forage Crops and Amenity Grasses" of Eucarpia*,
DOI 10.1007/978-3-319-28932-8_45

important applications of plant phenotyping: (1) the establishment of causal links between genotypes, environmental factors and phenotypes (expression of plant traits and features) in basic plant science; (2) the high-throughput and high-precision assessment of plant trait expression in the context of breeding and agronomy. Linking this with multi-level 'omics' analyses can provide a mechanistic understanding of critical plant processes. Therefore, HTPP is the tool to assist gene discovery, marker assisted selection and mode-of-action analysis.

Bruno Pollet (BASF Plant Science, Belgium) highlighted that, especially in a transgenic trait discovery - breeding approach, HTPP in a controlled environment is useful for pre-screening and detection of hits and is a good alternative for expensive field trials with transgenics. Multi-location and multi-season field trials remain, however, necessary as they are the reference environment for new introductions. Transferability studies show that correspondence between (semi-) controlled environment phenotyping and field performance is not perfect, but still rather satisfactory. Good ecophysiological knowledge is necessary for correct data acquisition and analysis. Modelling approaches help to summarise and interpret the large amount of data generated in a typical phenotyping experiment.

From a breeder's perspective, Alan Lovatt (IBERS, UK) argued that for practical breeding, it is not always necessary to break-down traits into physiological components. This can however be very valuable for the development of robust molecular selection tools. Field phenotyping, allowing traits to be assessed objectively, in a fast and cheap fasion and ideally as a single value per plant/plot/genotype would really be supportive for practical breeders. Susanne Barth (Teagasc, Ireland) suggested that airborne or tractor-mounted imagery (for example, for mobile NIRS on wet samples) can become very valuable, if cost-effective.

In his consideration of root phenotyping, Ulf Feuerstein (Deutsche Saatveredelung Ag) acknowledged that through selection for aboveground traits, breeders have been indirectly breeding for well-developed root systems in the past. Nonetheless, as aspects related to resource use efficiency become more important, the need for root phenotyping is increasing. Current strategies range from those allowing rapid assessment of large numbers of plants at the seedling stage, to the time-consuming evaluation of adult root systems in long-term experiments. Compared to above ground phenotyping, experimental conditions for root phenotyping should be controlled even better because the environmental influence of soil properties (e.g., bulk density, water availability, temperature,…) on root phenotypes is even stronger than the aboveground environment on aerial plant parts.

In summary, the workshop highlighted that: (i) Phenotyping in controlled environments enables achieving mechanistic understanding and performing pre-screening but 'the truth' can only be found in the field evaluations; (ii) Breeders' field evaluations cannot be replaced by phenotypic/physiological measurements in controlled environments, but in the near future field-derived data will become valuable upon proper calibration and conversion into measures of overall performance of the plant; (iii) The breeders are expecting integrated values (also of visually hidden properties) derived from cost-efficient, fast, and objective measurements.

Reference

Johannson,W. (1909). Elemente der Exakten Erblichkeitslehre. Jena: Gustav Fischer Verlag.

Workshop "Genomic Selection and Association Mapping"

46

P. Barre, H. Muylle, and T. Ruttink

Abstract

The aim of forage crop breeding is to create improved material by selection based on repeated phenotypic observations in one or more environments. High-throughput NGS applications such as genotyping-by-sequencing (GBS) or SNP chips facilitate genotyping of a large number of plants with ten thousands of markers for an "affordable" price. Simultaneously, advanced statistical methods have been developed to find correlations between genotyping and phenotyping data. The question now is how to optimize the use of molecular markers in breeding programs to go further, faster, and cheaper than classical breeding schemes. Here, we report on the discussions held at the workshop on association studies and genomic selection, held during the 31st symposium of the Eucarpia section on Forage Crops and Amenity Grasses.

Keywords

Forage • Genomics • Breeding • Amenity grasses

Association Mapping

Association studies (AS) detect the genetic loci associated with variation of quantitative traits (QTL) and identify alleles with favorable effects. The frequency of such favorable alleles can then be increased in elite material by selection or introgression. Association mapping can be based either on candidate genes or on genome wide molecular markers (GWAS). Limitations of QTL analysis include the detection of false QTL (e.g., due to the high number of non-independent tests, the structure of the population, and the relatedness between individuals) and the lack of detection power for QTL with small effects. The workshop presenters focussed on how QTL found in association studies can be validated and used for plant breeding.

P. Barre (✉)
Biology, Plant Breeding, INRA UR4 Unité de Recherche Pluridisciplinaire, Prairies et Plantes Fourragères C80006, Lusignan 86600, France
e-mail: philippe.barre@lusignan.inra.fr

H. Muylle • T. Ruttink
Plant Sciences Unit, Institute for Agricultural and Fisheries Research (ILVO), Melle, Belgium

I. Roldán-Ruiz et al. (eds.), *Breeding in a World of Scarcity: Proceedings of the 2015 Meeting of the Section "Forage Crops and Amenity Grasses" of Eucarpia*,
DOI 10.1007/978-3-319-28932-8_46

Bruno Studer (ETH, Zurich, Switzerland) illustrated how two complementary approaches, based on two types of populations, both identified the same genomics regions that control self-incompatibility (SI) in perennial ryegrass. First, the SI locus was identified in a bi-parental mapping population (e.g., Manzanares 2013). This type of populations, in which linkage disequilibrium (LD) declines slowly, allow QTL detection with a moderate number of markers, but the genetic resolution is relatively low and only a few alleles are surveyed. Second, the SI locus was also identified by a candidate gene approach in a genetically broad association mapping population. This type of population, which includes mainly unrelated individuals, allows surveying of a large number of alleles, and the location of QTL with high resolution thanks to the fast decline of LD. It is highly efficient when candidate genes are known but would require several tens of millions of markers in perennial ryegrass to survey the entire genome. Another type of population can be envisioned such as a "multi-family based population" in which the crossing scheme (number of parents, combinations, and generations) can be optimized depending on the required resolution and on the genotyping budget. Other approaches for the validation of the effect of QTL include genetic transformation, tilling or eco-tilling (screening for induced or natural mutations) and creation of selected populations where the favourable alleles have been fixed.

Leif Skøt (IBERS, Aberystwyth, UK) has presented the validation and exploitation of the gene LpFT3, which is associated to heading date (HD) and seed yield in perennial ryegrass. A major QTL for HD was identified in a bi-parental population. A candidate gene for HD, LpFT3, is situated in the QTL, and allelic diversity in the promoter region explains variation in heading date in a core collection (Skøt et al. 2011). Markers in this gene are currently used in a breeding scheme to discard undesired alleles.

Genomic Selection

The principle of genomic selection (GS) is to create a statistical model (prediction equation) for the prediction of the breeding value, based on the genomic information of genome wide markers. This equation is calibrated in a training population, which is both phenotyped and genotyped. This equation is then used to predict the breeding value of an individual or population from the breeding population, which is only genotyped (genomic prediction). Individuals or populations of the following generations are then ranked and selected based on this predicted breeding value (genomic selection).

High quality prediction equations are crucial. This is assessed by a Pearson correlation coefficient between the predicted and the true breeding values by cross validation with individuals not included in the training population, but which are phenotyped and genotyped. This accuracy depends on many parameters such as the heritability of the trait, the size of the training population, the number of markers, the relationship between the training and the breeding population and to a lesser extent on the statistical model used.

Torben Asp (Univ. Aarhus, Slagelse, Denmark) has reported on genomic selection in ryegrass (e.g., Fè 2015) and underlined the importance of choosing a training population closely related to the breeding population and the necessity to update the prediction equation regularly during the cycles of selection. When using different populations or families, it is essential that each population is represented in the training population. Furthermore, the LD in the training population defines the number of markers and, in turn, the genotyping strategy.

Paolo Annicchiarico (CRA-FLC, Lodi, Italy) gave an example of GS in alfalfa for forage yield on two unrelated populations. The accuracy of prediction was about 0.3, which should give a better genetic gain per time unit as compared to phenotypic selection, because GS allows to increase the selection intensity and to decrease the duration of a cycle of selection. The accuracy was half as high when a calibration based on one population was used on the other population.

Two main points were highlighted in a general discussion. First, there is a great variability of the accuracy in cross validation depending on the set of individuals selected for validation. Second, there is a trade-off between the number of plants in the training population and the quality of phenotyping. A high number of plants in the training population are required to ensure good accuracy of the prediction equation but high quality phenotyping is only feasible on relatively small populations.

Conclusion

AS and GS give good results in forage species, even on alfalfa, an auto-tetraploid species. The choice of the population is crucial. Family based populations from a few founders could be a good compromise to detect QTL and to create varieties, possibly by GS, since they contain enough diversity but have a low LD decay. We suggest that the advantages of the two approaches can be combined by including in a mixed model the phenotype as the dependent variable, the QTL detected by AS as fixed effects, and the rest of the markers as random effects. This should increase the accuracy of predicted breeding values.

References

Fè D. (2015) Establishing the basis for genomic prediction in perennial ryegrass. Ph.D. thesis, University of Aarhus. 124 p.

Manzanares C. (2013) Genetics of self-incompatibility in perennial ryegrass (*Lolium perenne* L.). Ph.D. thesis, University of Birmingham. 319 p.

Skøt L, Sanderson R, Thomas A, Skøt K, Thorogood D, Latypova G, Asp T, Armstead I. (2011) Allelic variation in the perennial ryegrass FLOWERING LOCUS T gene is associated with changes in flowering time across a range of populations. Plant Physiology 155(2): 1013–22.

47 Feeding Value Evaluation in Grass and Legume Breeding and Variety Testing: Report of a Debate

J. Baert and H. Muylle

Abstract

Because of the growing interest for the implementation of feed evaluation in grass/legume breeding and variety testing, we organized a survey among breeders, variety testing authorities and animal nutritionists. The survey dealt with 4 topics: the relevance of feed value in breeding and testing, the feed value parameters that should be assessed, a recommended testing method and the impact of the feeding value parameter on the evaluation of a variety. The outcome of the survey was the basis of a debate. Although very contrasting opinions the 3 parties agreed that information on feeding value of varieties is valuable for the farmer esp. when he has to select among high yielding and persistent varieties. Digestibility, protein content and the balance of nutrients in the rumen are the most important feeding value parameters. NIRS is a very useful method to test the feeding value but sample preparation is still labour consuming and differences in phenology have to be taken into account. Concerning the impact of feeding value on the variety evaluation the use of an index, threshold levels or just a description was discussed.

Keywords

Feeding value • Fodder grass • Legumes • Breeding • Variety testing

Introduction

There is growing interest in the implementation of feed evaluation in grass/legume breeding and variety testing. Some countries (e.g., UK, Denmark, Ireland, France) have already implemented feeding value parameters in their variety testing programme, others not (e.g., The Netherlands, Germany). The purpose of this feeding value debate was to discuss the

J. Baert (✉) H. Muylle
Plant Sciences Unit, Institute for Agricultural and Fisheries Research (ILVO),
Caritasstraat 21, Melle 9090, Belgium
e-mail: joost.baert@ilvo.vlaanderen.be

I. Roldán-Ruiz et al. (eds.), *Breeding in a World of Scarcity: Proceedings of the 2015 Meeting of the Section "Forage Crops and Amenity Grasses" of Eucarpia*,
DOI 10.1007/978-3-319-28932-8_47

relevance of feeding value in the context of breeding and variety testing and the parameter(s) that is (are) most important and realistic to implement.

For the preparation of the debate we organized a survey among breeders, variety testing authorities and animal nutritionists. Four main questions were formulated to each participant. We got answers from 6 breeders (Piet Arts (NL), Christof Boehm (DE), Petter Marum (NO), Joseph Robins (US), Franz Xavier Schubiger (CH), Lukas Wolters (NL)), 7 variety testers (Vincent Gensollen (FR), Dermot Grogan (IR), Tomas Mezlik (CZ), Liv Ostrem (NO), Uta Schnock (DE), Jan Rinze van der Schoot (NL), Daniel Suter (CH)) and 4 animal nutritionists (Johan De Boever (BE), Michael Lee (UK), Irene Mueller-Harvey (UK), Marketta Rinne (FI)). Here we summarize their answers and the outcome of the discussion during the meeting.

Further on, the following abbreviations are used: dry matter (DM), crude protein (CP), water soluble carbohydrate (WSC), organic matter (OM), neutral detergent fibre (NDF), acid detergent fibre (ADF), neutral detergent fibre digestibility (NDFD).

Is Feed Evaluation Relevant in Testing/Breeding Grasses/Legumes?

The opinions of breeders and variety testing authorities were divided among yes and no, while animal nutritionists unanimously answered positively.

Yes, because:
- It is essential for sustainable intensification.
- Farmers know the feeding value of maize but not of grasses.
- Biomass yield evaluations often work to distinguish elite from poor material, but fail to distinguish among the elite varieties.

Yes but:
- One needs to combine quality and yield.
- It depends on the species.

No, because:
- Feeding value of perennial ryegrass is already very good.
- Differences among varieties are small and management is more important than variety.
- Varieties are used in mixtures.
- It is difficult to measure because of its dependence on growth stage.
- It is expensive to determine.
- DM yield and persistence are more important in some production systems.

Discussion:

Is feed evaluation more relevant for tall fescue, timothy and cocksfoot than for perennial ryegrass?

- Depending on the region you are selecting for. In regions such as USA & Canada or Nordic countries other traits are more important. In these regions the major aim are high DM yield and persistency.
- If feed quality is evaluated during breeding or variety testing, farmers tend to use this information to select varieties or mixtures.

Which Feeding Value Parameter(s) Should Be Assessed in VCU Trials/Improved by Breeders? For Which Grass/Legume Species? Which Parameter(s) Would You Evaluate if Cheap and Rapid Methods Were Available?

Breeders and variety testing authorities ranked the parameters in order of importance as follows:

1. Digestibility (OM and NDF) in grasses and legumes.
2. Crude protein in legumes and grasses.
3. WSC in grasses.

   Other: ADF in grasses and lucerne, cyanogenic glucosides in *Trifolium repens*, formononetin in *Trifolium pratense*, condensed tannins in *Onobrychis viciifolii*. In species such as tall fescue and cocksfoot with high NDF content, improvement of NDF quality can be interesting, not the decrease of NDF content.

(b) If efficient methods would be available they would like to evaluate.

Animal nutritionists replied:

> The most important characteristics are digestibility and the balance of nutrients in the rumen to improve microbial protein synthesis. Therefore breeding for higher digestible NDF and WSC, and reduced protein solubility (increased rumen escape protein) are desirable.
>
> Characteristics of secondary relevance are higher lipid content (non-fermentable energy), higher dry matter content, improved vitamin and mineral composition, use of legumes as nutraceutical feeds to enhance nutrition and animal health (polyphenol oxidase, isoflavone, tannin).
>
> It would be worth to consider cereals and grain legumes used as whole crop feeds and species with potential for reducing greenhouse gas emissions from ruminants.

Discussion:

What kind of digestibility: *OM*, *NDF*, *high WSC*, *low ADF*?

- OM digestibility is the most important, followed by WSC, NDFD and CP. Parameters, such as fat content, are neglected but they can contribute to the reduction of methane emission.

What Do You Recommend as a Testing Method? When Should It Be Carried Out? Under Which Management?

Variety testing authorities:

- NIRS (near infrared reflectance spectroscopy)
- Mainly in first utilization year, on all locations, particularly in spring cuts. There is a lot of variation in number of years, cuts, locations, replicates among testing authorities.

Answers of breeders can be summarized as follows:

- We need faster, more efficient and less expensive phenotyping methodologies. NIRS is great, but the grinding of samples remains a major bottleneck.
- There is no consensus about the cuts that should be tested. The digestibility and the cell wall constituents should be determined at least during the first growth cycle (generative, main part of the yield) and during a summer growth (vegetative growth).
- At least one full harvest year should be tested; all varieties should be sampled at the same stage of development.

Animal nutritionists:

- NIRS is very useful (e.g., for digestibility preferably based on rumen fluid *in vitro* digestibility) but should be calibrated carefully to get accurate results. NIRS holds much promise for plant secondary metabolites, such as tannins. Indirect methods such as SPAD (chlorophyll) could be used for digestibility, lipid content and WSC testing. Leaf to stem ratio is also a good indicator of digestibility and CP content. Visual observation of crown rust infestation is valuable in the frame of ingestibility.
- Typically grazed species should be evaluated over the grazing season; typically mown species should be evaluated during the optimal growth stage. Testing should include ensiling and hay drying. Drying (wilting) has a positive effect on ingestibility as well as on the % rumen escape protein; ensiling strongly lowers the % rumen escape protein and sugar content.

Discussion:

All VCU (value for cultivation and utilization) tests are at present carried out in a harvest-all-varieties-on-one-date system. In such a system there can never be a good comparison between varieties with different heading dates?

- A correction for differences in phenology at harvest could be introduced on the basis of the DUS (distinction, uniformity, stability) testing

results, as DUS results aim to describe the phenology of a variety.
- NIRS is useful especially if on-line systems or hand-held systems prove to be accurate. A higher throughput phenotyping system would be useful. With the upcoming UAV technology, hyperspectral images can be taken every second day levelling the difference in earliness. Yield and quality can be monitored by remote sensing technology simultaneously.

What Is/Should Be the Impact of the Feeding Value Parameter on the Evaluation of the Variety?

Variety testing authorities:
- Feeding value parameters are included in an index. Their coefficient in the index depends on the variation: less variation (e.g., DMD up to 5 %) higher coefficient, more variation (e.g., WSC up to 30 %) lower coefficient.
- Varieties are excluded if their content of antinutrional compounds exceed that of standard varieties (e.g., cyanogenic glucides in *Trifolium repens*).
- Feeding value is just additional information.

Breeders:

- Feeding value parameters can be weighted in an index. Several options were suggested:
  - If feed value is a criterion then it should be weighted heavily (25–50 %).
  - The distribution of the weights in an index could be 40 % for yield components, 30 % for feeding quality and 30 % for other characters.
  - There should be an equal weight for both, feeding value and yield. The impact of feeding value of species such as tall fescue or cocksfoot, with a low feeding value, should be twice the weight of the yield.
  - Ideally one should use economic weights (based on dairy or beef production).
- There should be threshold levels for feed value to ensure no declines in quality.
- Feeding value should be used just as additional information.

Animal nutritionists:
- A percent-unit increase in digestibility results in a increase of potential roughage milk production based on energy and protein intake of 0.5 kg for wilted grass silages.
- An increase of 3 % lipid in grasses could supply an extra 1.1 MJ of gross energy per kg DM. This would result in a significant increase in animal performance as for dairy cows 5 MJ ME corresponds to approximately 1 L of milk.
- Nutraceutical feeds can make significant progress, reduction of bloat alone can improve farm profitability

Discussion:

A simple index would be to multiply energy content /kg DM by total DM yield?

- Animal nutrition (especially ruminants) is about designing the optimal, not the maximum, quality. WSC, CP, NDF, digestibility etc. should be combined in a way that makes sense for a well-balanced diet. This might lead to an index in which for example too high sugar contents get a negative weight, while NDF (effective fibre) gets a positive weight. However, each individual farm has its own desired grass quality (based on feeding system, milk production, available forages etc.) which makes it impossible to evaluate quality in a positive or negative way with an index.
- Variety testing should provide a description of the varieties. The choice of variety on farm should be farmer-driven. Depending on the farming system, a set of varieties can be chosen to fulfil the needs, which can be quite diverse among the season.
- Are we sure to see the same quality levels if we use the varieties in mixtures? If used in mixtures, the quality effect at mixture level is relevant. This implies that the quality traits of a variety should be used as descriptors, but not assigned positive or negative weights in an index during testing. This would allow different possible uses of each variety.

Any Further Comments Concerning Feed Evaluation?

Variety testing authorities:

- Breeders are strongly involved and informed at all stages of the introduction and evolution of feeding value parameters included in the VCU system.
- Breeders pay for the VCU and they decide what and how assessments should be done.
- It would be much easier to find a common European protocol for quality parameters than for the complete agronomical testing protocol, since testing of feeding quality parameters is much less dependent on the forage production system of a country than agronomical parameters.
- A common protocol will be very difficult due to the use of different quality measurement systems in different countries, as well as local requirements for climatically suitable and sustainable varieties. Only general principles for a small number of parameters may be suitable for a common protocol.

Breeders:

- When breeding for feed quality there may be drawbacks in DM yield, in tolerance and resistance to abiotic and biotic stresses and in animal health (e.g., subacute ruminal acidosis when WSC content is too high). However one has to breed for both, better yield and quality, and there is not necessarily a positive correlation between feed value and disease susceptibility.
- Selection for quality is expensive. There has to be return on investment for companies and customers.
- We need faster, more efficient, and less expensive phenotyping methodologies.

Conclusions

Before the meeting Ulf Feuerstein (Deutsche Saatveredelung Ag) formulated three questions. At the end of the debate he summarized the answers obtained as follows:

Is knowledge on the quality of forage varieties an advantage for the farmer?

Yes, when this information is available, farmers make use of it.

What are the prerequisites for a (good) quality testing?

All varieties should be examined in a similar way irrespective of earliness. VCU testing officials are aware of this problem and take into account these biases.

Does the Eucarpia section 'Forage crops and amenity grasses' recommend a quality evaluation in official tests?

Yes, the section recommends a quality evaluation and is prepared to discuss this with the variety testing authorities. Being aware of the importance of feeding value and starting a discussion, as done during this conference, is already a first great step.

Report of the Breeding Debate

48

D. Reheul, J. Aper, H. Amdahl, P. Annichiarico, P. Arts, J. Baert, P. Barre, V. Beguier, B. Boller, J.L. Durand, U. Feuerstein, A. Ghesquiere, M. Ghesquière, C. Grieder, K. Gylstrom, B. Julier, R. Kölliker, I. Litrico, P. Marum, L. Pecetti, H. Riday, J.P. Sampoux, F. Schubiger, C. Scotti, T. Vleugels, and L. Wolters

Abstract

Based on a questionnaire with 11 questions, 5 breeding institutes and 4 breeding companies defined their 2035 horizon for grass and forage crops breeding. Visions and opinions differed a lot regarding targeted species, breeding goals, importance of plant physiology, breeding techniques, testing environments, the use of molecular tools and the influence of regulations and sustainability drivers. The report can be considered as a joint reference document for future debates.

D. Reheul (✉)
Faculty of Bioscience Engineering, Department of Plant Production, Ghent University, Ghent, Belgium
e-mail: dirk.reheul@ugent.be

J. Aper • J. Baert • A. Ghesquiere • T. Vleugels
Plant Sciences Unit, Applied Genetics and Breeding, Institute for Agricultural and Fisheries Research (ILVO), Melle, Belgium

H. Amdahl • K. Gylstrom • P. Marum
Graminor, Ridabu, Norway

P. Annichiarico • L. Pecetti • C. Scotti
Centre for Fodder Crops and Dairy Productions, Centro di Ricerca per le Produzioni Foraggere e Lattiero-Casearie (CRA-FLC), Lodi, Italy

P. Arts
Barenbrug Holland, Wolfheze, The Netherlands

P. Barre • J.L. Durand • M. Ghesquière • B. Julier
I. Litrico • J.P. Sampoux
Unité de Recherche Pluridisciplinaire Prairies et Plantes Fourragères, Institut National de la Recherche Agronomique (INRA), Lusignan, France

V. Beguier
Jouffray-Drillaud, Saint-Sauvant, France

B. Boller • C. Grieder • R. Kölliker • F. Schubiger
Institute for Plant Production Sciences, Agroscope, Zurich, Switzerland

U. Feuerstein
Deutsche Saatveredelung Ag, Asendorf, Germany

H. Riday
Plant Genetic Resources, Genomics and Genetic Improvement, U.S. Dairy Forage Research Center (USDA-ARS), Madison, WI, USA

L. Wolters
DSV zaden Nederland BV, Ven-Zelderheide, The Netherlands

I. Roldán-Ruiz et al. (eds.), *Breeding in a World of Scarcity: Proceedings of the 2015 Meeting of the Section "Forage Crops and Amenity Grasses" of Eucarpia*,
DOI 10.1007/978-3-319-28932-8_48

Keywords

Horizon 2035 • Forage and turf species • Breeding goals • Breeding techniques

Introduction

The aim of the breeding debate was to discuss the vision of breeding companies and institutes regarding forage crops and amenity grasses breeding for the following decades (horizon 2035). A standardised questionnaire was sent to more than 20 companies and institutes asking for clear, affirmative, quantified answers to 11 questions. We would like to thank 4 companies and 5 breeding institutes that have provided answers. Based on their contributions the moderator of the debate launched a number of statements and questions that were discussed with the representatives of the companies/institutes and the whole audience during the 31th International EUCARPIA Symposium of the Section Fodder Crops and Amenity Grasses.

The questionnaire contained the following questions:

1. What is the vision of your company/institute regarding forage crops/amenity grasses breeding with horizon 2035 (20 years from now)?
2. What species will be bred; do you expect a shift in focus species?
3. What will be the most important breeding goals?
4. What breeding techniques and testing environments will be used; to what extent genomic selection will be part of the breeding programme?
5. What scientific qualifications will be most important in your breeding team?
6. What is the influence of the growing world population for the breeding strategy?
7. How does the knowledge from plant physiology influence the breeding work?
8. How do climate change and the "renewable carbon based economy" influence your breeding strategy?
9. How does consumers' attitude influence your breeding strategy?
10. How do regulations and sustainability actions influence your breeding strategy?
11. Any other important issues

Report

Vision

All respondents emphasize an increased importance of feeding value, low input, tolerances, legumes and mixtures in 2035. Some respondents aim for a leading role in the future.

The intention of **Agroscope (Plant Breeding Institute, CH)** is to provide adapted germplasm for sustainable production systems, to be part of the Swiss grassland success story with an international impact and to have a 50 % domestic market share of forage seed. Their aim is also to have at least one organically bred variety for each important species on the national list.

CRA-FLC (Research Centre for Forage and Dairy Production, IT) has an increasing interest in species that fix nitrogen and are drought tolerant, highly digestible (intensive systems), grazing tolerant (extensive systems) and that are suitable for mixed cropping.

ILVO (Institute for Agricultural and Fisheries Research, BE) aims to breed ryegrasses with an added value (stress tolerances, feeding value) and will have more attention to legumes and grass/legume mixtures. ILVO wants also to play a role in breeding "new" crops and guiding them to the market and to fill the gap between field and lab using molecular tools.

INRA (National Institute for Agricultural Research, FR) emphasizes (i) the growing role of grasslands for delivering ecosystem services in response to climate change and (ii) the expected continuous decrease of artificial inputs, nitrogen fertilization in particular. As this might be completed through a better use of legumes, in pure or mixed stands, it is essential that forage breeding incorporates the genetic complexity and diversity of the grasslands in the future.

USDFRC (Dairy Forage Research Centre, USA) wants to lead the world in integrated dairy forage systems research and wants to provide the dairy industry with solutions that comply with food security, environmental sustainability and economic viability.

Barenbrug (Plant Breeding Company, NL) wants to breed for high quality productive forage varieties adapted to lower inputs (good land might become scarce for forage crops) and tolerant to abiotic stresses. They foresee that turf will be less irrigated; when irrigation will be available more and more effluent and saline water will be used. The cutting frequency should decline to use less energy and to decrease the amount of clippings to be processed.

DSV (Plant Breeding Company, DE) aims for a leading role in 2035 and will reduce its activities in turf grass breeding and in breeding energy crops. The decrease of the size of turf grass breeding is mainly driven by the lower number of big buyers of turf grass (e.g., governments), the existing set of high performing varieties and the rather slow progress in turf grass breeding.

Graminor (Plant Breeding Company, NO) wants to put more emphasis on forage quality. They expect an increased importance of perennial ryegrass and alfalfa in Norway.

Jouffray-Drillaud (Plant Breeding Company, FR) will have more attention to forage varieties for hay and conservation purposes and less for grazing. They will focus on improved tolerances to abiotic stresses and varieties that are performant both in pure stands and in mixtures (min 3 species of minimum 3 families). For turf grasses, they will pay extra attention to seed yield and stress tolerances.

Focus on What Species

In general, companies and institutes focus on forage species adapted to low inputs, drought and heat tolerance with more utilization for conservation at the expense of grazing. They expect a growing role of legumes. *Lolium* spp. are dominant, but 6 out of 9 respondents mention *Festuca arundinacea* as becoming more important in the future (Table 48.1). CRA-FLC favours *F. arundinacea*, because it has a better adaptation to a wider range of environments in comparison with *Dactylis glomerata*. Agroscope will give less attention to white clover due to its greater drought susceptibility when compared to red clover. ILVO and INRA both emphasize breeding species for mixtures. For turf, species tolerant to harsh conditions, adapted to warmth and a slower growth will become important.

Breeding Goals

For forage crops, no real surprises in breeding goals were reported. Both companies and institutes mentioned: yield, yield stability, persistence, (new) disease resistances, seed yield, nitrogen use efficiency, drought tolerance and forage quality.

Jouffray-Drillaud declares yield under a cutting regime as important, while CRA-FLC focuses on grazing tolerance and competitive ability. Graminor emphasizes the interaction between a higher temperature and an unchanged day length as a consequence of climate change. INRA mentions ecosystem services such as carbon sequestration, water quality and the impact on biodiversity. Although ecosystem services are very difficult to quantify, collaboration with e.g., soil sciences can identify indirect traits (like roots) for better ecosystem services.

For turf species, seed yield and performance under suboptimal conditions are reported.

Breeding Techniques and Testing Environments

Phenotyping will remain very important: the respondents see a shift from less field testing and less Haldrup harvesting to more image analyses by camera or drone or the use of rhizotrons and more experiments in controlled conditions. Barenbrug considers controlled conditions as necessary for

Table 48.1 Overview of species considered to be important by 2035 by breeding companies and institutes

	Bar	DSV	JfD	Gram	Agr	CRA	ILVO	INRA	USD
Forage species									
Avena strigosa			×						
Bromus		×							
D. glomerata			×		+			+	
F. arundinacea	+	×	×		+	×		+	
F. pratensis				×	–			+	
L. x hybridum			×		–		×		
L. multiflorum		×	×		=		×		
L. perenne		×	×	+	=		×	×	
P. pratensis					–				
Phleum pratense		×		×			×		
Sorghum		×							
Plantago lanceolata					(×)				
Turf species									
F. arundinacea	+	+	+						
L. perenne		+	+						
F. longifolia	(+)								
F. rubra		+	+						
Koeleria	+								
Legumes									
Medicago sativa		×	×	+	(×)	●		+	●
T. alexandrinum			×						
T. ambiguum									•
T. incarnatum			×						
T. repens		×		–	–		×		
T. pratensis		×	×	×	=		×		●
Lotus corniculatus						•			•
O. viciifolia					(×)				
Pisum sativum						•			
Vicia sp.			×			•			
Legumes for mixtures									
								+	

Abbreviations: *Bar* Barenbrug, *JfD* Jouffray-Drillaud, *Gram* Graminor, *Agr* Agroscope, *CRA* CRA-FLC, *USD* USDFRC, *D. Dactylis*, *F. Festuca*, *L. Lolium*, *T. Trifolium*, *O. Onobrychis*
Symbols: × present and future breeding activities, + increase in the future, – decrease in the future, = equal efforts in the future, ● major effort, • minor effort, () new species to consider in the future

the simulation of certain stresses because companies don't have all stresses in their direct area.

Another trend will be the growing importance of molecular breeding and more particularly genomic selection. Agroscope foresees 2/3 phenotypic work and 1/3 genotypic work in 2035. There is a difference point of view at what stage genomic selection will be used. CRA-FLC thinks it will play a dominant role in the first selection stages, while Agroscope positions it further in the breeding program: "genomic selection may be used to select elite clones for synthetics and thus make clonal progeny tests redundant." ILVO sees genomic selection playing a role in a way NIRS is used today. USDFRC argues that genomic selection activities in alfalfa will take place with elite material at the end of the variety development, since phenotyping continues to be cheap. DLF is going to implement genomic selection in their perennial ryegrass program: so far it looks promising and perennial ryegrass has enough potential to pay back the efforts. The cost of genomic selection (e.g., processing of one sample) should be balanced with the cost of running one field plot (but there are calibration costs: phenotyping will still be necessary). According to Agroscope the growing role of genomic selection may result in a lesser use of newly collected ecotypes. Establishing the specific marker-trait associations in the active breeding pool, which has been developed using a certain set of ecotypes, is a great effort. This effort would have to be repeated when new ecotypes are introduced, because the diversity of these populations as a whole is most probably too large to expect universally valid marker-trait associations.

INRA points towards the importance of better statistical techniques of 'modelling' genotype x environment interactions e.g., in changing species mixtures. DSV argues that feeding value analysis will become standard. Jouffray-Drillaud points towards the use of semi-hybrids.

In general, the respondents want more testing in environments with extreme climate conditions and aim for a greater use of international networks for testing (and selection). ILVO wants to give more attention to a participatory approach (as testing in farmers' environments) and using cows as selectors. In Ireland variety performance is evaluated at farms (Pasturebase): this helps to obtain new breeding targets. Graminor notices that testing facilities are disappearing in Norway.

USDFRC aims for an open-ended breeding program pursuing multiple breeding strategies simultaneously: they don't want to lock themselves into a single strategy or program.

Qualifications of the Breeding Team

The respondents mention field breeders, molecular breeders, bio-informatics, statisticians, modelling experts, quantitative geneticists, seed researchers, experts in feeding value, chemists, ecophysiologists, experts in robot techniques, IP experts (DSV) and qualifications like curiosity and persistence (ILVO). USDFRC mentions not to forget unskilled labour and Agroscope projects that they will need 50 % more work force in the next 10–20 years.

Integration of Knowledge of Plant Physiology

CRA-FLC considers the knowledge of gene actions to become more important than knowledge on plant physiology; the knowledge related to adaptations of plants to contrasting environments will continue to be important. In case genomic selection is the focus, Agroscope considers an integration of the knowledge of plant physiology in the breeding work not immediately necessary in the absence of new breakthroughs. USDFRC puts no active integration in the breeding program other than a general understanding of bred species; exploratory projects will determine high throughput phenotyping procedures. According to Jouffray-Drillaud, ecology should be the focus and not physiology. Graminor considers plant physiology important for the adaptation of crops to changing growing seasons and Barenbrug puts forward that an understanding of physiology will be necessary for an effective application of new techniques like marker-assisted selection. DSV mentions four important physiological domains, related to rooting, feeding value and mechanisms of disease resistance and wear tolerances. ILVO calls for effective methods to measure physiological parameters related to their effect on yield and INRA emphasizes the importance of gene based modelling: "we will need both plant physiology and genotypic information to feed virtual plant models. These models can deliver a better insight in plant interaction-competition in the light of multi-species conditions". INRA also mentions that studies in plant physiology help at identifying new traits or functions that show genetic variation; interaction between geneticists and ecophysiologists helps defining adapted experimental designs for studying the effect of environmental stresses on genotypes. There was a general remark that there are a lot of public funding opportunities for companies and institutes to get started with the integration of physiological and genotypic information in current breeding programs.

Influence of Growing World Population

The respondents see the influence of the growing world population as relatively modest. According to Graminor the work and strategy of breeders is mainly driven by farm policy; according to

USDFRC cost effectiveness of the breeding work and the demand of the end-user are the most important drivers. Surprisingly, nobody mentioned that "without breeding the world will collapse". We have to better emphasize that breeding forage crops will continue to be necessary since the demand for grassland-based products is expected to increase. The SCAR foresight predicts the need for an additional 200 million kg of protein in 2035. In contrast, the public awareness of our sector is low because "nobody eats our crops". Grass and legume breeding companies and research institutes should increase the promotion of their activities and their link with sustainable agricultural systems (low input conditions, local production chains, ecosystem services). In that perspective, seed companies of major staple crops are far better in advertising and in hiding how they "sustain unsustainable systems".

Influence of Climate Change and Renewable Carbon Based Economy

The respondents emphasize the need for adapted varieties to abiotic and biotic stresses produced by a changing climate. The main focus lies on improved performance under harsh conditions (temperature extremes, scarcity of water, shortage or excess of nutrients) and more predators. Another theme is to be dependent on fewer inputs to have a positive impact on carbon footprint. Agroscope will breed for resilience and put an increased focus on minor species that are more tolerant to extreme conditions. ILVO will breed in artificial dry conditions, while CRA-FLC puts more emphasis on N-fixing species.

For INRA, breeding research should assess how populations react on climate change; in line with this they mention the strategic importance of collecting and preserving genetic resources of forage crops. ILVO emphasizes the relation between performance under "normal" conditions and under reduced input of nutrients.

There is not much enthusiasm to breed for products fitting into a renewable, carbon based economy. CRA-FLC mentions the dual use of crops: stems of alfalfa for energy, leaves for protein.

Influence of Consumer's Attitude

For turf, Barenbrug mentions that new species may have a role owing to reduced nutrient inputs.

For forage crops, the future situation will probably be not very different from the current situation. A dual influence of consumers' attitude is apparent: if demand for red meat decreases, the importance of forage crops drops. On the other hand, forage crops are perceived as rather environmental friendly and forage fed animals as producing healthy products. Improving forage quality keeps this good image upright. According to INRA, breeders have to be careful that future innovations will not ruin this picture. Jouffray-Drillaud mentions in this perspective that consumers won't allow breeding GM fodder crops.

Both Agroscope and USDFRC see breeding for organic agriculture as an opportunity. Nowadays, the organic sector feels that companies are not breeding for them, although there are a lot of funding possibilities available. Agroscope runs an organic breeding program for the main grass and clover species; red clover in particular is very important for organic farms as they need nitrogen fixing crops in their rotation. INRA puts forward that varieties adapted to mixtures are already very close to organic agriculture, because pesticides and fertilisers are not necessary. CRA-FLC states that weed competitiveness and coping with nutrient deficiency are two major goals in organic breeding programs.

Influence of Regulations and Sustainability Drivers

Regulations

Regulations reduce the access to gene banks: Graminor found quite some difficulties to get certain accessions in their trials. INRA states that free access should be preserved and gene patents should not restrict the free use for breeding. USDFRC fears that if royalties have to be paid to genebanks, no alfalfa seed company in the US will use genebank accessions in their program.

Agroscope emphasizes that selection decisions continue to be influenced by regulations and practices for variety testing, variety protection and recommendation. ILVO supports a more straightforward use of VCU criteria to allow for distinguishing varieties while INRA wonders if molecular data could be included in DUS testing.

Sustainability Drivers

The reduction of nutrient inputs will lead to a more important role for legumes. According to INRA their use and breeding should drastically increase both for application in pure stands and in mixtures. Barenbrug and DSV mention increased attention for the relations between plants, fungi and bacteria e.g., to reduce pesticides (endophytes, mycorrhizae). CRA-FLC states that systems rewarding sustainability have the capacity to support perennial species, multispecies systems, and legumes. VCU trials offer opportunities to include sustainability criteria (ILVO) or criteria for specific practises (e.g., mixtures) (INRA).

Any Other Issues

Jouffray-Drillaud sees opportunities for seed technology to decrease seeding rate when installing grassland and DSV calls for adequate IT and bioinformatics support to handle big data (the massive increase in genotypic and phenotypic information).

49 Breeding Ideas Presented During the Meeting

B. Julier, P. Barre, F. Gastal, M. Ghesquière, I. Litrico, G. Louarn, J.-P. Sampoux, A. Aleliūnas, G. Brazauskas, T. Carita, J.P. Carneiro, N. Simões, J.P. Fragoso de Almeida, I. Duarte, M. Cougnon, and H. Riday

Breeding Idea: Breeding of Lucerne for Better Adaptation to Growth in Mixtures

B. Julier, Ph. Barre, F. Gastal, M. Ghesquière, I. Litrico, G. Louarn, J.P. Sampoux

Problem

Lucerne (alfalfa; *Medicago sativa*) is often used as monoculture but may also be used in mixture with tall fescue (*Festuca arundinacea*) or cocksfoot (*Dactylis glomerata*) in order to balance protein and energy contents of the forage, to stabilize yield production over cuts and to limit weed development. However, lucerne varieties are bred and evaluated for registration in monoculture. We have shown that the value of a lucerne variety in mixture was not predictable from its value in monoculture. Breeding schemes that take the value in mixture into account should be elaborated. This question of breeding for monoculture *vs.* mixture is probably a concern for most forage species. Here, we assume a goal of creating a new lucerne variety that performs well in both pure and mixed stands.

Idea

Up to now, breeding criteria are not substantially different for pure stand or mixture. The first step, i.e., selection in nursery for heritable traits such as plant height, plant vigor and lodging resistance, could be identical for pure or mixed stands. Current breeding methods with isolated

B. Julier • P. Barre • F. Gastal • M. Ghesquière
I. Litrico • G. Louarn • J.-P. Sampoux
INRA, UR4, URP3F, Lusignan F-86600, France

A. Aleliūnas • G. Brazauskas
Institute of Agriculture, Lithuanian Research Centre for Agriculture and Forestry,
Instituto av. 1, Akademija, Kėdainiai District
LT-58344, Lithuania

T. Carita • J.P. Carneiro • N. Simões
J.P. Fragoso de Almeida • I. Duarte
Instituto Nacional de Investigação Agrária e Veterinária, I.P., Apartado 6,
Elvas 7351-901, Portugal

M. Cougnon
Department of Plant Production, Ghent University,
Proefhoevestraat 22, Melle B-9090, Belgium

H. Riday
U.S. Dairy Forage Research Center (USDA-ARS),
Madison, WI, USA

I. Roldán-Ruiz et al. (eds.), *Breeding in a World of Scarcity: Proceedings of the 2015 Meeting of the Section "Forage Crops and Amenity Grasses" of Eucarpia*,
DOI 10.1007/978-3-319-28932-8_49

plant design could alternatively be planted in denser stands of either lucerne or grass. Once promising plants are identified in the nursery, their half-sib progenies would be evaluated in both pure and mixed stands. Forage yield, lodging resistance and if possible, forage quality and legume proportion assessed using NIRS, would be recorded. To limit costs, the number of repetitions should be the same as before with half of the repetitions in pure stand and half of the repetitions in mixed stands. The mother plants of the best performing families would be then intercrossed to form potential varieties or to be introduced in a next breeding cycle. These progeny would be tested in larger plots in a multisite trial, again with both pure and mixed stands.

Scheme

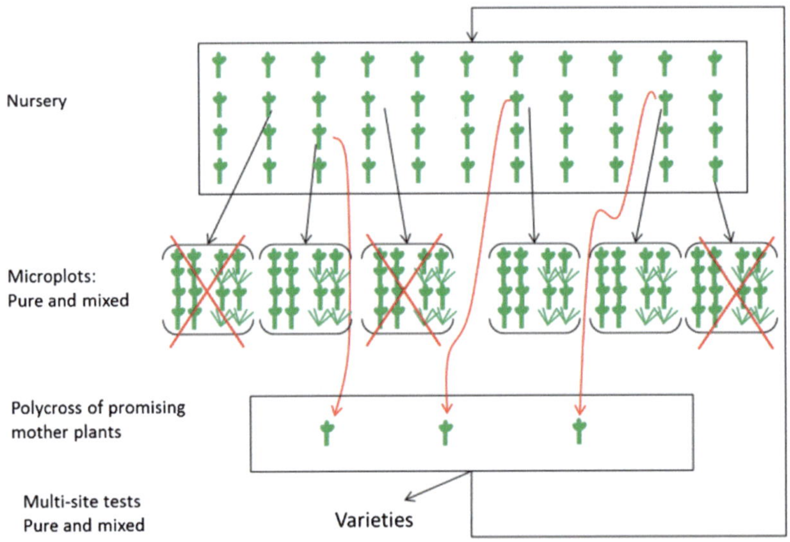

Summary

The benefit of the application of this breeding idea lies in the release of lucerne varieties adapted to all types of current uses, both pure and mixed stands.

Breeding Idea: Population-Wise Evaluation for Freezing Tolerance in Perennial Ryegrass Under Controlled-Environment Conditions

A. Aleliūnas, G. Brazauskas

Problem

Perennial ryegrass is a species of paramount economic importance used both for turf and forage. However, the species is prone to winterkill. Freezing tolerance is a major component of winter hardiness, yet it is difficult to breed for, because overwintering in the field is highly variable due to yearly fluctuations in overwintering conditions or the differences between breeding environments and target environments. Testing for freezing tolerance under controlled-environment conditions have so far relied on identification of single genotypes with higher freezing tolerance. Population-wise screening, however, would enable evaluation of freezing tolerance in perennial ryegrass populations, which could be further extended to the identification of causal SNP markers, conferring higher freezing tolerance in populations if combined with population-wise genotyping such as GWAFF.

Idea

Large-scale screening of multiple populations for freezing tolerance under controlled-environment conditions would not only help to indicate freezing tolerant populations by calculating LT_{50} value, but would also help to select freezing tolerant genotypes from multiple populations. The screening could be beneficial at early breeding stages for freezing tolerance evaluation of wild accessions, as well as for the improvement of the existing cultivars. Perennial ryegrass is an outcrossing species, resulting in genetically diverse populations. Freezing tests at the most extreme temperatures for this species would identify extremely freezing-tolerant genotypes in multiple perennial ryegrass populations. Such plants could be used to form a panmictic, cross-pollinated population. As freezing-tolerance is a multigenic trait, cross-pollination of genetically diverse freezing-tolerant genotypes would encourage the accumulation of new beneficial allele combinations, while the gain in freezing tolerance could be evaluated in subsequent freezing tests. All offspring would be subjected to selective freezing temperatures to identify if beneficial allele accumulation has occurred and to select the surviving offspring. This cycle could be repeated for several generations. Such freezing-tolerant genotypes would represent sufficient genetic diversity and would cover multiple beneficial alleles due to the large and diverse primary pool of alleles present in the initial collection of perennial ryegrass populations. This enrichment in beneficial alleles could be tracked by applying population-wise genotyping methods, such as GWAFF, and would lead to the identification of causal SNP markers conferring higher freezing tolerance.

Scheme

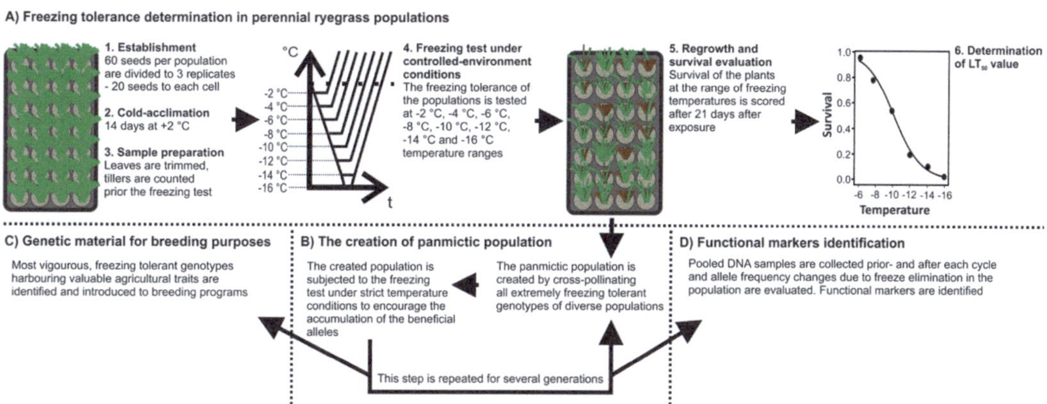

Summary

Freezing tolerance evaluation in multiple perennial ryegrass populations would lead to the identification of the freezing tolerant genotypes. Further creation of panmictic populations using diverse freezing tolerant genotypes would encourage accumulation of beneficial alleles, providing genetic material for breeding purposes and identification of causal SNP markers through allele frequency changes in these populations.

Breeding Idea: Pasture Legume Breeding for Phosphorus Use Efficiency

T. Carita, J.P. Carneiro, N. Simões, J.P. Fragoso de Almeida

Problem

Portuguese pasture farming systems based on annual legumes rely on phosphoric fertilizers to assure persistence and productivity. Portuguese soil characteristics are acidic, with low P content available for plant uptake, with a need for high P application due to a very low efficiency of phosphorus uptake by plants (around 15 %). The phosphorus used in fertilizers is obtained from mineral phosphate, a non-renewable resource. It is anticipated that phosphate resources will be depleted within 50–100 years. Therefore, it is essential to identify genetic lines with higher efficiency of P assimilation, combined with more efficient use of phosphate fertilizers (new forms of fertilizer; more and application of adjusted cultural pasture nutrient management itineraries) to reach agro-ecosystem sustainability, optimize production of seed and biomass, and increase the persistence of pasture legume varieties.

Idea

For the benefit of all relevant "Mediterranean pasture stakeholder groups" (e.g., farmers, seed producers, crop breeders and conservation authorities) the intention is to identify *Trifolium subterraneum*, *Ornithopus compressus* and *Astragalus pelecinus* genotypes with high ability to acquire, translocate, accumulate and utilize phosphorus. INIAV has a large germplasm collection of these species. We intend to start a pioneering research to evaluate these accessions (in a state of seedling and vegetative phase) for their phosphorus "acquisition efficiency" and "use efficiency". After this preliminary assessment, it should be studied which of these mechanisms has the major influence in the efficiency of phosphorus use so the selection criteria can be adjusted. Several field trials will be installed with the most promising genotypes to evaluate phosphorus use efficiency throughout the growing cycle. The selected genotypes of each species will be proposed to integrate into biodiverse mixtures.

At the same time, chromosome mapping (determining QTL for phosphorus use efficiency) will be investigated with the aim to develop molecular markers to assist selection.

Scheme

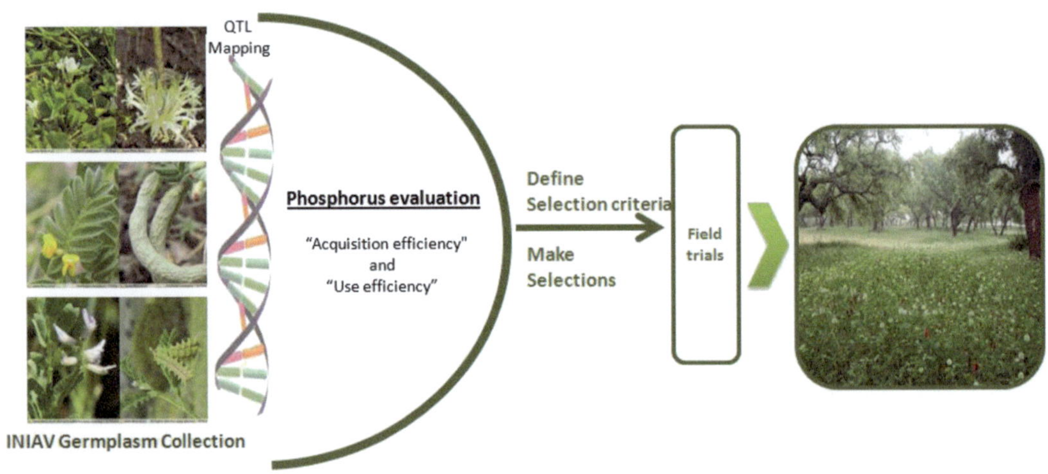

Summary

The implementation of this research may allow the increase of sustainability and persistence of permanent pasture on Mediterranean dryland areas. Moreover, on a global scale it may contribute to reducing the exploitation of non–renewable phosphate resources.

Breeding Idea: Adapted Vetches for Mediterranean Rainfed Conditions

F. P. Carneiro, N. Simões, T. Carita, I. Duarte

Problem

Under Mediterranean cultural conditions, common vetch (*Vicia sativa*) and hairy vetch (*Vicia villosa*) are important crops that produce low cost feed resources with good quality value in extensive systems. Frequently, the occurrence of dry years or water stress at the end of the vegetative growth and during reproductive growth imposes a reduction in biomass and in seed production. Therefore it is important to study genetically diverse germplasm to identify genotypes with high tolerance to dry conditions.

Idea

<u>Phase A – Productivity evaluation</u>

For several years a high number of vetch accessions should be evaluated under rainfed conditions to select the plants with the best performance when drought occurs, expressed as:

1. dry matter yield per plant, measured 90 days after emergence, and at the end of flowering;
2. seed yield.

<u>Phase B – water status with limiting water</u>

With the accessions selected on phase A, the water status (*relative water content – RWC*, *predawn leaf water potential* – ψ_{pd} and *midday leaf water potential* – ψ_{md}) should be determined to estimate the plants' ability to remain hydrated, with two treatments:

- W80 – watered conditions: irrigation fills 80 % of *available water content – AWC*;
- RF – rainfed conditions.

The data of RWC, ψ_{pd}, ψ_{md} and forage yield with the abovementioned treatments allow to identify the most promising vetch accessions for water-limited conditions.

At the same time, under drought conditions, the water use efficiency (WUE) should be assessed to confirm the accessions which, despite being grown under water-limited conditions, have high dry matter yield and high seed yield. These accessions can be the base of new commercial varieties for water-limited conditions.

Scheme

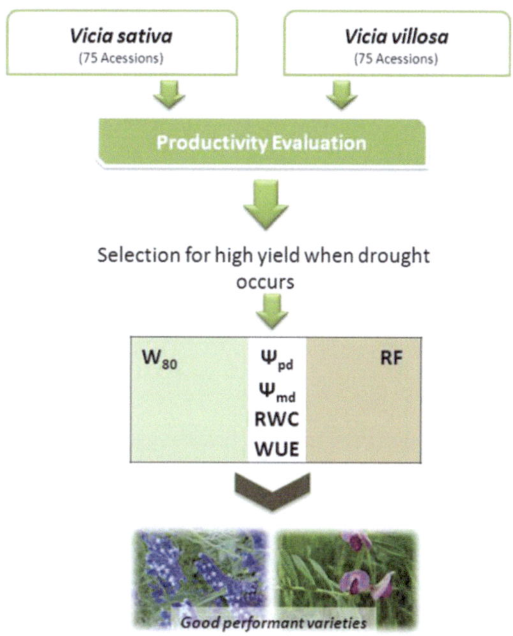

Summary

The benefit of the application of this breeding idea lies in selecting vetch accessions for high yield under water-limited conditions, by using water status parameters obtained under dry conditions to confirm which vetch accessions are most valuable.

Breeding Idea: Low-Silica Tall Fescue (LSTF)

M. Cougnon

Problem

Tall fescue has low palatability and lower organic matter digestibility compared to ryegrasses, resulting in lower animal performances. Selection for soft leaved genotypes has improved animal preference in early tall fescue breeding. Softness, however, is a very subjective trait. To make further progress, **quantifiable** traits that allow selection of tall fescue varieties with an improved animal preference are needed.

Idea

This breeding idea can be understood in three steps:

1. Plants accumulate silica to protect themselves against herbivory. Ecologists have studied the growth and defense investment in 18 grass species and compared them against vole feeding preference. Silica was the most influential defensive factor that determined vole feeding preference.
2. We found that tall fescue has a significantly higher silica content compared to other forage grass species and that there is a large genotypic variance for silica content within the species.
3. Silica content of plants can be measured using NIRS.

LSTF can be selected as follows: In the clonal nursery, plant material is harvested, dried and ground. The NIRS spectra of these samples allows to predict both silica content and digestibility of the organic matter (DOM), allowing to select clones with a low silica content and a high digestibility. This can lead to a new generation of tall fescue varieties: **LSTF**.

Scheme

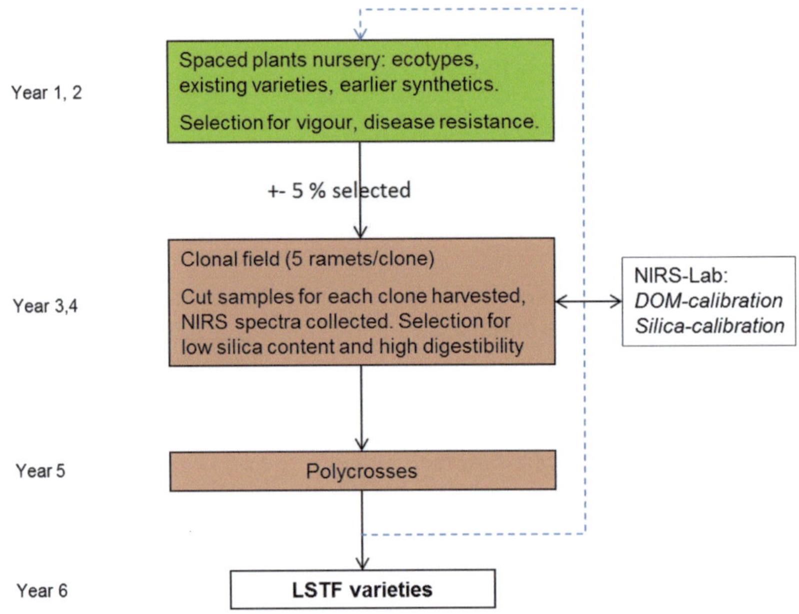

Summary

The benefit of the application of this breeding idea is to breed tall fescue varieties with an improved animal preference due to lower silica content.

Breeding Idea: Breeding Nursery Tissue Collection for Future Genomic Analysis

H. Riday

Problem

Phenotyping is a major bottleneck in breeding programs. With new genomic technologies, high-throughput genotype schemes are constantly being developed. However, every genomic technology requires phenotypic data to inform the generated prediction models. Forage breeders continue to phenotype and breed various forage species, often without using any molecular genetic approaches in their breeding programs. The phenotyping accomplished in these traditional field breeding situations is essentially wasted in terms of contribution to current or future prediction models developed based on genomic technologies.

Idea

Field breeders engaged in phenotyping should retain tissue samples from every selection unit in the breeding programs, even if they are not currently using "genomic" technology. If selection is based on family structure (e.g., half-sib selection), then parental tissue of the "family" should be retained. If selection is based on individual plants (e.g. phenotypic selection), tissue of every individual plant entering the breeding nursery should be retained. This scheme is particularly relevant to expensive phenotyped traits or those requiring a long time for measurement (e.g., biomass yield and persistence). Tissue for DNA extraction is usually frozen for storage. Breeding programs have the potential to generate thousands of tissue samples, requiring a well-thought-out system of storing and tracking frozen tissue samples. A low-cost solution would be to collect tissue samples in inexpensive paper envelopes, with identification information (and associated barcode) laser-printed onto the envelopes. Small quantities of tissue collected from selection units in such envelopes could be then put in a lyophilizer (freeze drier) to dehydrate the tissue while preserving DNA integrity. The tissue-containing envelopes can then be stored at ambient temperature under low humidity (<15 %). The shelf-life of tissue stored this way could last decades. Capital costs for this scheme are low (i.e., laser printer, benchtop lyophilizer, desiccant dehumidifier, and standard air conditioning unit). This scheme would allow breeders to access their tissue sample collection for DNA samples even decades later. Routine and consistent use of this scheme would eventually build an extensive collection of tissue samples over a range of germplasms and locations. The associated phenotypic information and even associated remnant seed can be utilized for genomic model development or post-phenotyping selection based on genotypic information.

Scheme

Summary

This scheme allows *a posteriori* genomic use of phenotypic information gathered during a breeding program's plant evaluation phase, even years after termination of the phenotype evaluation phase.

Fifth Workshop of the Eucarpia Festulolium Working Group

50

M.W. Humphreys and M. Ghesquiere

Abstract

A trial in large dense plots is running currently across 9 locations and 7 European countries with diverse climates. In the trial, 15 Festulolium varieties, considered the elite choices currently commercially available are compared with 6 control varieties of pure *Lolium* and *Festuca* species; the trials were sown from spring 2012 to 2014. The report of the first year DM yield was presented an a submitted paper, "Enhancing the productivity of forage grasses on European scale using interspecific hybridization" at the Eucarpia Fodder Crops Section Meeting. The field locations used represented very diverse climatic zones and whilst certain Festulolium varieties performed better in specific climates, others showed good adaptive plasticity and field performance at locations of contrasting stress conditions. Professor Humphreys and Dr Kopechy provide updates on the latest Festulolium research and on its uses.

Keywords

Festulolium varieties • Multi-site field trials

Multi-location Trial

A trial in large dense plots is running currently across 9 locations and 7 European countries with diverse climates. In the trial, 15 Festulolium varieties are compared with 6 controls of pure *Lolium* and *Festuca* species; the trials were sown from spring 2012 to 2014 at several locations. The report of the first year DM yield was presented as a submitted paper, "Enhancing the productivity of forage grasses on European scale using interspecific hybridization". V. Černoch (DLF-Trifolium) who manages the database for the whole network, presented slides and results of his own trial at Hladke Zivotice in Czech Republic. There is no doubt that the group is collecting a unique data set, including forage quality traits, to document Festulolium's ability to grow and persist in a very large range of environments.

M.W. Humphreys (✉)
IBERS, Aberystwyth University,
Gogerddan Campus, Aberystwyth, Ceredigion,
Wales SY23 3EE, UK
e-mail: mkh@aber.ac.uk

M. Ghesquiere
URP3F, INRA, Le Chene, RD 150-BP 80006,
Lusignan 86600, France

I. Roldán-Ruiz et al. (eds.), *Breeding in a World of Scarcity: Proceedings of the 2015 Meeting of the Section "Forage Crops and Amenity Grasses" of Eucarpia*,
DOI 10.1007/978-3-319-28932-8_50

On-going Projects at IBERS (Aberystwyth, UK)

M.W. Humphreys presented on the BBSRC-LINK SUREROOT project (http://www.sureroot.uk/). This project focuses on use of various novel Festulolium hybrid combinations and clovers. The plants were screened initially for their enhanced root growth at the National Plant Phenomics Centre at IBERS; after selection, they were planted in either plots or fields (e.g., the Farm Platform at Rothamsted Research, North Wyke and 8 commercial development farms under diverse livestock management systems). The project aims to explore the possibilities for biological approaches for enhanced soil hydrology to reduce incidents of flooding and to safeguard grassland production during times of low rainfall. Another project (Climate Smart Grasslands) with Bangor University and the Centre for Ecology and Hydrology at Bangor (http://www.nrn-lcee.ac.uk/climate-smart-grass/) examines the impact of one particular stress on the resilience of grasslands to combat a subsequent period of stress conditions and to identify the "tipping-point" when grasslands can no longer recover production. It aims to find solutions that may enhance grassland persistence at times of stress conditions. M.W. Humphreys described also current development in the scope of Festulolium for reclamation of mine spoil and contaminated soils in South Wales and similar objectives in China.

Recent Results at IEB (Olomuc, Czech Republic)

D. Kopecky gave new taxonomic results within the Schedonorus group of *Festuca* sp. emphasizing altitudinal distribution among the 2x *F. pratensis* and the 4x *F. apennina* with evidence of recent 3x direct hybrids between the two cytotypes when both are present. Furthermore, new results of gene expression using the DArTseq Platform underlined that the *Festuca* genes appeared to be preferentially expressed in Festulolium hybrids while, interestingly, they appeared phenotypically much more *Lolium* sp. like 68 than *Festuca.* sp.

Report of the Eucarpia Multisite Rust Evaluation Working Group

51

T. Vleugels and F.X. Schubiger

Abstract

From 2001 to 2013, five EUCARPIA multisite rust evaluation trials were conducted at 21 to 32 sites in 10 to 12 European countries. The aim of the trials was to assess the rust resistance of 16 Italian (*Lolium multiflorum*), 4 Hybrid (*L. boucheanum*) and 34 perennial ryegrass (*L. perenne*) cultivars. During the meeting of the section "Forage Crops and Turfgrasses" of Eucarpia at Ghent in 2015, a workshop was organized to discuss the future of the trial. The starting position of the discussion was the fact, that the seed stock is almost used up and the germination rate of some varieties is now very poor. The group decided to start a new trial, which focuses on leaf spot and rust diseases of *Lolium* species.

Keywords

Crown rust • Stem rust • Leaf spot • Italian ryegrass • Hybrid ryegrass • Perennial ryegrass

Twelve Years of the Eucarpia Multisite Rust Evaluation

The workshop started with a presentation of the results from the past 12 years of rust evaluations. Some major results are:

- The incidence of crown rust on Italian and perennial ryegrass was very consistent among locations and among years.
- Stem rust was uncommon on Italian ryegrass, but not on perennial ryegrass. The infection pressure was comparable among years.
- Both in perennial and Italian ryegrass, there were significant differences in crown (and stem) rust susceptibility among cultivars. However, the ranking of the cultivars remained mostly the same accross years and locations.
- There was no evidence of a breakdown of resistance in any cultivars during the 12 years of experimentation.

T. Vleugels (✉)
Plant Sciences Unit, Institute for Agricultural and Fisheries Research (ILVO),
Caritasstraat 21, Melle 9090, Belgium
e-mail: tim.vleugels@ilvo.vlaanderen.be

F.X. Schubiger
Forage Plant Breeding, Agroscope INH,
Reckenholzstrasse 191, Zürich CH-8046, Switzerland

I. Roldán-Ruiz et al. (eds.), *Breeding in a World of Scarcity: Proceedings of the 2015 Meeting of the Section "Forage Crops and Amenity Grasses" of Eucarpia*,
DOI 10.1007/978-3-319-28932-8_51

Results of the Questionnaire

Before the symposium, Franz Schubiger sent a questionnaire to all participants in the 2013 trial. These answers were briefly presented before the start of the open discussion.

Do you think that we should continue with the Eucarpia multisite rust evaluation? Do you consider a further trial useful?

The interest varied from low to high.

If yes:

Should we use new cultivars or the same as before?

- Most participants said yes, but complemented by some cultivars from the previous trials as controls.
- Accessions from gene banks should be included.

Some participants suggested to use spaced plants (genotypes or clones) instead of cultivars established as sown rows.

Do you recommend a change of the methods used?

- Most participants said no.
- Application of artificial rust inoculations in the field and recording of meteorological data was suggested.

What would be the benefits of continuing the trial and what would be the cost and effort?

- Overall, most participants see few benefits in a continuation.
- Continuing the trials could yield information on rust resistance of (new) cultivars in different environments, as well as evidence about a potential increase of the disease pressure in the areas where the trials are performed.
- Continuing the trial may allow the identification of new pathotypes of crown or stem rust.

Should we focus on forage grasses and/or forage legumes? Which species?

- Legume species (alfalfa, white clover, red clover): evaluation of different diseases in different environments.
- Forage grasses (*Lolium perenne*, *Dactylis glomerata*, *Poa pratensis*, *Phleum pratense*): rust and leaf spot diseases

Is rust disease relevant in breeding grasses/legumes?

- Yes, for specific species such as *Lolium* sp. and *Poa pratensis*.
- Resistance to rust diseases is important to get cultivars registered on the list of recommended varieties.

Which information is most important for a better understanding of resistance to rust of forage grasses?

- Information about the genetic variation of the pathogen.
- Allelic diversity at specific QTLs

Do you think there are other diseases worth focusing on?

- Soilborne pathogens
- Rust on *Poa pratensis* and *Dactylis glomerata*
- Stem rust in forage grasses
- Legumes: southern anthracnose, clover rot, mildew, foliar diseases

What is the knowledge most lacking in the host-pathogen relationship?

- Sources of rust resistance
- Mechanism of rust resistance
- Stability of the rust resistance in different environments
- How many genes confer rust resistance

Open Discussion

An open discussion was held between all participants to determine the future of the Eucarpia multisite rust trials. The group decided to start a new trial on both *Drechslera* leaf spot and crown/stem rust. The advantages of this trial is that multisite rust evaluations will be carried out over the next 15 years, so that each breeder can get information on the rust susceptibility of his/her material in various European environments. At the same time we will get novel information on the susceptibility of different varieties to leaf spot on a European scale.

Conclusion

- The group agrees to continue with *Lolium* species. Leaf spot diseases will be the focus. Crown and stem rust will also be scored.
- Five varieties (of Italian and perennial ryegrass) from the previous rust trials will be integrated in the new trial. Romanian crown rust resistant genebank accessions will be assessed.
- Information on leaf spot is available from the Dutch variety trials.
- Plants will be grown for two years: autumn sowing in Italy and spring sowing in other regions. Leaf spot will be scored in autumn. Another scoring will be done in the second year.

Index

I. Roldán-Ruiz et al. (eds.), *Breeding in a World of Scarcity: Proceedings of the 2015 Meeting of the Section "Forage Crops and Amenity Grasses" of Eucarpia*,
DOI 10.1007/978-3-319-28932-8

MIX
Papier aus verantwortungsvollen Quellen
Paper from responsible sources
FSC® C105338

If you have any concerns about our products,
you can contact us on
ProductSafety@springernature.com

In case Publisher is established outside the EU,
the EU authorized representative is:
Springer Nature Customer Service Center GmbH
Europaplatz 3, 69115 Heidelberg, Germany

Printed by Libri Plureos GmbH
in Hamburg, Germany